T0317463

Beef Practice:
Cow–Calf Production Medicine

Beef Practice:
Cow–Calf Production Medicine

Peter J. Chenoweth and Michael W. Sanderson

Blackwell Publishing Professional
2121 State Avenue, Ames, Iowa 50014, USA

Orders: 1-800-862-6657
Office: 1-515-292-0140
Fax: 1-515-292-3348
Web site: www.blackwellprofessional.com

Blackwell Publishing Ltd
9600 Garsington Road, Oxford OX4 2DQ, UK
Tel.: +44 (0)1865 776868

Blackwell Publishing Asia
550 Swanston Street, Carlton, Victoria 3053, Australia
Tel.: +61 (0)3 8359 1011

First edition, 2005

Library of Congress Cataloging-in-Publication Data

Chenoweth, Peter J.
Beef practice : cow-calf production medicine / Peter J. Chenoweth and Michael W. Sanderson.
p. cm.
Includes bibliographical references and index.
ISBN 13: 978-0-8138-0402-6
ISBN 10: 0-8138-0402-7 (alk. paper)
1. Cows—Diseases. 2. Cows—Health. 3. Calves—Diseases. 4. Calves—Health. 5. Veterinary medicine. I. Sanderson, Michael W. II. Title.

SF961.C48 2004
636.2′0896—dc22

2004017606

The last digit is the print number: 9 8 7 6 5 4 3 2 1

Contents

Editors

Peter J. Chenoweth, BVSc, PhD, Diplomate, American College of Theriogenology, Member, Australian College of Veterinary Scientist (Animal Reproduction). Professor, Coleman Chair of Food Animal Production Medicine, Department of Clinical Sciences, College of Veterinary Medicine, Kansas State University, Manhattan, Kansas.

Michael W. Sanderson, DVM, MS, Diplomate, American College of Veterinary Preventive Medicine, Epidemiology Specialty, Diplomate, American College of Theriogenology, Associate Professor, Beef Production and Epidemiology, Department of Clinical Sciences, College of Veterinary Medicine, Kansas State University, Manhattan, Kansas.

Contributors

Grant Dewell, DVM, MS, Postdoctoral Fellow, Animal Population Health Institute, College of Veterinary Medicine and Biomedical Science, Colorado State University, Ft. Collins, Colorado.

Temple Grandin, PhD, Associate Professor, Department of Animal Sciences, Colorado State University, Ft. Collins, Colorado.

D. Dee Griffin, DVM, MS, Professor, University of Nebraska, Director, Great Plains Veterinary Educational Center, Clay Center, Nebraska.

Thomas R. Kasari, DVM, MVSc, Diplomate, American College of Veterinary Internal Medicine, Associate Professor of Medicine, College of Veterinary Medicine, Texas A&M University, College Station, Texas.

Robert Larson, DVM, PhD, Diplomate, American College of Theriogenology, Clinical Associate Professor of Veterinary Medical Extension, College of Veterinary Medicine, University of Missouri, Columbia, Missouri.

Twig Marston, PhD, Associate Professor, Animal Sciences and Industry, Kansas State University, Manhattan, Kansas.

Glenn Nader, **MS**, Certified Range Manager (CRM #55) Area Livestock & Natural Resources Advisor, Yuba City, California.

Gary Rupp, DVM, MS, Diplomate, American College of Theriogenology, Professor, University of Nebraska, Director, Great Plains Veterinary Educational Center, Clay Center, Nebraska.

David R. Smith, DVM, PhD, Diplomate, American College of Veterinary Preventive Medicine, Epidemiology Specialty, Associate Professor, Veterinary and Biomedical Sciences, University of Nebraska–Lincoln, Lincoln, Nebraska.

Gary Veserat, MS, American Registry of Professional Animal Scientists, Veserat Consulting, Woodland, California.

Valerie Veserat, MBA, Veserat Consulting, Woodland, California.

Preface

The inspiration for, and foundations of, this book may both be traced to an innovative educational program provided for beef cattle veterinarians by the University of Nebraska Great Plains Veterinary Educational Center. The Beef Cattle Production Management Certificate Program, which has been in operation for 10 years, is designed to train veterinarians to develop new skills to better serve their beef clients. It does this via a series of educational modules that provide in-depth training in areas related to beef cattle production, management, and economic strategies. This is achieved with a unique multidisciplinary approach in which emphasis is placed upon nontraditional veterinary areas including critical thinking, business and financial management, statistics, risk analysis, epidemiology, applied animal breeding, nutrition, data management and analysis, and communication skills. Veterinarians who are oriented toward production animal agriculture are exposed to new concepts, disciplines, and opportunities. Approximately 130 veterinarians and several other specialists, representing an eclectic range of interests and career status, have graduated from this course. Many of these "graduates" have used their training to assume leadership roles in both the profession and industry, and most, if not all, have been both stimulated and challenged in the process.

The teachers, or "mentors," who participate in this program also represent a diverse group of key industry specialists, prominent scientists, and educators. We have unashamedly drawn heavily on this group in requesting contributions to this book, as well as in providing other input and advice.

Thus this book represents an attempt to build a bridge between traditional beef cattle, cow/calf practice, and a number of emerging opportunities that can enable veterinarians to become more effective at the level of whole-farm management. In pursuing this goal, the underlying premise has been that the information provided should be both easily accessible and useful, although occasional forays into the esoteric were not discouraged.

PJC MWS

Acknowledgments

Many have contributed to this book, in both tangible and intangible ways. The senior editor was most fortunate to have been mentored by H.G. (George) Osborne of the University of Queensland at an early stage of his career, and was thereby thoroughly drenched with the principles of the herd/flock approach to veterinary medicine. A later stint as Director of the University of Queensland's Pastoral Veterinary Center (Goondiwindi, Queensland), which in itself was the realization of a vision by George Osborne, helped to forge a deeper understanding of—and empathy with—rural producers and their needs. The sometimes contradictory urges to both question and create can be attributed to genetics and upbringing, with due credit (or blame) being apportioned accordingly.

The junior editor was trained and mentored in epidemiology and the value of data in herd decision making by Drs. John Gay and Dale Hancock of Washington State University. Their influence is reflected in his chapters, and a debt of gratitude is owed and gladly offered to them.

Early in the development of this book, the messages of prophets such as Doug Blood (Australia) and Otto Radostits (Canada) were striking a responsive chord in relation to the future of food animal veterinary medicine. Other leaders who helped shape current awareness of veterinary opportunities in animal agriculture include the late A. Leman (Swine Consultant, formerly with the University of Minnesota), Larry Rice (Oklahoma State University), and Gary Rupp (Great Plains Veterinary Educational Center). The latter provided the vision and energy for the Beef Production Management Certificate Series—a program that directly inspired this book.

A number of mentors for this course (Grant Dewell, Dee Griffin, Bob Larson, Gary Rupp, and David Smith, as well as the two editors) are represented here as contributing authors, and others have assisted with inputs and inspiration. Of particular mention here is John Spitzer (Clemson), whose friendship and valuable, uncompromising critiques have been of immense help and support for the senior editor. The response from contributing authors, all of whom are exceptionally busy people, has been universally enthusiastic and dedicated. For this we are immensely grateful.

Individuals who have assisted in reviewing chapters, and in making constructive suggestions, include Brad DeGroot (Kansas State University), Bob Gentry (veterinary practitioner, Beloit, Kansas), Janice Swanson and Sandy Johnson (Kansas State University), and Greg Ologun (National Technical University, Akure, Nigeria),To these individuals, and to others who helped in many ways, we extend our sincere gratitude.

In particular, we would like to acknowledge the assistance of Lisa Sisley, who edited, organized, exhorted, and nudged until the job was completed to her high standards.

To the staff at Blackwell Publishing (formerly Iowa State Press), who have kept faith during an extended gestation period, we are also most grateful. We also acknowledge, with gratitude, the infrastructure and resources generously provided by Kansas State University, particularly the College of Veterinary Medicine, whose people do, indeed, make a difference. The junior editor has been providentially blessed with a consistent, dependable source of guidance and instruction by his parents, Reg and Deanna, and support from his wife Barb and two daughters, Bethany and Bailey. Their contribution to this book is heartily proclaimed.

Beef Practice:
Cow–Calf Production Medicine

1
Introduction

Peter J. Chenoweth and Gary Rupp

The domestication of cattle can be traced back more than 10,000 years (Dunlop and Williams 1996), a period over which cattle have played a pivotal role in sustaining the physical, social, and spiritual needs of many cultures. Cattle played an important role in facilitating civilization by helping early humans to make the transition from small bands of nomadic hunters to larger, self-sustaining settlements. Today, the significant contribution of cattle to human history and human development is reflected in many aspects of our everyday culture and vocabulary.

In turn, the major impetus for the establishment of veterinary colleges in most countries has been their perceived potential to assist and protect animal agriculture (Prichard 1993) as well as other animals of utilitarian value to society (Rivas et al. 1996). Indeed, the stated purpose for the establishment of the first modern veterinary college, at Lyon, France, in 1762 was, "to open a school where would be taught publicly the principles and the method to cure the diseases of livestock" (Dunlop and Williams 1996, p. 321). It is perhaps ironic that the disease threats of the time were major epizootics such as rinderpest and cattle plague, whereas today we are grappling with major livestock disease concerns such as bovine spongiform encephalopathy and foot-and-mouth disease. In our rapidly changing world, the future of the livestock industry and of the veterinary profession are even more interdependent. World population predictions indicate that the demand for animal protein will double or triple in the next 40–50 years. Growing numbers of human beings, and their relatively increasing wealth, will drive this demand. Veterinarians will play a pivotal role in helping livestock industries meet this challenge while safeguarding both health (animal and human) and the environment. This role was dramatically illustrated by efforts of the beef industry to maintain consumer confidence in the wake of the recent (2003) discovery of bovine spongiform encephalopathy cases in both Canada and the United States, in which veterinarians figured prominently.

Veterinary roles within the beef industry are as diverse as the industry itself. Throughout the world, and even within the continental United States, great variation exists in beef cattle breeding operations in terms of management, environmental setting, degree of intensification (and capitalization), types of cattle used, productivity, and profitability. Not surprisingly, great variation also occurs in those health and production strategies that practitioners and producers employ. Such variation poses difficulties in defining universally applicable practical health and management programs. However, commonalities imposed by cattle behavior, diseases, physiology, and ontogeny allow a generic framework to be established, on which certain health and production principles may be imposed.

For cow–calf producers to survive and thrive, they must achieve optimal production per unit asset, resource, or input within an increasingly complicated and economically challenging context. The role of veterinarians in helping producers to achieve this aim is multifaceted and evolving, just as the animal industry itself is also evolving. Current veterinary contributions to cow–calf production range from care for, and treatment of, individual animals to responsibility for the entire production unit. For examples of dramatic recent changes in livestock industries—and their associated veterinary consequences—one might look at the modern swine and poultry industries, which have become very intensive, concentrated, and efficient within a relatively short time span (Henry 1997). In fact, swine practitioners may well have been the pioneers in species-specific practice (Schultz 2000), at least in the food animal realm.

Economic implications in cow–calf production are paramount, yet they must be considered within the context of animal and environmental well-being. The veterinarian must reconcile the biological requirements of livestock with the production and economic goals (as well as financial survival) of producers and must place these goals within the contexts of cyclical market changes, environmental vagaries, and societal perceptions. It is fortunate that good production, animal welfare, and environmental stewardship have more in common than in conflict.

Veterinarians are particularly well equipped to help producers navigate the many challenges they face. For example, veterinarians generally have high credibility with producers, have wide access to different operations and operators, and are able to coordinate with a network of appropriate professionals and services. Veterinarians interact with individual producers more than do all other providers of ranch management advice, they represent the most reliable source of information on health and production matters, and they are the second most consulted source of nutritional advice (National Animal Health Monitoring System 1994).

In addition, veterinary education is comparative and typically runs the gamut from biology and pathology to production. Veterinarians are trained to "gather information and solve problems" (DeGroff in Avery et al. 1998, p. 1681). They have a unique, analytical perspective that is useful in synthesizing and evaluating knowledge while bridging the divide between basic science and animal systems. In addition, because of their biological training and public credibility, they have a valuable role to play as the "honest brokers" in resolving production versus welfare issues.

Despite such favorable indicators, many rural veterinarians are involved only marginally in production management consultation (Wikse 1995).

There will always be a need for conventional rural mixed veterinary practice; however, such practices are decreasing in number. Demographic and economic forces are at work on many existing practices, which are responding either by closing their doors altogether or by evolving into strictly companion animal practices. Other practices, however, are changing direction to provide a range of services that is more in tune with production management concepts. Veterinarians today have a tremendous opportunity to affect herd profitability, especially with larger herds (more than 500 head of cattle; Leverett 1996). An evolutionary process has been at work whereby the focus of veterinary services has moved from the individual animal, to the herd, and subsequently to the whole-farm or whole-ranch enterprise. This last category of service has been described as consultative and information-based (DeGroff 1997). Similarly, the marketing of services is moving from those that are "procedure based" (otherwise termed "marketable skills" [Chenoweth 1989]) to those that are "package-pricing based" (Lechtenberg in Avery et al. 1998, p. 1681). Newly graduated veterinarians usually do not have the full repertoire of knowledge and skills to successfully make such transitions without extra training. However, a newly graduated veterinarian does possess a comprehensive, comparative education, which, with the additional development of appropriate marketable skills, should provide an effective base from which to evolve into a "species specialist" who offers production-oriented services within a particular livestock industry (Blood 1985).

This extra training is often within areas and disciplines that have not been regarded as part of the traditional veterinary realm. Modern production-oriented veterinarians need to understand economics as well as epidemiology, and they must be able to apply these disciplines to matters of health, management, nutrition, and genetics. Veterinarians have been urged to take leadership in areas such as animal welfare, food safety, and environmental effects (McGrann in Avery et al. 1998). Related areas of potentially valuable training for veterinarians include communication skills, employee training, production and financial analysis, record keeping, problem solving, business planning, and monitoring (DeGroff in Avery et al. 1998, p. 1681), as well as risk assessment.

A brief list of qualifications for a professional service provider should include focused expertise, good problem-solving capabilities, and useful relationships with other resource providers and specialists. He or she should elicit a good sense of the producer's personal goals and exhibit a sincere desire to help the producer achieve them. As a group, veterinarians currently working with the beef cattle industry generally possess many appropriate qualifications and have provided excellent health management advice and service in the past.

More recently, some practitioners have progressed to supplying a broader range of services in reaction to industry changes and increasing demand. In some cases, practitioners have been through well-rounded educational programs, which are essential for providing specialized services. Here, focused and relevant continuing education is necessary because production-oriented veterinarians generally do not obtain all of their essential tools from the professional DVM curriculum. Rather, they must augment those skills obtained through the DVM program with a combination of preprofessional, professional, and postprofessional training, as well as with practical experience. Beef cattle veterinarians cannot be specialists or experts in all disciplines, and they should not become discouraged because of this. Rather, with the motivation of accruing advanced training in important and relevant specialty areas, they should provide an invaluable resource in assisting producers to resolve current and future challenges.

INTEGRATING NEEDS AND SERVICES

Future needs in the beef cattle industry provide an opportunity for the veterinary profession to expand its valuable services beyond traditional individual animal care, and even beyond "production medicine," which implies a "health problem-oriented" approach. Today, there is an impetus toward "production management" within a challenging environment, driven not only by economics but also by additional consumer demands for a safe, wholesome, humanely raised product. A veterinarian who aspires to provide extended services today must make a lifelong commitment to intensive interdisciplinary education, information management, managerial skills, and problem solving on a whole-ranch or whole-farm basis.

In addition, it is very important for veterinarians to understand and combine animal performance and business principles within a production management approach. To succeed, veterinarians must employ initiative, leadership, and interpersonal skills, as well as strong networking relationships with professionals in complementary disciplines. The bottom line should be improved management/production practices, performance, and economic sustainability for beef cattle producers in accord with societal values.

MARKETING PRODUCTION MANAGEMENT SERVICES

Successful marketing of extended services to producers will require the veterinarian to answer two critically important questions: What motivates a producer to use specific services, and what motivates the veterinarian to offer them?

First, why would a producer want to take advantage of extended veterinary services? An obvious answer is increased profit for the producer, although the reason is usually much more complicated than profit alone. Producer motives may include independence, sustainability, recognition, respect, and a lifestyle that contributes to fulfillment and happiness. Other factors, such as familial relationships and dynamics, as well as "off-farm" goals and priorities may further complicate the picture. Thus, veterinarians must identify the reasons why clients might desire the services they offer and offer services that allow the producer to meet his or her goals. This calls for the veterinarian to have marketing skills, particularly when new and innovative services and concepts are involved (DeGroff 1997).

However, we should also consider what motivates the veterinarian—or supplier—to offer extended or enhanced services. His or her motivation may be even more difficult to pinpoint because it may not be related directly to income. Possible factors include the personal satisfaction of problem solving and making a positive contribution and of working with producers and their cattle in a desirable environment, as well as continuing to offer a viable benefit within a changing industry.

The four "P's" of marketing should be kept in mind: product (in this case, service), performance, place, and price. To fulfill these basic marketing requirements, veterinarians should start by designing a "product" that the beef producer perceives as being of value. Here, success depends on frequent and open communication between the supplier and the user. Such communication should allow the veterinarian to become aware of the producer's needs and priorities. In addition, the producer should gain an enhanced knowledge of the cost benefits of the veterinarian's services. The important message underlying this process may be summarized in the following quotation:

> Never expect anyone to engage in a behavior that serves your values unless you give them adequate reason to do so. (Dwyer 1996)

The second "P" is performance. The services offered by the veterinarian must be at or above the level the client determines to be acceptable. This is crucial. If the service does not meet the expected value, the client will look elsewhere for help.

The third "P" is place or, in this case, convenience in how the service is delivered. If the veterinarian can deliver a service within the confines of daily routines and operational management, then the service will be perceived as having greater value and a better chance to succeed. For example, if a veterinarian provides an information service such as herd records, he or she should ensure that the owner receives relevant herd information in a timely and convenient manner, such that the producer will benefit from the information.

The last "P" is price. The cost of a service is often a key factor in its initial marketing, and its value over time will determine long-term demand. The veterinarian will greatly benefit by researching the "going rate" for the service he or she wishes to offer and by "marketing" it as an investment rather than a cost. Pricing a service either too high or too low may cause the producer to perceive it as either impossible to afford or lacking in worth. This presents real challenges as well as opportunities when a "new" market is being created.

Successful marketing of services involves convincing and persuading others. This, in turn, depends greatly on the provider's leadership and interpersonal (communication) skills. Marketing experts describe this as "power." Power in this case is the ability to influence people to do what one would like them to do. From a professional service standpoint, two essential prerequisites for the possession of power are competence and confidence. A veterinarian must first be highly competent in the services he or she plans to offer. This leads to confidence. If one lacks expertise and experience (i.e., competence), it is difficult, if not impossible, to act with confidence. In turn, the veterinarian's competence and confidence will serve to build producer trust; an important factor for initiating extended services.

The last, and possibly most important, concept of marketing extended veterinary services is related to risk. Veterinarians who are motivated to provide production management services must take the necessary steps to be knowledgeable, competent, and confident. This represents a risk because of the personal time, expense, and sacrifice that must be committed. An additional risk lies in the real possibility of failure, rejection, or embarrassment. These personal and financial risks may outweigh the desire to offer extended services. However, such challenges may stimulate a veterinarian to new levels of commitment and service.

In this book, we attempt to describe the essential building blocks in developing enhanced and extended services for cow–calf producers. This involves bringing together a number of different threads, many of which receive scant attention in the traditional veterinary curriculum. The concept of expanded and enhanced veterinary services to animal agriculture relies heavily on a team approach. No single individual can possibly hope to accrue all of the necessary knowledge to appropriately advise on all aspects of a modern agricultural enterprise. Veterinarians have a unique knowledge base on which to build, so that they can become an invaluable resource for producers. In this book, we hope to season this wealth of knowledge with both an awareness of veterinary opportunities in a dynamic and changing world and wisdom in knowing that these opportunities are often best tackled via collaboration and networking with other professionals. Although directed primarily toward production management concepts, this book aims to include aspects of the full spectrum of services to cow–calf operations, from those that fall under the mantle of traditional services to those that are more theoretical and possibly futuristic. Throughout the text, however, we have attempted to maintain a theme based on good principles of livestock production that have stood the test of time. Here, the objective has been to integrate a wealth of information (production, health, genetic, nutritional, welfare, environmental, economic, and marketing) so that it is readily digestible, understandable, and usable by students, veterinarians, producers, decision makers, and advisors. Hopefully, the collective enthusiasm and optimism displayed by the different contributors will act as a "catalyst" for expanded and enhanced veterinary services for cow–calf producers.

REFERENCES

Avery, A., T. DeGroff, J. McGrann, G.R. Monfore. 1998. Summary, Symposium on Opportunities for Veterinarians in Agribusiness, Oct 17–18, 1998. *Journal of the American Veterinary Medical Association* 213:1679–1687.

Blood, D.C. 1985. The role of the species specialist in relation to the future of veterinary preventive medicine and animal production. *Proceedings, International Conference on Veterinary Preventative Medicine and Animal Production,* Melbourne: University of Melbourne. pp. 118–120.

Chenoweth, P.J. 1989. Marketable skills: Beef cattle services that produce economic gain. *Large Animal Veterinarian* May/June:26–28.

DeGroff, T. 1997. Beef cattle practice: Are veterinarians headed in the right direction? *Food Animal Compendium* August:S162–S165.

Dunlop, R.H., D.J. Williams. 1996. *Veterinary Medicine: An Illustrated History*. St. Louis: Mosby.

Dwyer, C. 1996. *Managing People*, 2nd ed. Dubuque, IA: Kendall/Hunt.

Henry, S. 1997. Missed areas of consultation in the beef cattle production cycle. *The Bovine Proceedings* 30:79–82.

Leverett, G. 1996. The role of food animal practitioners with grazing beef cattle. *Journal of the American Veterinary Medical Association* 209:2006–2009.

National Animal Health Monitoring System. 1994. Part II: Beef Cow/Calf Reproductive and Nutritional Management Practices. Beef CHAPA. *Cow/Calf Health and Productivity Audit*. Fort Collins, CO: U.S. Department of Agriculture.

Prichard, W.R. 1993. Comments on the evolution of veterinary medical education in the US and Canada: Some lessons from history. *Journal of Veterinary Medical Education* 20:53–55.

Rivas, A.L., M.T. Correa, M. Memon, et al. 1996. Functions of veterinary colleges and orientations of professional practice in the Americas. *Journal of the Veterinary Medical Association* 208:1630–1635.

Schultz, R. 2000. Preparing yourself and your clients for the 21st century. *Large Animal Practice* 21:32–37.

Wikse, S.E. 1995. Getting started in beef cow/calf production management consultation. *Compendium in Veterinary Medicine for the Practicing Veterinarian* August:S15–S19.

2
Cow/Calf Production Principles

Peter J. Chenoweth

INTRODUCTION

Cow/calf producers generally sell their product in a competitive market, where price is largely beyond their control (Dewell et al. 1999). This is because the selling price is affected not only by product weight and quality but also by supply and demand. If demand is high and supply is low, prices tend to be higher, and producers try to boost production accordingly. This does not happen immediately, and the resultant surge in supply may not occur for several years. In the meantime, other factors can intervene to change demand, including consumer preferences, prices of alternate goods, exports, or health concerns. Interruptions to predicted supply can occur as a result of adverse climatic conditions, disease outbreaks, and so on. The end result is that the cattle industry appears to be doomed to cyclical fluctuations in production and price (see chapter 12).

This recurring cattle cycle is fueled by the attempts of producers to take advantage of a market that pays for pounds (or kilograms) of calf produced, and it is exacerbated by a general lack of financial and economic acumen among producers. An effective advisor for cow/calf producers should thus understand both good financial and good production principles, as well as the relationships between them (McGrann and Wikse 1996).

The average costs of producing a calf in the United States (U.S. Department of Agriculture estimates) tripled in the 22 years from 1972 to 1994 (from $114 per breeding, or exposed, cow to $412, respectively). Differences occurred in geographic and environmental conditions and in their effects on breed types, feed types, weaning weights, and the resultant cost of gain. Differences in efficiency also occurred between high- and low-cost producers, as reflected in cow maintenance costs (Table 2.1).

Market forces are mainly beyond local control. What is under local control is producer efficiency, which includes reducing those costs associated with optimal production levels. In turn, optimum production implies a balance between production levels, input costs, welfare considerations, and sustainability.

A useful starting point is to attempt to define threshold production levels for a given region or environment. As an example, Table 2.2 provides production indicators and risk categories appropriate for the Midwest in the United States, a temperate zone populated primarily by *Bos taurus* cattle, particularly those of British breeds. In other regions and countries, such indicators will need to be modified to accommodate those economic and environmental considerations that apply.

Table 2.1. Cow maintenance costs for high- and low-cost producers in different regions of the United States.

Region	Northwest	Southwest	Midwest	Southern plains	Southeast
Low-cost producer	$273	$254	$239	$244	$224
High-cost producer	$399	$358	$340	$365	$342

Cattle-Fax Annual Cow-Calf Survey Data for 1997–1998. Available at http://www.cattlefax.com

Table 2.2. Production indicators and risk categories.

Risk Level and Category	Low	Acceptable	Unacceptable (action indicated)
Confirmed pregnant			
Pastured	>95	90–95	<90
Rangeland	>90	80–90	<80
Abortions (per cow pregnant)	<1	1–3	>3
Calving % (live calves/cows pregnant)			
Heifers	>98	94–98	<94
	<5	5–15	>15
Adult cows	<2	2–5	>5
	>95	85–95	<85
Dystocia	<2	2–7	>7
% Calving first 60 days			
Pastured		0.5–3	>3
Rangeland	<0.5		
Calf death loss (<14 days)	>90	85–90	<85
Calf death loss (>14 days)	>85	70–85	<70
% Weaned calf crop			

Adapted from Nebraska Veterinary Medical Association Profitability Guidelines for Livestock Production 1990.

Table 2.3. Standardized performance analysis measurements for low-, medium-, and high-profit herds.

Criterion	Low Profit	Medium Profit	High Profit
Pounds weaned/cow exposed	413	455	455
$ Total income per breeding female	391[g]	423[g]	495[h]
$ Total cost per breeding female	638[d]	387[e]	270[f]
$ Net income per breeding female	–247[a]	36[b]	225[c]
$ Total investment per breeding female	1538[g]	2308[h]	1397[g]
% Return on assets	–15.5[a]	2.88[b]	18.16[c]

Adapted from Dunn 2002.
[abc]Means in same row differ ($P < .01$).
[def]Means in same row differ ($P < .05$).
[gh]Means in same row trend ($P < .10$).

CHARACTERISTICS OF HIGH-PROFIT/LOW-COST PRODUCERS

Production is not synonymous with profit. Unfortunately, the majority of cow/calf enterprises in the United States do not make a profit (McGrann and Parker 1999). Here, a logical starting point for producers—and advisors—is to identify costs of production (i.e., the actual cost of producing a calf, or dollars per kilogram of beef). This allows an estimate to be made of the economic consequences of production-related recommendations (Larson and Pierce 1999). Those practices associated with high profitability are identifiable, as are those associated with low profitability. In the United States, the use of standardized performance analysis data has facilitated the generation of such comparative data for cow/calf operations. For example, Table 2.3 illustrates differences in production and financial characteristics of U.S. herds that were categorized into low, medium, and high levels of profitability (Dunn 2002).

Such figures are most revealing. For example, although operations differed in profitability, this variation was not reflected in differences in weaning

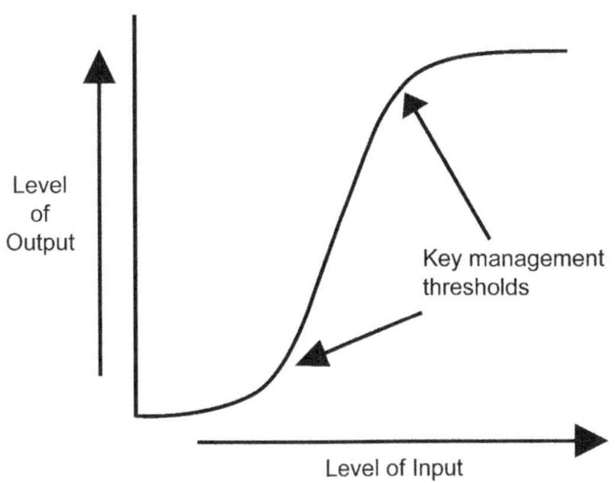

Figure 2.1. Classic input/output relationships.

Table 2.4. Factors associated with increased profitability in cow/calf operations.

Factors	Profitability Relationship
Feeding harvested feeds	–
Capital invested/cow	–
Business management	+
Size of operation	+
Pounds weaned/exposed female	+
Age at first calf	+
Crossbreeding	+
"Efficiency" (especially technical)	+
Product "quality"	+

Adapted from Larson and Pierce 1999.

weights, a result similar to those found in other studies, which have shown little difference in "production" criteria between low-profit and high-profit operations (Sprott et al. 1998). In addition, total investment per breeding female did not differ between low-profit and high-profit operations. However, to achieve equivalent weaning weights, low-profit operations spent over twice as much per breeding female as did high-profit operations ($638 vs. $270, respectively). In turn, net income per breeding female differed by nearly $500 in the two groups, and percentage return on assets varied by approximately 33%. In Texas, profitable operations emphasized overall production and financial goals, whereas unprofitable operations tended to concentrate on singular aspects such as stocking rates and calving season length (Sprott et al. 1998). The conclusion must be that efficiency and sound economic management are more important in determining profitability in cow/calf operations than is production per se.

Input and output relationships may be depicted as a production function curve (Figure 2.1). Here, different relationships can alter the shape of the curve. For example, the slope of the curve representing a production function could be fairly steep for reproductive traits, or relatively flat for a trait such as weaning weight. Identifying relevant traits and their relationships for different geographical regions is important and necessary.

This indicates that there is an optimal level of input to achieve a desired output, beyond which further input ceases to be cost effective, supporting the view that maximum production levels are not necessarily associated with positive returns. Factors that have been positively associated with profitability (positive or negative) in cow/calf operations in the United States include those shown in Table 2.4.

In Kansas, a number of additional management practices associated with increased beef herd profitability were identified by Engelken et al. (1991) as follows: fly control, calf creep feeding and growth-promotant usage, precalving parasite control in cows, and improved calf management in general. As a word of caution, it should be noted that improvements in the production and profitability of beef cow/calf operations often do not occur until 2–4 years after changes are implemented in managerial practices, although exceptions occur (Wikse et al. 2002).

"GOOD" COW/CALF PRODUCTION PRINCIPLES

As indicated above, a number of managerial practices are known to increase productivity in cow/calf herds when they are implemented in a practical and economical manner. These include a restricted breeding season, an "optimal" calving season, and a breeding season evaluation. A good heifer replacement program is crucial, as are proper nutrition and overall good herd health. Prior evaluation of bulls, an effective crossbreeding program, and good record keeping are also all essential.

Although such management tools are recognized as beneficial to herd productivity and profitability, many producers are not taking advantage of them (Dewell et al. 1999). In fact, the low level of utilization of such knowledge has been identified as the most important constraint to the productivity of the

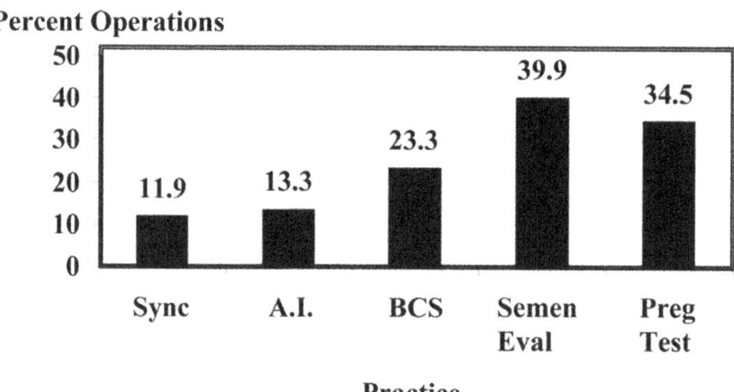

Figure 2.2. Adoption of management practices by U.S. cow/calf operations.

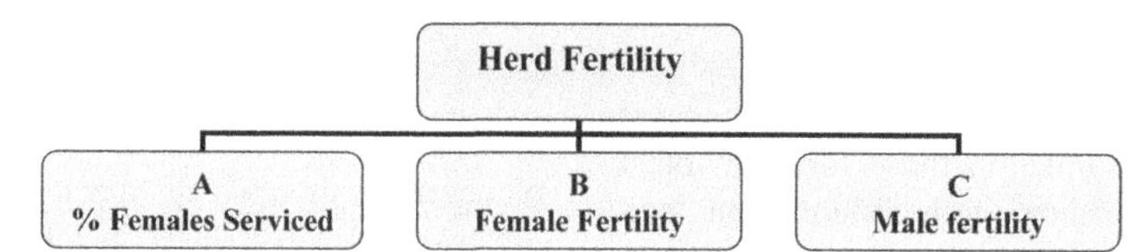

Figure 2.3. Herd fertility calculation. Percentage of females serviced—influenced by cyclicity, nutrition, puberty, postpartum interval, heat detection/bull libido, social/behavioral factors, and management (bull ratios, pasture management, breeding-season length, etc.). Female fertility—influenced by genetics, disease, pathology, postpartum interval, and management. Male fertility—influenced by sperm production, semen quality, bull ratios, libido, soundness, social interactions, inseminator skill, and semen handling. The final pregnancy rate, or herd fertility, is a product of all three factors (i.e., $A \times B \times C = \%$ pregnancy).

U.S. beef cow/calf industry (see Figure 2.2; Wikse et al. 1998). This perhaps explains why the weaning rate of calves has remained virtually static in the United States over the last several decades.

Examples of poor producer adoption rates of good management principles and practices in the United States are evident in the relevant National Animal Health Monitoring System (NAHMS) reports. For example, in 1993, nearly 53% of operations did not have a "set" breeding season, a figure that apparently did not change over the next 4 years (NAHMS 1998b). The adoption rate increased with herd size, and it was also higher in operations that were primarily dependent on cattle for income. Nearly 80% of cattle enterprises relied on handwritten record systems, and approximately half of the operations employed individual cow identification. The use of artificial insemination (A.I.) in beef herds remained relatively static (<10%) between 1993 and 1997.

The reasons for such relatively low adoption rates are many and complex. Beef producers tend to be conservative. Many are relatively small or part-time operators or have mixed enterprises in which the cattle operation shares priorities with other goals and activities. In the United States, approximately two-thirds of beef-producing operations do not rely on cattle as their primary income source. In addition, regional differences occur in outlook, social pressures, and economics.

Despite such considerations, veterinarians and advisors should be in a unique position to convince clients to adopt cost-effective managerial changes, improve efficiency, and assist clients in implementing these changes.

A number of principles and practices associated with good cow/calf production are discussed in this chapter (including restricted breeding seasons and whole-herd pregnancy diagnosis), and other aspects (e.g., bull evaluation and management, nutrition, and so forth) will be discussed in subsequent chapters. As many of these principles are associated with reproductive measures, it is useful to consider the reproductive process as one that has many contributing components. Here, herd fertility is recognized as a factorial that is influenced by a number of factors, as illustrated in Figure 2.3.

REPRODUCTIVE PERFORMANCE

Some definitions are necessary at this juncture. The descriptions of measures of reproductive performance given in Table 2.5 are used for the standardized performance analysis program within the United States and were reached by collaboration between the National Cattlemen's Beef Association and Texas A&M University.

RESTRICTED BREEDING

In general, any beef production system in which animals are required to be grouped for managerial purposes (e.g., weaning, vaccination, and marketing) will benefit from a restricted breeding season. Mother Nature imposes her own prerogatives such that, even in feral herds, a considerable degree of seasonality occurs in both breeding and calving (Spitzer 1986). Thus, in most regions of the world, there is an optimal period for females to calve, suckle, and rebreed. This period is mostly related to nutritional opportunity, although other environmental factors such as heat stress and parasite populations may play roles of varying importance. Producers have traditionally planned for females to calve at this optimum period, as the cows then tend to breed back faster—and their calves grow better—than those that calve at less opportune times.

It is recognized that in some parts of the world, such as Northern Australia, the level of managerial input required to implement a severely curtailed breeding season would be logistically difficult to implement, and most probably uneconomic. However, where restricted breeding seasons are feasible, their implementation is a major factor in achieving optimum production levels. Arguments in favor of restricted breeding seasons (<80 days) include enhanced production potential, the opportunity to work in concert with favorable environmental factors, and a more homogeneous calf crop. In addition, a restricted breeding season allows the producer to perform prebreeding management procedures, concentrate the

Table 2.5. Selected definitions of reproduction and performance.

Term	Definition
Pregnancy percentage	Number of females exposed diagnosed as pregnant/number of females exposed that are pregnancy tested × 100[1]
Pregnancy loss percentage	Number of females diagnosed as pregnant which fail to calve/number of females diagnosed as pregnant × 100[2]
Calving percentage	Number of calves born/number of exposed females × 100
Calving distribution	Cumulative number of calves born by 21, 42, and 63 days of the calving season, and those born after 63 days[3]
Calf death loss, based on exposed females	Difference between pregnancy percentage and total live calves based on exposed females[4]
Calf death loss, based on calves born	Number of calves dead/number of live calves born
Calf crop, or weaning percent	Number of calves weaned/number of exposed females × 100
Average weaning weight	Total weight of weaned calves/total number of calves weaned
Pounds weaned per exposed female	Total pounds of calf weaned/total number of exposed females
Female replacement rate	Raised replacement heifers exposed for first calf + purchased replacement heifers and exposed breeding cows/number of exposed females × 100

Adapted from McGrann and Parker 1999.

[1] Adjustments to exposed female numbers are made as follows: Exposed animals that are sold or otherwise removed between breeding and pregnancy diagnosis are subtracted; exposed animals purchased during that period are added. All death losses of exposed females should remain within that group.

[2] The numerator should include females that abort between pregnancy test and calving. Purchased pregnant females should be added to the divisor, whereas pregnant females sold should be subtracted.

[3] The assumption is made that the first 21-day calving period should start 285 days following introduction of bulls to the mature cow herd. If this date is unavailable, then counting starts at the time of calving of the third mature (3 years and older) cow. All calves delivered, dead or alive, are included in this analysis.

[4] Calf death loss should include calves lost at birth plus calves that die before weaning.

calving season, plan for strategic nutrition enhancement, and improve monitoring. Other general advantages include early problem detection, an improved female replacement program, programmed herd health, and improved culling procedures.

Where a survey indicated that most U.S. calves (64%) were born within a 3-month period (February, March, and April) during spring (NAHMS 1998a), these combined advantages were estimated to provide an increased return of $50/cow (Toombs 1996). Including January and May increased the percentage of calves born to 80%, reflecting the outcome from a breeding season that occurs predominantly in late spring and early summer. This survey showed, however, that there was no month of the year in which fewer than 10% of operations had calves born.

Although this indicates room for considerable improvement, it should not be assumed that the adoption of a severely restricted breeding season is advantageous in all situations and environments, as was already noted for Northern Australia. For example, in Nebraska, comparison of breeding season lengths of 30, 45, and 70 days for crossbred females showed that their highest productivity coincided with the longer breeding season (Deutscher et al. 1990).

In some situations, short breeding seasons may result in unacceptably low calving rates (Larsen et al. 1994). The reasons for this include the extra pressure that such systems impose on "normal" physiologic events. For example, if a beef female is pregnant for 284 days, then she has 81 days to recycle and breed to achieve a 12-month calving interval. If she has no problems at parturition and nutrition is good, she will take approximately 35–70 days to return to cyclicity. If cyclicity is resumed at, for example, 40 days, this leaves 41 days (or two to three estrus cycles) to become pregnant. Under optimal conditions, she will have an approximately 60% chance of achieving pregnancy at each estrus. Thus, her chance of becoming pregnant in time to fulfill a 12-month calving interval is either 84% (two cycles) or 94% (three cycles). The fact that beef females undergo a variable, and often extended, period of anestrus following parturition can represent a major constraint on conception rates during the breeding season. Even greater constraints occur in tropical environments, where genotype and environment interact to further delay postpartum return to estrus. Producers in such areas often compensate for this by employing extended breeding seasons. Other managerial options to reduce the postpartum return-to-estrus period are discussed in the section on weaning strategies.

The relationship between breeding season length and pregnancy rate is affected by one or more of the following:

- Postpartum return to estrus (period from calving to first estrus)
  Conception rate per estrus exposed (pregnancy rate per cycle)
- Gestation length (length of pregnancy).

Gestation length is also influenced by calf sex, with male calves generally having a longer gestation (by several days) than female calves. Breed type can also cause considerable variation (up to 19 days) in gestation length (Larsen et al. 1994). A longer gestation length in certain breeds (particularly *Bos indicus*) reduces their window of opportunity to rebreed in a timely enough manner to achieve a 12-month calving interval. If both gestation length and postpartum return to estrus are sufficiently extended, then even greatly increasing the breeding season length will not compensate for lowered reproductive rates (Larsen et al. 1994). There are a number of potential disadvantages of restricted breeding seasons, including lost production, difficulties in coordinating calendar and season, the need for close monitoring, increased importance of good herd health, and the fact that late calvers, if not culled, will often miss years.

Moving cattle breeding operations from an extended breeding season to a restricted one can present real challenges. If done too rapidly, a severe drop in production may result. Associated changes in culling patterns can alter herd dynamics such that it may take years for optimum production levels to be established. Prior consultation with clients should include appropriate advice on realistic time frames for equilibrium to reoccur. Use of computer-generated "what-if" scenarios (e.g., the Decision Evaluator for the Cattle Industry, or DECI; http://www.marc.usda.gov/) can be useful in mapping different strategies and predicting outcomes. A relatively safe strategy is to reduce bull presence in the herd in increments over a number of years, so that the desired breeding season is gradually achieved. This should coincide with the steady removal of late-bred females from the female herd and with an emphasis on heifer breeding within a tight schedule (possibly using estrus synchronization; see chapter 10). For year-round breeding herds, the initial strategy should be to first remove bulls during the period when experience and records indicate that little breeding activity occurs. Bull removal can then be extended on a realistic timetable.

Strategic Pregnancy Testing

Pregnancy testing represents an important management tool in beef breeding herds. It is a "marketable skill" that allows veterinarians to gain access to, and credibility with, cow/calf operations. To be most effective, it should be much more than a simple exercise of sorting female cattle into pregnant and nonpregnant groups (Hamilton 1997). Additional information such as projected calving dates and patterns (as described below) can render this service more attractive to producers (Toombs 1996, Australian Association of Cattle Veterinarians 1998).

The herd pregnancy diagnosis represents an important starting point for beef herd diagnostics and advice and should provide an invaluable basis for informed managerial decision making. For instance, analysis of group patterns ("breeding season evaluation") and sorting of animals into groups for specific purposes such as strategic feeding, calving supervision, culling, or rebreeding are valuable services the veterinarian can provide. Other advantages of herd pregnancy diagnosis include selection of replacement females, culling against infertility, and identification and treatment of other problems.

To make the best use of pregnancy test information, it is helpful to determine the distribution of pregnancies within the herd (and its subgroups) and to compare this distribution with herd averages, both actual and expected. This helps determine whether or not a problem exists and, if so, where to seek the cause. Such analyses can also help to indicate potential solutions. Further discussion of this process occurs in the section on breeding season evaluation that is found later in this chapter.

When used in conjunction with a restricted breeding season, the herd pregnancy diagnosis is best done relatively early after bull removal (e.g., 6–8 weeks) but late enough so that early pregnancies are not missed. Veterinarians usually perform this task via transrectal palpation of the female reproductive tract, a technique that has been adequately described elsewhere (BonDurant 1986, Australian Association of Cattle Veterinarians 1998). With this method, the greatest accuracy in aging the conceptus occurs between days 35 and 65 of pregnancy, with precision decreasing as gestation proceeds. However, useful accuracy is obtainable at most stages of pregnancy, with the exception of the very early stage (less than approximately day 35).

It is helpful to emphasize the importance of good, safe, working facilities for this task (Grandin 1993, Australian Association of Cattle Veterinarians 1998). These conditions help to safeguard both the operator and animal and to facilitate accurate results. In addition, awareness of the biomechanical principles involved with this task (Australian Association of Cattle Veterinarians 1998) can prevent chronic injuries to palpators. In a U.S. survey, a significant number of bovine practitioners reported traumatic injuries (especially in the shoulder, elbow, and wrist areas) that were most probably associated with transrectal palpation of cattle and that could probably be alleviated or avoided by improved technique or facilities (Cattell 2000). Recommendations to reduce the prevalence of cumulative trauma disorders in veterinarians who are regularly palpating cattle are available (Australian Association of Cattle Veterinarians 1998).

The Australian Association of Cattle Veterinarians coordinates a national accreditation scheme for cattle pregnancy diagnosis (Australian Association of Cattle Veterinarians 1998) that aims to promote excellence in the skills of pregnancy testing and to provide accountability. Accredited veterinarians are issued plastic, wrap-around tail tags that are used to identify four classes of tested cattle. These categories are NDP (not detectably pregnant), CATA (open, or up to 3 months pregnant; this tag is only used in Victoria, and it indicates category "A"), U4 (up to 4 months pregnant), and O4 (>4 months pregnant).

Ultrasonography—particularly real-time B-mode technology—has shown rapid development and acceptance for pregnancy diagnosis within the beef industry (see chapter 10). Ultrasound has advantages beyond those of transrectal palpation, including the ability to accurately (and conclusively) diagnose early pregnancies (before day 30), to determine fetal sex (after day 55), and to help diagnose problems in the female reproductive tract. In addition, ultrasound can assess fetal viability through heartbeat, and the operator is less apt to "miss" detecting twin fetuses than when employing transrectal palpation. Ultrasound images of actual pregnancies can also be useful as a marketing tool for purebred and specialty cattle operations. Conversely, the equipment is expensive (although the price is steadily dropping) and susceptible to damage, and extra training is necessary for proficiency to be obtained.

Breeding Season Evaluation

Breeding season evaluation is a useful concept for the assessment of the reproductive performance of cow herds (Spire 1986). This evaluation includes assembling relevant information, analyzing

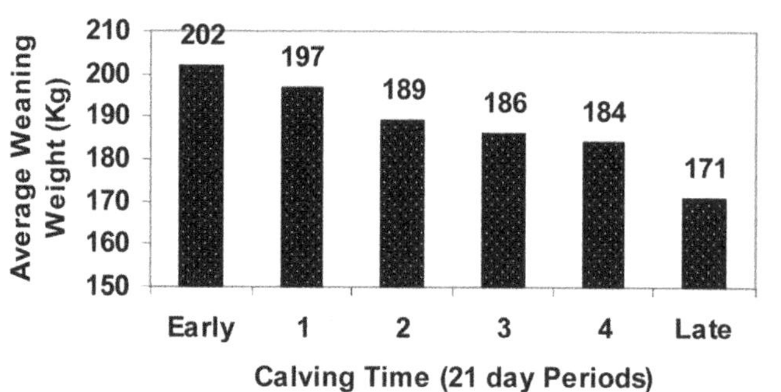

Figure 2.4. Effect of cow calving time on average weaning weights (in kilograms).

Table 2.6. Hypothetical pregnancy patterns per 21-day breeding period in three herds.

Periods	Herd A Pregnancy (%) per Period	Herd B Pregnancy (%) per Period	Herd C Pregnancy (%) per Period
First 21 days (0–21)	62	15	23
Second 21 days (22–42)	24	25	12
Third 21 days (43–63)	9	28	15
Fourth 21 days (64–84)	—	14	21
Fifth 21 days (85–105)	—	8	19
Sixth 21 days (106–126)	—	5	5

Dash indicates no more pregnancies.

and interpreting these data, and making recommendations for improvement.

Breeding seasons may be evaluated in a number of ways. Pregnancy (or calving) data may be analyzed on the basis of timing (e.g., 21-day estrus periods), breeding groups, female age or parity, nutritional opportunity (e.g., body composition score, or BCS), pasture, bull group, or any other identifiable category. Accuracy of input data is important if such analysis is to be valid (Mossman and Hanly 1977, Traffas and Engelken 1999). Useful information that can facilitate an accurate breeding season evaluation includes breeding herd inventory, sires used, pastures available, bull-to-female ratio, breeding season dates, and pregnancy rates. In addition, it is helpful to know expected calving dates and distribution, herd averages and variations, and female factors such as age/parity, breed, body weight, frame score, BCS, and lactation status.

Calving distribution is also an important concept, as the timing of calf births influences their weaning weights if mass weaning is done at a predetermined date (Figure 2.4).

The interpretation of pregnancy (or calving) patterns may be illustrated with reference to pregnancy patterns in three herds (Table 2.6). In this diagrammatic representation of the breeding performance of three herds (Figure 2.5), all herds ultimately achieved the same percentage of pregnant females (95%). However, differences in the paths taken to achieve this result provide not only considerable differences in economic returns but also valuable diagnostic clues.

Herd A

The pattern of pregnancies occurring with herd A (with 62%, 24%, and 9% of pregnancies occurring in the first, second, and third 21-day periods of the breeding season, respectively) is close to ideal. To achieve such a pattern requires a conception rate (i.e., one cycle pregnancy rate) of approximately 60%. In turn, this requires that most females are cycling at the beginning of breeding and that they are exposed to an adequate number of active, fertile bulls.

Herd B

Herd B illustrates a problem of delayed conceptions, in concert with an inconsistent pattern. Delayed conception may occur for several reasons, as follows.

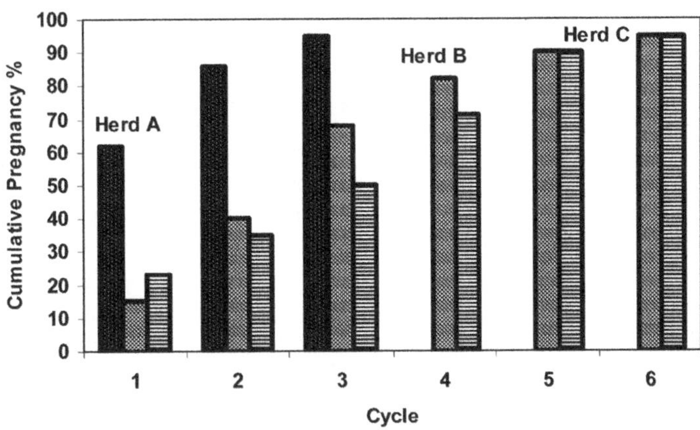

Figure 2.5. Hypothetical pregnancy patterns in three herds.

Females not Cycling at Commencement of Breeding

A common cause of delayed conception is lack of sufficient time postpartum for females to resume fertile cycles. The time required for this to occur is affected by a number of factors including uterine involution time, nutrition, interactions between growth and lactation demands, and suckling stimuli. Beef cows often reach peak lactation between 6 and 8 weeks postpartum. Younger females, particularly first-calf heifers, require longer postpartum intervals (PPIs) than do older, multiparous females. Some breed types (e.g., those that have *Bos indicus* background) require longer PPIs than others. Female parity, calving dates, size and lactational state of the udder ("wet" or "dry"), and size of calves all provide valuable clues in relation to possible causes of lack of cyclicity.

Inadequate Nutrition

Herd B's pattern of delayed conceptions may also occur when nutrition is inadequate, although continuation of this situation would probably lead to lower final pregnancy rates than those shown. Where overall nutrition is the problem, BCS and its relationships can be useful in providing clues, even though some months may have elapsed between breeding and assessment.

Bull Infertility

Another possible cause of delayed conception in the herd is subfertile (or infertile) bulls. Here, the final pregnancy rate would most likely also be depressed. Bull problems are usually identified with a specific breeding group where either nonpregnancy or late pregnancy is encountered in females that otherwise should have become pregnant relatively early. The risk of bull problems occurring can be reduced by subjecting the bull "team" to breeding soundness evaluations before commencement of the breeding season (see chapter 9).

Infertility Disease

The pattern of delayed conceptions in herd B may be a result of the effects of infertility diseases such as vibriosis or trichomoniasis, although this would not represent a typical pattern for these diseases when introduced into a naive female population. Such a pattern could, however, occur if the population contained females of differing susceptibility to either of these diseases at the time of initial disease introduction. If this were the case, lower final pregnancy rates than those shown in this example (Table 2.6; Fig. 2.5) could be expected. If there is any suspicion that infertility disease might be a contributing factor to the delayed conceptions, then appropriate disease testing should be considered.

Herd C

Herd C displays a picture of strung-out conceptions that lack a consistent, discernable pattern. This is more suggestive of infertility disease than is the scenario with herd B, although lower final pregnancy rates than those obtained in herd B again might be expected. Elimination of possible causes such as inadequate PPI (very young calves at foot), nutrition (BCS), and bull infertility would strengthen the tentative diagnosis of infertility disease in this case.

Table 2.7. Effect of calving pattern on weaning weights and returns.

Calving Period	Average Calf Weight (lb)	Herd A: Total Weaning Weight/Period	Herd B: Total WWT/Period	Herd C: Total WWT/Period
1	500	31,000	7500	11,500
2	458	10,992	11,450	5496
3	416	3744	11,648	6240
4	374	—	5236	7854
5	332	—	2656	6308
6	290	—	1450	1450
Total pounds weaned	—	45,736	39,940	38,848
Return ($)	—	38,589	31,952	31,078

Dash indicates no relevant figures.

Table 2.8. Management procedures employed in "package" programs.

Procedure	Program				
	Mossman & Hanly	PMP[1]	"O'Connor"[2]	Prince et al.	PEP[3]
Restricted breeding season	X	X	X	X	X
Heifer target weights	X	X	—	X	—
Early breed heifers	X	X	—	X	—
Prior bull evaluation	X	X	X	X	X
Cow BCS	X	X	X	X	X
Pregnancy test	X	X	X	X	X
Herd health	X	X	X	X	X

Dash indicates not applicable. "X" indicates action included
[1]PMP = production management program (Sykes and Stafford 1984).
[2]Cows only.
[3]Texas Beef Partnership in Extension Program (Wikse et al. 2002)

COST OF DELAYED CALVING

An economic estimate may be made of the effects of delayed calving based on calf weaning weights and growth rates and on calf price. In the examples shown above (Table 2.7), based on the three 100-cow example herds described previously, the average weaning weight for calves born in the first 21 days of calving is assumed to be 500 lb (227 kg). If birth weights averaged 80 lb (36 kg), then growth from birth to weaning (at approximately 7 months) averaged approximately 2 lb (0.9 kg)/day. A calf price of $0.80/lb is used in this example.

Here, an estimated 20% difference occurred in total calf receipts in cow herds that were similar in ultimate reproductive performance (on the basis of number of cows pregnant) and differing only in the time or rate at which most pregnancies occurred. The difference in total return between a herd that calved in 63 days and one in which calving extended for twice that period was approximately 20%. For a 1000-cow herd, this difference, based on calves at $0.80/lb, could be as great as $75,000. Managerial remedies for extended and erratic calving patterns are available, as discussed elsewhere in this chapter. The costs of implementation of these remedies need to be weighed against the estimated economic penalties for nonimplementation.

A number of the recommendations discussed above have been incorporated into "packages" (Mossman and Hanly 1977, Sykes and Stafford 1984, Withers 1984, Anderson et al. 1986, Prince et al. 1987, Wikse et al. 2002), with considerable agreement, as shown in Table 2.8. In general, such programs are regarded as cost effective, although hard economic data are often lacking.

In Texas, encouraging results are reported from collaborative educational programs for producers (Beef PEP, or Texas Beef Partnership in Extension Program), in which veterinarians, extension agents, and a major pharmaceutical company have combined forces to implement and monitor changes in beef herds (Wikse et al. 2002), with the goal of improving profitability and sustainability of cow/calf operations in Texas (Tables 2.9 and 2.10).

GENETIC CONSIDERATIONS

Despite the genetic tools and associated knowledge available to cattle producers today, cow/calf producers have lagged behind other industries in adoption (NAHMS 1996, 1998b) for reasons that include lack of knowledge, economics, logistics, and tradition, as well as producer priorities and goals. A reasonable genetic goal for the cow/calf industry would be to produce low-cost, high-profit cattle while conserving and improving the resources used (Field and Taylor 2002). The systems and technologies employed to achieve this objective should be cost effective and user friendly (Field 2003).

Three primary genetic tools are available to cow/calf producers; namely, selection pressure, breed differences, and mating systems (Field 2003). In turn, the success of purebred breeders is dependent on the long-term success of the commercial cow/calf producers who purchase and use their genetic product.

Selection of breeding females should be made within the context of environmental and managerial constraints, market dictates, economic realities, and conflicting priorities. Genetic influences should be evaluated in terms of the quantity and quality of the product, the associated costs of production, and the product's market value (Field 2003). The related genetic traits to match with these production parameters include

- Volume and quality of production (per resource unit): market weight, yield, and quality
- Units of production (per enterprise): reproductive rate, calf survival, and cow survival
- Cost of production: cow maintenance, longevity, calving difficulty, and feed efficiency
- Market value: retail yield, retail quality, consistency, and conformity to market specifications.

These and related traits, may be assigned values for heritability and heterosis, as shown below (Table 2.11).

Such genetic information is useful to producers in helping them to align genetic goals with production and other priorities. However, difficulties often occur with the practical application of this knowledge. For example, many of the desired traits are not easily or accurately measurable, and their economic consequences in different environments are often poorly understood. Some traits are antagonistic. For example, selection for high milk and high growth has been associated with lowered female reproductive rates. In other words, high milk and yearling weight EPDs (estimated progeny differences) tend to be linked with delayed rebreeding of younger females. Models have been developed using multiple traits to predict the profitability of sire offspring within a given production situation. For example, in comparing 352 Angus sires using a multiple-trait model, M.D. MacNeil and W.O. Herring (unpublished data) found a range in profitability of $41.65 between sire offspring. The implication was that bulls with a desirable balance of traits may be more profitable than those that excel in just one or two traits.

Although such considerations imply that overemphasis on individual traits should be avoided, and also that a certain degree of genetic flexibility is necessary, it is apparent that many beef females are not productive. Here, genetic and phenotypic characteristics are both important, and some generalizations apply. For example, there are several general female traits that are considered to be important in diverse environments, including high fertility, early puberty, longevity, low maintenance, milk and size compatible with ranch constraints, calving ease, and the ability to produce and raise healthy calves.

Selection and maintenance of such traits in the cow herd require good records and strict attention to a good replacement heifer program (see chapter 8). As production and reproduction traits do not always go hand in hand, it becomes a challenge for females to excel in all requirements. An ideal breeding system would allow some separation of lines that emphasize the reproductive and maternal traits in breeding females from those that emphasize the production traits of nonretained terminal offspring, as practiced in the swine industry. It is also difficult for a given genotype to excel in all environments, with this consideration being particularly relevant for temperate breeds raised in subtropical or tropical environments (Hauser 1994).

Table 2.9. Commencing reproductive management practices in Texas Beef Partnership in Extension Program study herds.

Study Herd	Breeding Season	Pregnancy Exam	Bull Breeding Soundness Evaluation	Cow Vaccinations	Calf Vaccinations	Cow Deworming (Times/Year)	Calf Deworming	Calf Implant
1	12 mo	Yes	No	+	+	2x	No	Steers only
2	12 mo	Yes	Yes	+	+++	1x	No	Steers only
3	6 mo	No	Yes	++	–	1x	No	No
4	3 mo	No	Yes	++	+	1x	No	All[a]
5	6 mo	No	Yes	+	+	2x	No	All
6	4 mo	Yes	Yes	+	+++	1x	No	No
7	7 mo	Yes	Yes	+++	+++	2x	Yes	All
8	7 mo	No	No	–	+++	1x	No	All
9	6 mo	No	No	+	++	2x	No	All
10	3 mo (× 2)	Yes	Yes	++	+++	1x	No	No

Pluses and minuses represent relative degrees of emphasis.
[a]Not replacement heifers

Table 2.10. Examples of changes in performance measures (1999–2000) in standardized performance analysis of Texas Beef Partnership in Extension Program study herds.

Study Herd	Calf Crop (%)	Wean Weight (lbs)	Cost/100 lbs Calf	Price/100 lbs Cow	Net Income
1	3.6	104	−$12.60	$8.57	$107.46
2	29.8	−87	−$73.44	−$15.51	$118.61
4	6.1	69	−$62.34	−$2.52	$341.87
5	−3.3	9	−$9.56	$15.30	$93.53
6	12.1	−60	−$10.13	$15.72	$93.73
7	4.1	44	−$9.97	$13.77	$99.26
9	−0.1	81	−$0.50	$8.75	−$41.41

Wikse et al. 2002.

Table 2.11. Heritability and heterosis of various traits and their effects.

Traits/Classes	Heritability	Heterosis	Increase vs. Costs	Increase vs. production	Increase vs. market value
Offspring market weight	40%	Moderate	Variable	+	Neutral
Cull cows market weight	50%	Moderate	Variable	+	Neutral
Reproductive rate	<20%	High	Variable	+	Neutral
Offspring survival rate	20%	High	+	+	Neutral
Parent survival rate	20%	High	+	+	Neutral
Milk production	20%	Moderate	Variable	+	Neutral
Calving difficulty	15%	Moderate	−	−	Neutral
Fleshing ability	40%	Moderate	+	Variable	Variable
Feed efficiency	45%	Moderate	+	+	Neutral
Retail yield	25%	Low	Neutral	Variable	+

Field 2003.

Fertility is poorly defined and measured in cattle. It represents a complex trait, influenced by male, female, and environmental factors. Genotype–environment interactions further cloud the issue, with some evidence that selection for fertility, at least in females, might be more rewarding in stressful environments (Deese and Koger 1967). Despite these considerations, it is clear that selection against low fertility, or subfertility, in cattle breeding herds is both important and rewarding (Koger 1967, Bellows and Staigmiller 1994). This presents a major incentive to identify and remove individuals that are below herd norms or goals in reproductive performance. Such selection should include not only "open" females but, if feasible, those that have suffered from dystocia and those with substandard calves.

Despite the relatively low heritability of cattle fertility, selection for certain traits associated with fertility can be advantageous. For example, scrotal circumference in bulls is favorably associated with fertility traits in both males and females (see chapter 9).

CROSSBREEDING

Although within-breed selection is useful, greatest genetic benefit is usually obtained by exploiting breed differences and heterosis in some form of crossbreeding program. No single breed has all of the characteristics needed to excel in all beef production traits in all environments. Heterosis has been shown to favorably affect the survival and growth of crossbred calves by increasing both the reproductive rate and longevity of crossbred cows and the weaning weight of their calves (Cundiff et al. 1992, Koch et al. 1994). In general, the performance of crossbreds exceeds that of either parental breed, especially for composite traits such as lifetime productivity. Particular advantages in pounds of calf weaned occur with *Bos indicus–Bos taurus* crosses (Cundiff

Table 2.12. Benefits and disadvantages of different crossbreeding systems.

System	Benefits	Requirements	Disadvantages
Two-breed rotational	WWT + 16%	Moderate management intensity Minimum of two breeding pastures Herd size 50+	Possible large variation between generations Replacement heifers identified by sire breed
Three-breed rotational	WWT + 20%	High management intensity Minimum of three breeding pastures. Herd size 75+	Possible large variation between generations Replacement heifers identified by sire breed
Rotation terminal sire (two-breed)	WWT + 21% Can target specific marketing goals	High management intensity Minimum of three breeding pastures Herd size 100+	Replacement heifers identified by sire breed plus year of birth
Terminal sire × purchased F1 females	WWT + 21% Average herd size Can target specific marketing goals	Moderate management intensity Purchase females	Purchased females identified by source Biosecurity risk
Four-breed composite	WWT + 17.5% Minimum of one breeding pasture Any herd size Reduce intergenerational variation	Low management intensity	Availability Low EPD accuracy in bulls resulting from population size(s)
Composite-terminal sire	WWT + 21% Minimum of one breeding pasture Any herd size	Moderate management intensity	Availability of composite

WWT = weaning weight; F1 = first generation; EPD = estimated progeny difference.
Adapted from Field 2003.

and Gregory 1999), with this advantage being greatest in subtropical and tropical regions. A number of crossing systems are available, including two- and three-breed rotational, terminal sire, and composite methods. Each system has particular advantages and disadvantages (Table 2.12).

Tools have been developed that allow producers to identify the most profitable sires and systems, using economic and management descriptions of their operation. One example is the terminal sire profitability index that has been developed by the American Charolais Association (http://www.Charolaisusa.com/).

Complementarity is the process whereby breed types are crossed so that direct and maternal breed effects are combined with heterosis to optimize performance. This process can help match genetic potentials for factors such as growth, size, reproduction, and maternal ability, as well as carcass and meat characteristics, with external factors such as climate, nutrition, and market preferences. An example would be to use specialized male breeds as terminal sires with crossbred females to maximize favorable, and minimize unfavorable, traits.

Important interactions occur between genotype and environment, such that it is useful to identify those biological types of cattle, or crosses, that are best suited for both production and market environments. The genetic potential for mature size and productivity should be matched with both actual environment and economical, available feed. Large, important interactions can occur when environmental effects are large and genetic effects are wide (Nunn et al. 1978). For producers in areas such as Northern Australia and Argentina, southern Brazil, and the United States, this often means using crosses between *Bos taurus* and *Bos indicus* breeds, which have consistently outperformed purebred cattle in subtropical and tropical environments (Franke 1980).

WEANING STRATEGIES

Weaning strategies fall under two major banners: reducing the PPI in dams and minimizing weaning stress on calves, with considerable overlap between these objectives occurring with a number of strategies.

Reducing Postpartum Acyclicity

Prolonged PPIs in suckled beef cows represent a major source of economic loss to cow/calf producers (Yavas and Walton 2000). In beef cows, postpartum ovulatory cyclicity is usually not restored for 35–60 days or even longer. Prolonged acyclicity in the postpartum period can lead to lowered calf crops and be detrimental to lifetime reproductive efficiency (Bell et al. 1998)

The time of onset of postpartum cyclicity (or postpartum return to heat, PPR) is influenced by a number of factors, including nutrition and suckling. In turn, suckling less than two times a day permits an earlier return to cyclicity than does suckling twice or more a day. Thus, suckling appears to exert a threshold, rather than linear, effect on PPR (Senger 2003). Neither the act of suckling in itself nor mammary stimulation alone can account for all of the effects encountered, which appear to be influenced by visual, olfactory, and auditory inputs (Yavas and Walton 2000). In beef cows, this combination of effects can delay postpartum cyclicity for 60 or more days. Even further delay may be imposed by breed type (e.g., *Bos indicus*) and environmental effects (e.g., production in the tropics), and their interactions (Chenoweth 1994), such that herd fertility is consistently lowered (Larsen et al. 1994).

Thus achievement of a consistent 12-month calving interval in beef herds is often a challenge. Even under ideal conditions, females have relatively little time to recommence cycling and relatively few opportunities to become pregnant. When nutrition is limited, the reproductive process becomes secondary to maintenance, growth, and even lactation. A number of strategies are reported for manipulating weaning or suckling to promote earlier cyclicity. These include early weaning, creep feeding, temporary calf removal, and once-daily suckling.

Early Weaning

Early weaning has often been regarded as an emergency management decision, taken in the face of drought, poor body condition scores, and drylot situations (Marston 2002). However, there is growing acceptance of early weaning and other weaning manipulations as strategic management tools.

Advantages of early weaning include reduced forage consumption by the dam as well as lowered supplementation requirements (Lusby et al. 1981). Early weaning also reduces overall ranch nutritional demands (Marston 2002). As suckling is associated with postpartum acyclicity, early weaning can promote cyclicity if it is performed once luteinizing hormone reserves are replenished in the dam (Yavas and Walton 2000). Early weaning may considerably reduce PPIs. As such, it can represent a practical and profitable management tool, especially as it lowers nutritional demands for the cow/calf unit. Here, it is more efficient to feed the calf directly than via the dam. Early weaning also represents a valuable management tool in drought mitigation.

Calves may be early weaned at 3–6 months of age, although even earlier weaning (45 days) has been reported. Variations in acceptable weaning age will occur with different managerial and environmental situations. Nutritional interactions arise, with best effects occurring in first-calf heifers and young cows that calve in appropriate body condition (BCS of >4). In contrast, poor results may be seen when females are not in good body condition (BCS of 4 or less) at time of early weaning. Tactically, early weaning may be applied only to those females that could best benefit from it. Calves that are early weaned and immediately shipped are best handled as a group (not commingled) to minimize stress and to ensure consistently good nutrition. If the latter requirement cannot be provided via pasture, then drylot feeding may be indicated. Retained calves can obtain better prices if they are healthy and well fed. In general, early weaning necessitates extra calf management inputs, with greater attention to calves that are younger at weaning. The extra management effort associated with early weaning renders this procedure impractical for many cow/calf operations.

Creep Feeding

Creep feeding is the practice of providing supplemental feed to nursing calves. This is usually achieved by using a device that allows the calf, but not the dam, access to supplemental feed. Reasons for considering creep feeding may include

- Increasing calf weaning weights
- Compensating for insufficient milk production by the dam
- Reducing nursing pressure on the dam, thus reducing PPR and increasing early pregnancy rates.

Creep feeds vary considerably, from high-quality pastures ("green creeps") to grain-based ("energy creeps") or limit-fed, protein-based systems. The decision as to whether to attempt a creep-feeding program and which type should be based on a number of factors including calf and feed prices, efficiency of feed conversion, forage quality and quantity, and labor availability.

Here it should be stated that convincing evidence is lacking for the cost effectiveness of "energy creeps" in reducing nursing pressure. Calves tend to nurse to repletion before tackling the creep feed. Heifer replacements to the breeding herd that have been creep fed may, in turn, rear lighter calves. This trend has been associated with increased fat deposition in the udders of heifers grown rapidly in the preweaning phase (Lusby and Gill 1999).

Temporary Calf Removal

One alternative to early weaning is temporary calf removal, which increases gonadotropin-releasing hormone release in eligible dams (Yavas and Walton 2000). This technique was initially investigated by Wiltbank and coworkers as an adjunct to estrus synchronization programs using Syncro-Mate-B, where the fixed-time insemination protocol called for breeding at 48–54 hours after implant removal. Calf removal during this interim (i.e., between implant removal and artificial breeding) was shown to improve both estrus response and pregnancy rates. As expected, greater response occurs in younger females, especially first-calf heifers, and in cattle of moderate body condition at time of calf removal. If body condition is too low (e.g., BCS of 4 or less), then females will often not respond. If body condition is good (e.g., BCS of 6 or more), however, then females will often cycle early enough to obviate temporary calf removal (Warren et al. 1988). Trials confirmed that temporary calf removal was beneficial when used in concert with progestin regimes (Smith et al. 1979), and also with prostaglandin-induced estrus, although here it should be performed either before or at the time of initial prostaglandin injection (Chenoweth 1984). Other work has shown that temporary (48-hour) calf removal could also be advantageous for females entering a natural breeding program in which they are not subjected to estrus synchronization regimes (Chenoweth 1984, Odde et al. 1986).

Once-Daily Suckling

Another alternative to early weaning is once-daily suckling (ODS). Early research with ODS was conducted with Brahman cross first-calf females in Texas (Randel 1981). Calves were penned at 30 days of age, and their free-ranging dams were allowed to suckle their calves for one 30- to 45-minute period per day. After 45 days of ODS, calves and mothers were reunited. This regime significantly reduced PPR in the dams while having no adverse affect on calf weaning weights. Although nutrition levels in the dam also interact with ODS response, suckling manipulation was considered to exert the major influence.

Restricted suckling (once per day for 30–90 minutes), beginning on days 21–30 postpartum, has been shown to increase gonadotropin-releasing hormone release in the dam and shorten PPR, whereas twice-daily suckling did not reduce PPR (Yavas and Walton 2000). In one study, comparable reductions in PPR occurred in females subjected to either early weaning (17 days) or ODS (12 days). Interestingly, neither treatment influenced PPR in females that consistently calved early.

Most weaning manipulations require increased management inputs, as well as additional facilities. There is always also a risk of decreased calf growth rates (Yavas and Walton 2000) and increased health problems, although these hazards can usually be circumvented with good, preemptive management.

Reducing Weaning Stress on Calves

Weaning stress, especially when combined with other stressors such as transport, commingling, and environmental changes, can lower calf immunity and increase susceptibility to disease. It also contributes to decreased feed intake in both dam and calf. Different strategies, including preconditioning, are employed to reduce the cumulative adverse effects of these stressors. Weaning strategies to reduce cumulative stress include weaning calves onto grass while maintaining fenceline dam contact, as well as the use of antisucking devices.

Two-step weaning, first developed in the Veracruz region of Mexico, employs a simple antisucking device inserted into the calf's nostrils, thus preventing suckling, although calves can still eat and drink. The calf is allowed to run with its mother with the device in place for a period before their ultimate separation. Work in Canada indicates that installation of the device for as few as 4 days resulted in less behavioral evidence of anxiety at separation than evidenced by either controls or calves separated from their mothers by a fenceline. For cows, a slightly longer period (4 to 7 days) appeared to be necessary for optimal stress reduction. Benefits of two-step weaning include less distress and more time feeding in both mother and calf, as well as the capability to immediately ship calves on separation with less stress than is incurred with abrupt weaning. Information on the antisucking device can be obtained at http://www.quietwean.com.

BIOSTIMULATION

Biostimulation, or the "male effect," is effective in initiating female puberty and stimulating cyclicity in a number of species. With livestock, practical managerial applications of this phenomenon have mainly occurred with sheep and swine. However, there is also good evidence for biostimulatory effects in cattle.

For example, several studies have advanced the resumption of cyclic activity in mature postpartum cows by exposing them to bulls (Zalesky et al. 1984, Alberio et al. 1987, Naasz and Miller 1987). An interaction has been observed between bull exposure and nutrition level in postpartum cows, with cows in lower (although moderate) body condition being apparently more responsive to the biostimulatory effects (Stumpf et al. 1992). In a study with Angus first-calf heifers, contact with vasectomized bulls reduced the cows' PPR from 62 to 46 days, with more exposed heifers being cyclic by 40 days postpartum than were nonexposed heifers (Gifford et al. 1987). Use of either testosterone-treated cows or sterile bulls resulted in similar biostimulatory effects (Burns and Spitzer 1992).

Male-mediated acceleration of puberty in females has been reported in a wide variety of mammals, including domestic animals (Izard 1983). However, in cattle, studies of male effects on heifer puberty onset have been inconsistent. One study showed positive effects of bull urine on puberty and subsequent calving dates in crossbred beef heifers (Izard and Vandenbergh 1982), whereas another found that the placement of testosterone-treated cows with prostaglandin-treated, bull-bred beef heifers increased pregnancy rates as compared with controls (Chenoweth and Lennon 1984). It would appear that nutritional (Roberson et al. 1991) and probably social influences interact with the effects of biostimulation on puberty in heifers. More work is needed to establish the timing and other conditions necessary for optimal exploitation of biostimulation. However, in common with sheep, it is reasonable to assume that the male should be presented as a novel stimulus and that the females should be transitional at the time of introduction. (Chenoweth and Spitzer 1995).

REFERENCES

Alberio, R.H., G. Schiersmann, N. Carou, J. Mestre. 1987. Effect of a teaser bull on ovarian and behavioral activity of suckled beef cows. *Animal Reproduction Science* 14:163–272.

Anderson, R.S., H.L. Filmore, J.N. Wiltbank. 1986. Improving reproductive efficiency in range cattle: An application of the O'Connor management system. *Theriogenology* 26:251–260.

Australian Association of Cattle Veterinarians. 1998. *Pregnancy Testing of Cattle.* Indooroopilly: Australian Association of Cattle Veterinarians.

Bell, D.J., J.C. Spitzer, G.L. Burns. 1998. Comparative effects of early weaning or once-daily suckling on occurrence of postpartum estrus in primiparous beef cows. *Theriogenology* 50:707–715.

Bellows, R.A., R.B. Staigmiller. 1994. Selection for fertility. In *Factors Affecting Calf Crop*, edited by M.A. Fields, R.S. Sands. Boca Raton, FL: CRC.

BonDurant, R.H. 1986. Examination of the reproductive tract of the cow and heifer. In *Current Therapy in Theriogenology*, 2d ed., edited by D. Morrow. Philadelphia: WB Saunders.

Burns, P.D., J.C. Spitzer. 1992. Influence of biostimulation on reproduction in postpartum beef cows. *Journal of Animal Science* 70:358–362.

Cattell, M.B. 2000. Rectal palpation associated cumulative trauma disorders and acute traumatic injury affecting bovine practitioners. *Bovine Practitioner* 34:1–4.

Chenoweth, P.J. 1984. Reproductive management procedures in control of breeding. *Proceedings of the Australian Society of Animal Production* 15:28–31.

Chenoweth, P.J. 1994. Aspects of reproduction in female *Bos indicus* cattle: A review. *Australian Veterinary Journal* 71:422–426.

Chenoweth, P.J., P.E. Lennon. 1984. Natural breeding trials in beef cattle employing oestrus synchronization and biostimulation. *Proceedings of the Australian Society of Animal Production* 15:293–296.

Chenoweth, P.J., J.C. Spitzer. 1995. Biostimulation in livestock with particular reference to cattle. *Assisted Reproductive Technology/Andrology* 7:221–228.

Cundiff, L.V., R. Nuñntildeñez-Dominguez, G.E. Dickerson, et al. 1992. Heterosis for lifetime production in Hereford, Angus, Shorthorn, and crossbred cows. *Journal of Animal Science* 70:2397–2410.

Cundiff, L.V., K.E. Gregory. 1999. What is systematic crossbreeding? *Beef* 8–16.

Deese, R.E., M. Koger. 1967. Heritability of reproduction. In *Factors Affecting Calf Crop*, edited by T.J. Cunha, A.C. Warnick, M. Koger. Gainesville: University of Florida Press.

Deutscher, G., D. Colburn, M. Knott, M. Nielsen, J. Stotts, D. Clanton. 1990. Breeding and calving effects on cow productivity. *University of Nebraska Beef Cattle Report* 55:9–13

Dewell G., N.L. Dalstead, F.B. Garry. 1999. Ramifications of the down market for food animal veterinarians. *Large Animal Practice* May/June:22–27.

Dunn, B. 2002. Factors affecting profitability of the cow-calf enterprise. *Proceedings of the American Association of Bovine Practitioners*. 45–49. Rome, GA: American Association of Bovine Practitioners.

Engelken, T.J., M.F. Spire, D.D. Simms, et al. 1991. Management practices that increase beef herd profitability. *Veterinary Medicine* 86:851–857.

Field, T.G. 2003. Tools for making genetic change. Proceedings of the Beef Improvement Federation, Lexington, KY. Available at: http://www.bifconference.com/bif2003/symposiumpapers.html. Accessed July 10, 2004.

Field, T.G., R.E. Taylor, editors. 2002. *Beef Production and Management Decisions*, 4th ed. Saddle River, NJ: Prentice-Hall.

Franke, D.E. 1980. Breed and heterosis effects of American zebu cattle. *Journal of Animal Science* 50:129–137.

Gifford, D.R., M.J. D'Occhio, P.J. Sharpe, T. Weatherly, B.P. Setchell. 1987. Postpartum anoestrus in first-calf heifers exposed to bulls. *Proceedings of the Australian Society for Reproductive Biology* 32.

Grandin, T., editor. 1993. *Livestock Handling and Transport.* Cambridge: CAB International.

Hamilton, E.D. 1997. Is open/bred enough? *Proceedings of the Society for Theriogenology.* Montreal, Quebec, Canada. pp. 133–134. September 17–20. Montgomery, AL: American Society for Theriogenology.

Hauser, E.R. 1994. Importance of genotype X environment interactions in reproductive traits of cattle. In *Factors Affecting Calf Crop*, edited by M.A. Fields, R.S. Sands. Boca Raton, FL: CRC.

Izard, M.K. 1983. Pheromones and reproduction in domestic animals. In *Pheromones and Reproduction in Mammals*, edited by J.G. Vandenbergh. New York: Academic Press.

Izard, M.K., J.G. Vandenbergh. 1982. The effects of bull urine on puberty and calving date in crossbred beef heifers. *Journal of Animal Science* 55:1160–1168.

Koch, R.M., L.V. Cundiff, K.E. Gregory. 1994. Heterosis and breed effects on reproduction. In *Factors Affecting Calf Crop*, edited by M.A. Fields, R.S. Sands. Boca Raton, FL: CRC.

Koger, M. 1967. Selection and culling in Florida. In *Factors Affecting Calf Crop*, edited by T.J. Cunha, A.C. Warnick, M. Koger. Gainesville: University of Florida Press.

Larsen, R.E., S.C. Denham, J.F. Boucher, E.L. Adams. 1994. Breeding season length versus calving percentage in beef cattle herds. In *Factors Affecting Calf Crop*, edited by M.A. Fields, R.S. Sands. Boca Raton, FL: CRC.

Larson, R.L., V.L. Pierce. 1999. Agricultural economics for veterinarians: Partial budgets for beef cow herds. *Compendium on Continuing Education for the Practicing Veterinarian* 21(9):S210–S219.

Lusby, K.S., R.P. Wetteman, E.J. Turman. 1981. Effects of early weaning calves from first-calf heifers on calf and heifer performance. *Journal of Animal Science* 53:1193–1197.

Lusby, K.S., D.R. Gill. 1999. Creep feeding. *Beef Cattle Handbook*, BCH-5476. Madison: University of Wisconsin Extension.

Marston, T. 2002. Early weaning reduces demand on cows. *K-State Beef Tips*, July 2002, *Kansas State University Department of Animal Sciences and Industry Newsletter*, pp. 1–3. Available at: http://www.oznet.ksu.edu/ansi/nletter/bt/bt0702.pdf. Accessed July 22, 2004.

McGrann, J.M., J. Parker. 1999. Factors influencing profitability of the cow-calf operation. *IRM-SPA Handbook.* Standard Performance Analysis 50–1, pp. 1–36, College Station, TX: Texas Agricultural Extension Service, Texas A&M University.

McGrann, J.M., S.E. Wikse. 1996. Standardized performance analysis: Opportunities for beef cow-calf veterinarians. *Compendium on Continuing Education for the Practicing Veterinarian* 18:S199–S206.

Mossman, D.H., G.J. Hanly. 1977. A theory of beef production. *New Zealand Veterinary Journal* 25:96–100.

Naasz, C.D., H.L. Miller. 1987. Effects of bull exposure on postpartum interval and reproductive performance in beef cows. *Journal of Animal Science* 65(Suppl 1):426.

National Animal Health Monitoring System. 1996. *Management practices associated with profitable cow-calf herds.* USDA:APHIS:VS, CEAH. N197.796, pp. 1–2. Fort Collins, CO: National Animal Health Monitoring System.

National Animal Health Monitoring System. 1998a. *Beef '97 Part III: Reference of 1997 beef cow-calf production management and disease control.* USDA:APHIS:VS, CEAH. N247.198, pp. 1–42. Fort Collins, CO: National Animal Health Monitoring System.

National Animal Health Monitoring System. 1998b. *CHAPA Part IV: Changes in the U.S. beef cow-calf industry, 1993–1997.* USDA:APHIS:VS, CEAH. N238.598, pp. 1–48. Fort Collins, CO: National Animal Health Monitoring System.

Nebraska Veterinary Medical Association. 1990. *Profitability Guidelines for Livestock Producers.* Hastings: Nebraska Veterinary Medical Association.

Nunn, T.R., D.D. Kress, P.J. Burfening, D. Vaniman. 1978. Region by sire interaction for reproduction traits in beef cattle. *Journal of Animal Science* 46:957–964.

Odde, K.G., G.H. Kiracofe, R.R. Schalles. 1986. Effect of forty-eight-hour calf removal, once- or twice-daily suckling and norgestomet on beef cow and calf performance. *Theriogenology* 26:371–381.

Prince, D. K., W.D. Mickelsen, E.G. Prince. 1987. The economics of reproductive beef management. *Bovine Practitioner* 22:92–97.

Randel, R.D. 1981. Effects of once-daily suckling on postpartum interval and cow-calf performance of first-calf Brahman x Hereford heifers. *Journal of Animal Science* 53:755–757.

Roberson, M.S., M.W. Wolfe, T.T. Stumpf, et al. 1991. Influence of growth rate and exposure to bulls on age at puberty in beef heifers. *Journal of Animal Science* 69:2092–2098.

Senger, P.L. 2003. *Pathways to Pregnancy and Parturition*, 2d ed. Moscow, ID: Current Conceptions.

Smith, M.F., W.C. Burrell, L.D. Shipp, L.R. Sprott, W.N. Songster, J.N. Wiltbank. 1979. *Journal of Animal Science* 48:1285–1294.

Spire, M.F. 1986. Breeding season evaluation of beef herds. In *Current Veterinary Therapy 2: Food Animal Practice*, edited by J.L. Howard, 808–811. Philadelphia: W.B. Saunders.

Spitzer, J.C. 1986. Influences of nutrition on reproduction in beef cattle. In *Current Therapy in Theriogenology,* 2d ed., edited by D. A. Morrow, pp. 320–341. Philadelphia: W.B. Saunders.

Sprott, L.R., B.B. Carpenter, B.S. Robert, et al. 1998. Characterizing profitable East Texas cow-calf operations. *Compendium on Continuing Education for the Practicing Veterinarian* 20:S170–S181.

Stumpf, T.F., M.W. Wolfe, P.L. Wolfe, M.L. Day, R.J. Kittok, J.E. Kinder. 1992. Weight changes prepartum and presence of bulls postpartum interact to affect duration of postpartum anestrus in cows. *Journal of Animal Science* 70:3133–3137.

Sykes, W.E., R.W. Stafford. 1984. Productive management programs for beef breeding herds. In *Postgraduate Committee in Veterinary Science.* Proceedings 68, pp. 291–301. Sydney: Beef Cattle Production, University of Sydney.

Toombs, R.E. 1996. Setting fee schedules: How much are veterinarians worth to beef cattle operations? *Compendium on Continuing Education for the Practicing Veterinarian* 18:S85–S93.

Traffas, V., T.J. Engelken. 1999. Breeding season evaluation of beef herds. In *Current Veterinary Therapy 4: Food Animal Practice,* edited by J.L. Howard, R.A. Smith, pp. 98–104. Philadelphia: WB Saunders.

Warren, W.C., J.C. Spitzer, G.L. Burns. 1988. Beef cow nutrition as affected by postpartum nutrition and temporary calf removal. *Theriogenology* 29:997–1006.

Wikse, S.E., J.C. Paschal, D.B. Herd, et al. 1998. The Texas Beef Partnership in Extension (Beef PEP) Program: An innovative approach to increase the profitability of cow/calf operations. *Proceedings of the 20th World Buiatrics Congress* 1:147–151.

Wikse, S.E., J. Paschal, D.B. Herd, J.G. McGrann, P.S. Holland, D. Bade. 2002. The Texas Beef Partnership in Extension (Beef PEP) Program. *Proceedings of the Australian Association of Cattle Veterinarians,* Annual General Meeting, Adelaide, May 6–10, 2002. Brisbane: Australian Association of Cattle Veterinarians, pp. 81–89.

Withers, G.N. 1984. Productive management programme: A practitioner's perspective. *Proceedings of the Australian Society of Animal Production* 15:206–209.

Yavas, Y., J.S. Walton. 2000. Postpartum acyclicity in suckled beef cows: A review. *Theriogenology* 54:25–55.

Zalesky, D.D., M.L. Day, K. Imikawa, J.E. Kinder. 1984. Influence of exposure to bulls on resumption of estrous cycles following parturition in beef cows. *Journal of Animal Science* 59:1135–1139.

3
Records and Epidemiology for Production Medicine

Michael W. Sanderson

RECORDS COLLECTION AND MANAGEMENT

Production medicine is largely about using records, and proper analysis and interpretation of them, to make rational decisions based on production and economic reality. Of course, the first step in this plan is to identify and keep the appropriate records in a form that allows for analysis and interpretation. Production records can be valuable for measuring production levels and subclinical disease, identifying risks for increased disease and decreased performance, and identifying sources of revenue and cost. All of these data are needed for optimal decision making.

A significant proportion of cow/calf producers in the United States report that they keep some sort of records. However, when asked for specific production or health measures, the producers are generally only able to provide estimates, indicating that their records are not in an accessible form (Sanderson and Gay 1996). Records themselves are raw data on the workings of the ranch. Until they can be summarized, analyzed, and changed from "data" to "information," they are of limited usefulness for decision making. The summarization and analysis of production data are best performed with a computer, where results can be quickly obtained according to needed criteria. Pregnancy rates by body condition score (BCS) or parity, or average daily gain by sire, may be quickly generated from properly collected and stored computer records. Generation of multiple reports such as these can be quite time-consuming if done by hand. Daily operational data are most often written by hand in a pocket diary. With the recent advent of the handheld computer, records may be collected electronically for download to the primary computer where the records are maintained. Future improvements in data acquisition promise to increase the ability of the producers to directly input data to the computer at the time of the management event. A number of commercial cow/calf database record systems currently offer a handheld module for collecting "chuteside" data. This development promises to dramatically simplify the collection of real-time operational data and facilitate cow/calf record collection by bypassing the data entry step that follows collection.

PROGRAM OPTIONS FOR RECORD KEEPING

Spreadsheets

Spreadsheet programs such as Microsoft Excel may be used to collect and analyze cow/calf production data. A spreadsheet is essentially a list of data in a table. In general, columns identify variables in the data, such as cow identification (ID), BCS, or pregnancy status, for all animals. The rows in a spreadsheet identify individual records such as pregnancy status and BCS for an individual cow. A spreadsheet record of pregnancy outcomes for individual cows in the herd can be easily collected at the time of pregnancy examination and analyzed using built-in Excel functions. This analysis can provide an expected calving distribution for the herd as well as pregnancy percentage and expected distribution by cow age, breeding pasture, or BCS. Spreadsheets are less able to collect and analyze complex data over multiple years, but they can provide useful summaries within a year. For the producer without a commitment to keeping records, spreadsheets may serve to educate the client on the value of record systems and can provide some valuable information for decision making.

Relational Databases

For the producer with a desire to effectively track production, a well-designed relational database is much more suited to cow/calf production record keeping.

A relational database records the data from the herd in multiple tables that are related by common values. For example, one table may contain permanent data on the cow, such as date of birth, ID, and breed. This "cow" table will be linked to a "calf" table that contains information on all calves. An individual cow can be linked to multiple calves over multiple years. It is a relatively simple task, then, to get a report on all the calves a particular cow has produced or on her average calving date. All of these reports are considerably more difficult to generate from a spreadsheet program. A simple visual relationship model is given in Figure 3.1. Clearly, the way that the relationships are set up is critical to the efficient performance of the database in providing the needed information. The issues in proper database design are beyond the scope of this chapter, but careful consideration should be given to assessing the ability of a database to accurately collect the ranch data and provide the reports needed for management decisions.

Several databases are commercially available for cow/calf record management and analysis (e.g., CowCalf5, http://www.cowcalf.com; Cow Sense, http://www.midwestmicro.com). These are available to individual producers or veterinarians to assist them in keeping records on multiple herds. Some breed associations also offer record systems for purebred breeders that provide pedigree and expected progeny difference values in addition to performance records (e.g., Angus Herd Improvement Records, Hereford HerdMASTER). Integrated data companies also may provide record-keeping systems (e.g., eMerge Interactive, http://www.emergeinteractive.com; APEIS CattleTrax, http://www.cattletrax.com; AgInfoLink, http://www.aginfolink.com). These systems are designed with vertical coordination between industry sectors particularly in mind.

Significant opportunity exists for the appropriately trained veterinarian to be involved in the analysis and interpretation of cow/calf production records. In addition to the value of records on the individual herd, anonymous comparison of production data and costs among herds in a practice or region (benchmarking) can be useful in identifying production inefficiencies and opportunities. Summary data on production and economic measures in cow/calf herds are available (e.g., Cow Herd Appraisal Performance System [http://www.ag.ndsu.nodak.edu/dickinso/chaps/chaps.htm] or Standardized Performance Analysis [SPA]; Spire 1991). The cow/calf industry in the United States is moving toward production/marketing alliances that may provide an opportunity

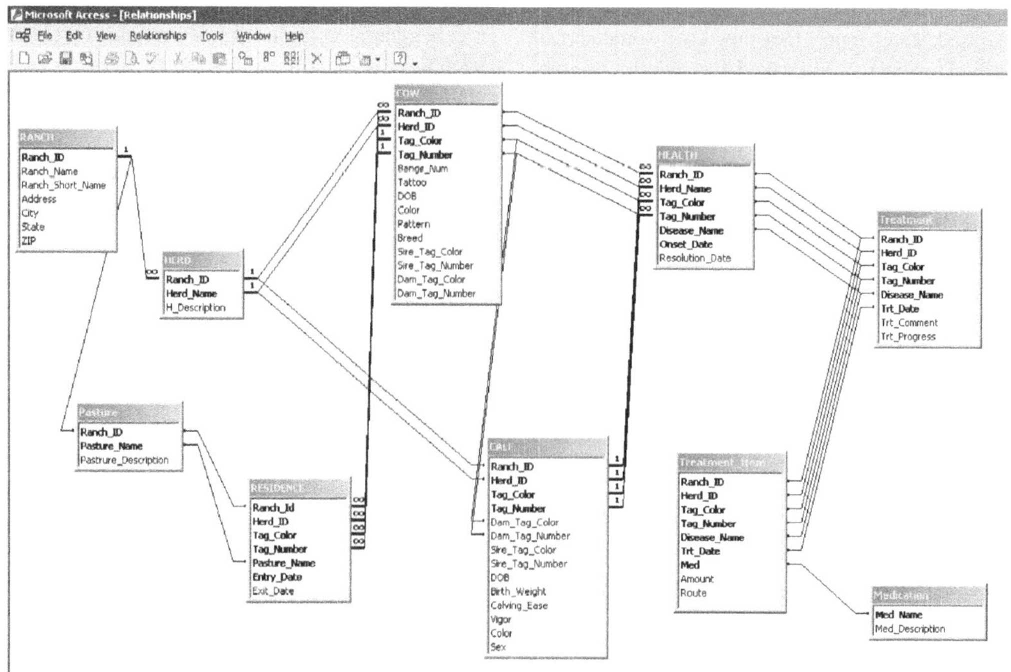

Figure 3.1. Database example.

for the comparison of production information from large numbers of herds. With additional training in database management and data analysis and interpretation, veterinarians could play an important role in this expanding arena.

Significant opportunity exists for veterinary involvement in SPA data, as described in chapter 12. SPA data have demonstrated the importance of controlling costs in beef cow/calf herds for profitability. Cost cutting without tracking production, however, has the potential to decrease production value by more than the amount of costs saved, decreasing net return. Similarly, a focus on production without an assessment of economic effects may elevate production but decrease net returns because of increased costs. Optimal decisions can only be made on the basis of records to identify input costs and production response for the individual herd.

The records that we keep are primary source data, or raw data, of the operation as it occurs. This primary source data would include, for example, the pregnancy status of the individual cow or the treatment of the individual calf. From this primary source data, we calculate useful indices of health and production that can inform decision making. These production indices would include pregnancy rate, morbidity rate, calf crop, and so forth. The primary source data must be collected in a reliable way to ensure their utility. This requires that the database information system be structured in a way that reflects the production process and minimizes data collection errors. Well-designed database systems are programmed to catch some data entry and logical errors. However, the database system itself may introduce errors if the design does not accurately reflect the production system it models in some way.

Goal of Record Systems

The goal of the record-keeping system should be to monitor production as well as identify and quantify sources of income and expense. The primary sources of revenue and expense for the operation should be considered, as should the suspected biggest sources of lost revenue (failure to conceive, poor performance, and marketing) and increased cost (nutritional supplementation and machinery costs). Quantifying answers to these questions is a major purpose of record-keeping systems. Producers should explicitly write out the answers to these questions and the components of the production system that relate

Table 3.1. Production targets.

	Target	Intervene
Reproduction		
Calf crop (%)	>90	<85
60-day pregnancy rate (%)	>95	<90
First 20-day pregnancy rate (%)	>65	<50
Calf		
Average daily gain (pounds)	>2.25	<2.0
Weaning weights (%)	>100	<90 (expected)
Cow		
Cow age	5–6	<4, >7
Cow body composition score		
Midgestation	4.5–6	<4, >7
Calving	5–6	<4.5
Health Targets (%)		
Dystocia		
Mature cows	<5	>8
Heifers	<15	>25
Gestational loss	<2	>3
Perinatal mortality	<5	>10
Cow death loss	<2%	>5%
Culling rate	15–20	<10, >25

Adapted from Spire, 1991.

to each question. With this information, they are ready to evaluate their record-keeping needs and decide which record-keeping systems will satisfy their specific requirements. Financial and economic record collection is covered in chapter 12 and will not be discussed here. This section will concentrate on production records and monitoring.

Decision Making

To be successful, the cow/calf production record-keeping system must collect data to aid decision making. Excessive, unnecessary data increase record-keeping time and costs without providing needed information for informed decision making. Producers

Table 3.2. Set of data for decision making.

Data Point	Description
Minimal production record requirements	
Number of cows and heifers in the herd at the beginning of the fiscal year	For any record system, an accurate inventory of production units throughout the year is essential. Without accurate inventories, the calculated production indices are meaningless.
Number of cows and heifers exposed to be bred	Number of cows and heifers you are supporting to produce the weaned calf crop.
Transfers of pairs or pregnant females in or out of the herd	Transfers must be accounted for to keep inventories correct and make calculated production indices meaningful.
Total number of calves weaned	Aggregate calf production of the herd.
Total pounds of calf weaned	Aggregate production of the herd in pounds.
Total price received for weaned calves	
Total acres used for grazing and raised feed	
Treatment and processing records of groups to facilitate quality assurance	These records are essential to ensuring that chemical and physical defects are avoided. These records should include the date of procedures performed, products used, lot number, administration route and site, and frequency of needle changes. Records kept here will be defined by the goals of the quality assurance program designed to meet the requirements of the customer.
Calculations of herd production levels allowed using minimum records	
Calf crop	Total number of calves weaned divided by the number of cows exposed to be bred. Calf crop includes losses from infertility, early embryonic death, abortion, stillbirth, and calf death from birth to weaning. It identifies depressed production without identifying the specific cause. Identification of the specific cause will require additional primary records targeted to each area of loss.
Pounds of calf weaned per cow exposed	Total pounds of calf weaned divided by the cows exposed to be bred, another important calculated production index. Pounds weaned per cow exposed include loss from all areas included in calf crop and losses caused by depressed weight gain in calves.
Pounds of calf weaned per acre of land used for grazing or feed production	Total pounds of calf weaned divided by total acres used for grazing or raised feed. Measure of the efficiency with which grazing and forage-producing land is used.

Chenoweth and Sanderson, 2001.

Table 3.3. Primary source data to allow tracking of production level and specific causes of production depression and inefficiency.

Data Point	Description
Individual cow and calf identification	The basis of tracking individual performance of cows in the herd and culling based on that performance.
Individual cow and heifer pregnancy status	Pregnancy percentage provides an index of the breeding efficiency of the herd. It includes failure to conceive, early embryonic mortality, and depending on the time of pregnancy examination, early- to midterm abortions. Along with body condition score at breeding and pregnancy, pregnancy percentage provides an estimate of nutritional effects on fertility.
Cow body condition scores or weights	Ideally taken at the time of pregnancy exam and before calving. Helps identify potential nutritional effects on reproductive performance and groups of cattle that may need separate nutritional management.
Calf birth date/calving date	To identify early-born calves and early-calving cows. When categorized by 21-day periods during the calving season to produce a calving histogram, it provides an index of fertility over the course of the breeding season.
Calf birth weight	
Calving ease score	Measure of the incidence and severity of dystocia. Along with calf birth weights, this score can effectively guide the use of expected progeny differences for sire selection to control dystocia rates and optimize calf growth.
Calf vigor score	Scored at birth; may reflect the effects of dystocia, nutritional deficiency, and infectious disease on the calf.
Date of all calf deaths	Along with calf birth date, this information allows assessment of when mortality is occurring. Analysis can assess rates of stillbirths and that of deaths from birth to 24 hours, from 1 to 28 days of age, and from 28 days of age to weaning. Provides details on when losses are occurring to guide intervention resources and efforts.
Individual calf weaning weight	Assesses the production of individual and cohorts of cows.

Chenoweth and Sanderson, 2001.

Table 3.4. Primary source data to allow complete tracking of disease cause and relation to production depression and inefficiency.

Data Point	Description
Date and diagnosis of morbidity in calves and cows	Allows calculation of disease-specific incidence rates. This will provide additional detail on the time of losses and the effect on production to effectively guide the rational allocation of resources to prevention. Record of disease-specific rates will require training of producers to validate case definitions to allow useful interpretation of the data. If rates are to be compared across herds, each producer will need to use the same case definition; that is, producers will have to all agree on what a "sick calf" is.
Date and treatment of morbidity in calves and cows	Useful in assessing response to treatment as well as providing quality assurance validation. If treatment response is to be measured, recovery, retreatment, and final disposition dates of all cases also must be tracked.

Chenoweth and Sanderson, 2001.

overburdened from data collection without a clear purpose and payoff can become frustrated and quit. For every proposed record, the use and value of the data for decision making should be clear. If we cannot articulate what the proposed data will be used for, it is probably not worth the time and expense to collect the record. In addition, the desire of the producer to keep and use records for decision making should be taken into account. Recording a small number of the "most important" records may be helpful and effective in showing the producer the value of records. Overwhelming the producer with too much "paperwork" will likely doom the record-keeping effort to failure.

Production Monitoring

Production records assess the productivity and efficiency of the enterprise and measure the attainment of production goals set in the planning phase. A general set of production targets are included in Table 3.1. These production targets are not universally applicable goals. Initial production targets should be established on the basis of the producer's perception of historic performance and acceptable production in the area. No generic, uniformly applicable production standards can be identified for all herds. As the record system produces data, actual production targets should be established for each herd on the basis of its own unique mix of inputs. The production targets should be those that result in the best economic return for the ranch. Production monitoring is also valuable as a sentinel monitoring system for subclinical disease. By definition, subclinical disease is not clinically detectable; however, it may commonly affect production before clinical disease is apparent. As such, production monitoring may serve as an early warning system for underlying disease.

A minimum set of data for decision making based on production records is listed in Table 3.2, along with the herd production measures that can be calculated. Collection and analysis of these records will allow assessment of the overall production of the herd. These data will identify depressed herd production but will not provide information on individual performance or specific causes of production depression.

Individual ID of cows and calves is necessary for the addition of the records listed in Table 3.3. This ID also will increase the ability of the record system to identify specific causes of production depression. If weaning weights are decreased, is it because of a change in calving distribution, or are the calves growing more slowly? If pregnancy rates are decreased, are they evenly decreased across different groups of cows, or is one group particularly affected? If the cause of the production depression can be more specifically identified, the appropriate response is more likely to be applied.

The collection and tracking of calf morbidity data as outlined in Table 3.4 allow assessment of the effects of morbidity on calf performance and the costs associated with each morbidity incident. Once these values are known, an appropriate amount of resources can be allocated to optimize net income. Record-keeping databases for cow/calf production must be specifically designed for keeping track of morbidity incidence correctly.

Commercially available record management programs provide standard reports of certain production measures. In addition, the program can usually be queried to provide additional reports on specific issues that may be useful in addressing production. The ability to query the database to obtain custom reports is necessary if the greatest value is to be gained from a record-keeping system. The data that can be obtained from the database may be limited by the database's design. The underlying data relationships in the record system must accurately reflect the production system they model. If not, the system will sometimes fail to allow the needed analysis. The program should also be able to write records data to an external data file that can be imported into statistical software for more sophisticated analysis.

Standard production measure definitions according to SPA are as follows.

Pregnancy percentage: indicator of breeding performance

$$\frac{\text{Number of females exposed diagnosed as pregnant}}{\text{Number of females exposed}}.$$

- Adjustments to number of females exposed:
  - Subtract:
    - Females sold or removed between the onset of breeding and pregnancy examination
  - Add:
    - Females purchased during breeding that were exposed to the bull
  - Retain:
    - All death losses of exposed females

Calving percentage: includes failure to conceive, early embryonic death, and abortion losses:

$$\frac{\text{Number of calves born}}{\text{Number of females exposed}}.$$

Pregnancy loss percentage: indicator of mid- to late-term abortion and palpation accuracy

$$\frac{\text{Number of females diagnosed pregnant that fail to calve}}{\text{Number of females diagnosed as pregnant}}.$$

Calf death loss: based on number of live calves born, death loss from birth to weaning. Include calves lost at birth and all death loss until weaning.

$$\frac{\text{Number of calves that died}}{\text{Number of live calves born}}.$$

Calf crop or weaning percentage: losses from conception to weaning

$$\frac{\text{Number of calves weaned}}{\text{Number of females exposed}}.$$

Female replacement rate

$$\frac{\text{Raised and purchased heifers or cows}}{\text{Number of females exposed}}.$$

Calving distribution: by 21-day periods

$$\frac{\text{Cumulative number of calves born by 21, 42, and 63 and after 63 days}}{\text{Total number of calves born}}$$

- Calving distribution assumptions
  - First 21-day period begins 285 days following introduction of the bull or at the time of calving of the third mature (>3 years old) cow.
  - All calves delivered dead or alive are included.

Pounds weaned per exposed female combined calf crop and weaning weight

$$\frac{\text{Total pounds of calf weaned}}{\text{Total number of females exposed}}.$$

Disease Monitoring

Production record systems may also monitor disease incidence within the herd. Aggregate count records of disease may be kept within cattle classes and ages and for specific disease syndromes. Calf morbidity may be monitored in the following categories: before 28 days of age, before weaning, and after weaning. Disease-specific data in each age category, such as for diarrhea, pneumonia, and pink-eye, would add value to the records. The proportion of cows and heifers that experience dystocia is another valuable aggregate morbidity measure. If data are collected for the specific date of morbidity, actual disease incidence can be calculated. This allows accurate targeting of management efforts at the highest-risk groups and times. Monitoring disease incidence at the individual level adds additional value in allowing tracking of the performance effect of morbidity. Knowing the performance effect of a morbidity incident along with the treatment and labor costs allows an accurate assessment of the resources to devote to disease prevention.

Quality Assurance

Record-keeping systems are also valuable for tracking compliance with quality assurance requirements. Data for validation of a Hazard Analysis Critical Control Point program can be integrated into a beef record system. Computerized records provide documentation of procedures and allow the rapid searching of records to ensure that withdrawal times are met. Beef quality assurance programs are covered extensively in chapter 13.

An additional quality issue is the herd disease status. Good production and disease records can provide significant assurance that the herd does not have a particular disease or does not have a high prevalence of disease. This may be an especially valuable tool for the registered producer selling seed stock to commercial ranches. Validated performance and disease records may be a useful marketing tool to enhance the biosecurity program of the commercial producer. Further discussion of these issues is offered in chapter 5.

Herd Performance and Target Assessment

At appropriate times during the year, the production and disease performance measures resulting from the record system should be evaluated. The end of calving season and after both weaning and pregnancy examination are opportune times for evaluation. Actual performance should be compared to the target goals for the year. If target goals are not met, then adjustments should be made. Both financial and production records are necessary to ensure that efforts to meet production target goals are cost-effective. At each time, a producer should set new and refined goals, monitor production and disease, and then reassess performance. The techniques and measures to assess production and disease are covered in the epidemiology section of this chapter.

Individual Animal ID

The need to identify cattle for trace-back in the event of an outbreak of foreign animal disease has raised animal ID to the top of the discussion list. The U.S. National Cattlemen's Beef Association has supported a voluntary national cattle ID system to support value-based marketing and rapid trace-back. At present, the

National Institute for Animal Agriculture task force on animal ID has put forth guidelines for a national ID system for livestock. Initially, all premises in the United States will be identified with a unique premise ID, followed by specification of all livestock by a unique group or individual ID. The goal is to provide trace-back of all contact animals within a 48-hour period. This system will have significant value in tracing contacts and quickly containing a contagious disease introduction. Further, it will facilitate consumer confidence in the beef supply by ensuring timely and effective trace-back of animals identified with noncontagious diseases such as bovine spongiform encephalopathy. In the case of bovine spongiform encephalopathy, a robust national ID system will allow rapid recognition of the herd of origin and of birth cohorts for investigation. Rapid and complete investigations will improve consumer confidence that all bovine spongiform encephalopathy–positive animals have been identified and excluded from the food chain.

The national ID system will likely be based on electronic ID (radio frequency ID). The electronic ID system will facilitate efficient collection and timely use of ID information. Such a system also has the potential to decrease data entry errors at the ranch level. The system will include an ear tag with an imbedded transponder that can be read by a transceiver and that will directly download the animal's ID to the computer system, avoiding data entry errors. Transponders and transceivers must comply with ISO 11784 and 11785 standards for data and technical specifications. Even so, mixing transceivers and transponders from different companies can sometimes cause problems with readability, so the compatibility of the equipment should be checked before its purchase.

Record systems promise to change significantly over the coming years. With the advent of a national individual animal ID system and the global marketing economy, the need to keep good records will only increase. The record system chosen should be based on the desire of the producer and the level of information needed to make decisions. Collection of primary source data on an individual animal basis allows much greater flexibility in identifying areas of production depression, as well as its likely causes. Financial and economic data must be kept as well to assess the effect of production changes on net income.

EPIDEMIOLOGY

The cow/calf production system is diverse and complex. Cattle thrive in a wide range of environments, and the production systems employed to raise cattle are equally varied. The individual cow is metabolically complex, and a herd of cows is part of a complex ecosystem. Cows live and reproduce in an environment that is not highly controlled and, as such, are greatly affected by it. Altogether, this makes rational, data-driven decision making for optimal herd production difficult.

Comparing production levels among herds is problematic because of differences in cow genetics and environment among herds. Even comparing production across years in the same herd is difficult. Effects of management intervention on production and disease are almost always confounded with yearly environmental differences. In spite of these difficulties, our clients need to make sound management decisions. In production medicine, we seek to make good decisions for the health and profitability of entire beef herds. Even though data from beef herds are challenging to interpret, we can improve these decisions through the use of such data. The purpose of this chapter is to examine the ability to look at herd data and analyze, interpret, and exploit them for management decision support. Epidemiology and biostatistics are the means of understanding herd-level data and using them to make rational herd-level decisions.

Epidemiology is a basic science for the study and understanding of the determinants of disease and production level in populations of animals. On the basis of its principles, we attempt to understand and quantify the myriad factors acting on a population of animals so that we can identify those factors associated with changing disease or production levels.

Why are some animals affected in a herd and others not? Why do some herds have high disease rates and other herds have low disease rates? The answers are rarely simple. Disease and production are complex, but epidemiology is the "tool kit" we must use to make progress in understanding disease and production determinants. Just as the techniques of immunology or pathology tell us about the processes going on within the individual animal, so the techniques of epidemiology can tell us about processes going on within a population of animals.

In beef cattle practice, the populations we are interested in are herds. Although epidemiology has historically been applied to the determinants of disease in a population, the principles are equally applicable to identifying factors associated with high or low production in a herd. Epidemiology and biostatistics provide the "tools" to collect and analyze data from the herd to discern why some animals are diseased and some remain well, or why some herds have high disease rates and others have low ones. Equally, these same tools allow us to discern why some herds have high and others low production. Indeed, without the ability to look at the numbers associated with a herd, we would fail to understand the factors affecting disease and production in the herd and would be relegated to surmising what is going on based on our intuition and biases.

This chapter explains why epidemiology and biostatistics are basic sciences for production medicine and how their methods improve our understanding of disease and production at the herd level, thereby improving our ability to make good herd-management decisions.

Multifactorial Nature of Disease and Production Efficiency

We understand that disease is multicausal. That is, single agents or factors do not "cause" disease, but a number of factors work together to result in disease. The concepts of necessary causes and sufficient causes enhance our understanding of disease.

A necessary cause is one without which we never have disease; it is necessary for disease. By definition, *Escherichia coli* is necessary for colibacillosis. In general, those diseases that are named for their etiologic agent by definition have necessary causes. This ability to attribute disease to a single "cause" is not as useful for problem solving as it appears at first glance. From a practical standpoint, neonatal diarrhea caused by differing etiologic agents is generally not clinically distinguishable. Therefore, a more useful approach is to consider neonatal calf diarrhea as a syndrome with no necessary cause; that is, no particular etiologic agent is necessary to cause disease.

A sufficient cause is one that can cause disease by itself; no other factor is required. If applied to infectious agents, then the agent must be capable of causing disease by itself and must do so in all cases. We commonly speak of, for example, coronavirus as a cause of neonatal diarrhea, but is coronavirus a sufficient cause? If we begin to look within the herd, we find sick calves shedding coronavirus, but we also find perfectly healthy calves shedding coronavirus. Clearly, the mere presence of coronavirus does not equal disease; that is, coronavirus by itself is not sufficient to cause disease. If both the sick calves and the healthy calves "have" corona, what is the difference that results in disease in one and continued health in the other? (See Figure 3.2.) Presumably, other factors are working in concert with coronavirus in the sick calves to "cause" disease. In almost all cases, it is this collection of component causes that together cause disease in the individual. This entire collection of component causes is considered a sufficient cause.

An individual disease syndrome may have more than one sufficient cause. Sufficient causes for neonatal calf diarrhea may include one of the etiological agents along with a sampling of other potential component causes such as failure of passive transfer; dystocia; a wet, muddy environment; poor nutrition; high population density; cold stress; or many other factors (see Figure 3.3).

Epidemiology is the science of identifying the component causes of a sufficient cause to modify them and reduce disease rates. If we can remove one component of a sufficient cause, it is no longer a sufficient cause, and though, for example, we still have corona virus, we see lower disease rates.

Epidemiology can be effective in this way even when we do not know the "etiologic" agent involved. For example, even though the etiologic agent associated with epizootic bovine abortion is unknown, risks for abortion have been identified that allow management changes to control the effects of the disease. The distribution of this disease has been shown to match the geographic distribution of the tick *Ornithodoros coriaceus*. The disease is controlled by managing the exposure of cattle to the foothills of California, Oregon, and Nevada, where the tick is common. Exposure of heifers to the ticks before breeding results in immunity, and exposure of cows in late gestation does not result in abortions.

Role of Epidemiology at the Herd Level: Herd-Level Diagnostic Kit

Epidemiology requires accurate observation and description of health and production. It does not necessarily require diagnostic laboratory support. Rather, its use is dependant on the collection of records about the health, disease, production, and risks of a herd and of groups of animals within the herd. We use these records to compare the disease risk or production level between different management (risk) groups. In these comparisons, we are as interested in healthy or unaffected animals as we are in diseased or affected animals. It is these healthy/unaffected animals that make up the comparison group with which we may discern what is different between the affected animals and the nonaffected animals. In addition, although sick animals provide the numerator in the rates of disease, numbers of well animals are required to provide correct denominators for the risks and rates we use to compare disease among groups.

Comparison of disease risk (or productivity) among different groups of animals allows us to identify risk factors and make decisions at the herd level. That is, we can determine which difference between

PATHOGEN DISEASE MODEL

- Non-Infected and all Healthy
- Infected and all Sick

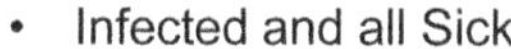

Agent **is Necessary** and **is Sufficient** – When the agent is absent there Is no disease, when the agent is present there is always disease. This model does not fit many diseases well and has little utility for solving problems in the real world of disease.

SINGLE AGENT DISEASE

- Non-Infected and all healthy
- Infected but only some sick

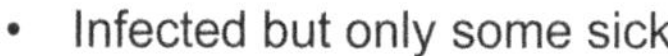

Agent **is Necessary** that is we never have disease in its absence, but it is **not Sufficient.** Even when the agent is present it does not always result in disease. This is most applicable to diseases that are named for their associated etiologic agent

MULTI-AGENT DISEASE

- Non-Infected but some sick
- Infected but only some sick

Agent is **Not Necessary** and **Not Sufficient.** So we have clinical disease in animals that do not have a particular agent (say neonatal diarrhea in calves that do not have Carona virus) and clinical health in some animals that do have the agent. Most applicable to disease syndromes such as neonatal diarrhea, bovine respiratory disease complex, or pinkeye. It is this view of disease that has the most utility for solving herd level disease problems.

Figure 3.2. Necessary and sufficient causes of disease.

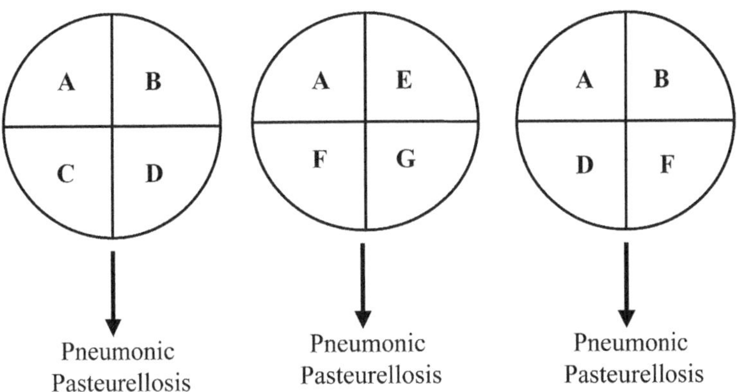

3 Sufficient causes for Pneumonic Pasteurelosis Component 'A' (Pasteurella) is necessary in each cause. By definition, if we call it Pasteurellosis, Pasteurella must be involved. Pasteurella is not sufficient by itself in any of the causes

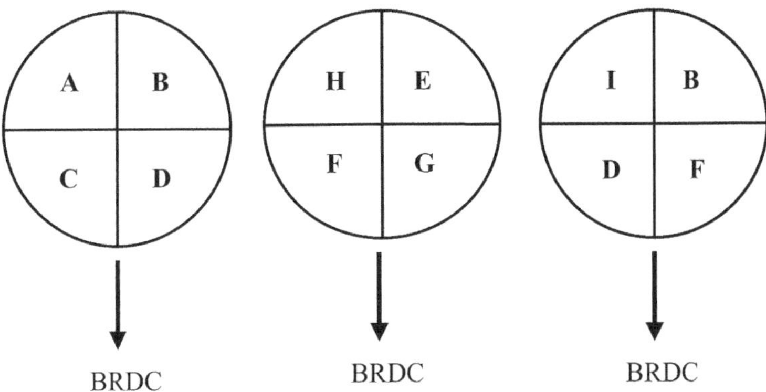

3 Sufficient causes for Bovine Respiratory Disease Complex. There is no necessary component for disease since no single component is included in each cause. This model of disease causation fits best with observation and has the most utility for solving problems. If we can remove one component from a sufficient cause it is no longer sufficient.

Figure 3.3. Component causes of disease. Disease is best understood as a group of component causes that together cause illness. If we can remove one of the components from a sufficient cause, it is no longer able to cause disease. This leads to a risk factor management approach to disease control. We do not have to completely eradicate particular disease agents to control disease rates in a herd, but rather, we must control individual components to reduce or eliminate sufficient causes.

two groups of animals is responsible for the difference in disease risk or production level. Epidemiology is the diagnostic toolbox that allows us to rationally and objectively compare these risks and to examine disease at the herd level.

The population at risk (PAR) is a key concept necessary for correctly calculating rates and comparing groups. The PAR is the number of animals available to become sick (or have a certain production outcome). Which animals are included in the PAR is dependent on the disease or production syndrome in which we are interested. If pregnancy is the outcome of interest, then cows exposed to the bull make up the PAR. If abortion is the outcome of interest, only those cows that are pregnant are included in the PAR. For infectious disease, we generally only count the first occurrence of disease in our calculation of a rate, so once a weaned calf gets pneumonia, he is removed from the PAR for pneumonia. Obviously, he could get pneumonia again at a later date, so we can allow for multiple occurrences by shortening the time period of the rate. We could, for example, calculate rates for each 3-week period after weaning.

Measuring Disease Frequency

To compare disease rates and risk factors between two groups of animals, we must collect data from both healthy and diseased animals in each group. We generally measure the effects of disease in terms of morbidity (sickness) and mortality (death). Depending on our objectives, we may need to measure each of these conditions in specific categories as well, such as respiratory or diarrhea morbidity and mortality.

To manage herd disease rates, we must measure disease in terms of prevalence or incidence. Prevalence is the proportion of animals affected by a particular disease, syndrome, or characteristic at one point in time and is the simplest measure of disease. For example, if we were to bleed cows at the time of pregnancy examination and assess their antibody levels to *Neospora caninum*, we would have the prevalence of serum antibodies, or seroprevalence, of *Neospora*. This does not necessarily tell us anything about current neosporosis in the cows, but it is a measure of what proportion of the cows have been exposed and responded to *Neospora* in the past. We can similarly find the prevalence of copper deficiency in calves at weaning time, or the prevalence of failure of passive transfer 24 hours after birth. The prevalence of neonatal diarrhea in a beef herd is the number of calves with diarrhea on the day we count them divided by the number of calves present that day. If 10 calves have diarrhea out of 100 calves, the prevalence is 10/100, or 10%.

Prevalence does not have to arise from sampling on a single day. Prevalence of failure of passive transfer could be taken over the course of the calving season; the point in time that it is taken could be 24 hours after each calf is born. Prevalence in a herd is a function both of how commonly a disease is transmitted and of the duration of disease signs. Thus, in the *Neospora* example above, if the duration or persistence of antibodies is long, the prevalence may be high even when exposure or transmission might be relatively uncommon.

Incidence measures of disease include a time component, so we measure the proportion of animals that become sick over a specified period of time. Incidence rates may come in either of two varieties: cumulative incidence rate (CIR), often referred to as a risk rate, and incidence density rate (IDR), also known as the true rate.

CIRs or risk rates report the total number of cases over a specified period of time. Pregnancy rate may be a true CIR if it is reported as the percentage of cows that conceived over a specified breeding interval (e.g., 21-day or 63-day pregnancy rates). The percentage of calves that have neonatal diarrhea before 3 weeks of age is also a cumulative incidence. The duration of the measured period is the same for each calf, but they occur at different times during the calving season for each calf depending on when the calf is born. CIR is a retrospective look at disease in the herd. That is, it is calculated at the end of the outbreak, when we count 50 of 100 calves that had pneumonia by 3 weeks postweaning. As such, CIR does not assist in prospectively assessing disease effect or altering management. It may, however, provide a useful measure of disease risk in the herd.

The CIR may be calculated over a long period of time; for example, calf morbidity from birth to weaning. If the risk for disease varies significantly during this period of time, important information regarding when disease occurs may be lost. For example, because morbidity risk for calves is concentrated in the first 3 weeks of life, combining this high-risk period with other lower-risk periods may obscure important information. Knowledge of the risk of disease in the herd can provide valuable information for decisions about the economics of different intervention plans based on their cost and the likely improvement in morbidity or mortality. CIR is similar to distance traveled at the end of the day: When the driving day (or the outbreak) is all over, how far have you gone (or what percentage of calves got sick)? (Table 3.5)

IDR is a current measure of disease. It measures the rate at which healthy animals are becoming sick animals. It is measured in terms of population time at risk: 100 weaned calves in a pen provide 100 calf-days of risk every day, or 700 calf-days of risk every week (100 calves × 7 days). Similarly, the same set of calves provides 100 calf-weeks of risk every week. If seven calves get sick the first week after weaning, the IDR is 7 per 100 calf-weeks. By using population time at risk, we can compare disease rates between populations of different sizes. A pen with 50 calves provides only 50 calf-weeks of risk each week, but by reporting the IDR for both the 50 calf pen and the 100 calf pen in terms of 100 calf-weeks of risk, we

Table 3.5. Cumulative incidence rate (risk) calculations.

The simplest way to consider incidence rate calculations is in a static population, where all animals enter at one time and there are no departures or additions to the group. In this case, we can calculate a CIR. For example, suppose you wean 100 calves and follow their respiratory morbidity over 4 weeks. Table 3.5 records the number of morbidities that occur each week.

The PAR for the 4-week cumulative morbidity is 100 animals—the number at risk at the beginning of the 4-week period. An individual calf in this population has a 30% risk of getting disease over the 4-week period.

The PAR for the weekly cumulative morbidity is the number at risk at the beginning of each week; that is, the beginning PAR (BPAR) for week 1 (100) minus the morbidities for week 1 (7) equals the BPAR for week 2 (93). The weekly risk varies, so an individual calf has a 7% risk of getting disease in week 1 and a 12.9% risk of getting disease in week 2. We can see that although the total risk of getting disease over 4 weeks is 30%, the risk appears to vary by week.

Week	BPAR[a]	Weekly Morbidity	Cumulative Morbidity	4-Week Cumulative Incidence (Risk)	Weekly Cumulative Incidence (Risk)
1	100	7	7	0.07 = 7% (7/100)	0.07 = 7% (7/100)
2	93	12	19	0.19 = 19% (19/100)	0.129 = 12.9% (12/93)
3	81	9	28	0.28 = 28% (28/100)	0.11 = 11% (9/81)
4	72	2	30	0.30 = 30% (30/100)	0.028 = 2.8% (2/72)

[a]BPAR = beginning population at risk.

can directly compare the rate of disease between the two pens.

The numerator of an IDR is the number of cases over the specified interval, and the denominator is the population time at risk. The defined time period may vary depending on the level of detail in the data or on the question to be answered. Human IDRs are commonly reported in person-years at risk. For uncommon diseases, it is usually reported in cases per 1000 person-years at risk (1000 people followed for 1 year gives 1000 person-years of risk). In production animal medicine, we are generally interested in IDRs that are specified to a shorter interval. Cows or calves move through significantly different risk periods over the course of a production year, so yearly incidence densities are often not useful (the number of calves with pneumonia per 1000 calf-years at risk does not seem to have great utility; however, the weekly number of calves with pneumonia per 1000 calf weeks at risk for the 6 weeks following weaning may be useful). IDR is similar to speed: It is a measure of how fast you are going. We might report our speed as 60 miles per hour, or 88 feet per second, as a measure of how fast we are going at a particular instant. IDR may be reported in different units as well (calf-days, weeks, or months at risk), and it measures how fast calves are changing from well to sick at a particular point. (Table 3.6)

As alluded to above, incidence and prevalence are related by the duration of signs

$$\text{Incidence} \propto \frac{\text{Prevalence}}{\text{Duration of signs}}.$$

For a given incidence of disease, prevalence is increased when duration of signs increases as a result of a buildup of historical cases that are detected in a one-time prevalence sampling.

Validating Measurement of Disease

We still need to make sure that we are measuring the right things when we calculate disease rates in the herd. That is, what criteria do we use to decide who is sick and who is well? It might seem like a trivial question at first, but on further reflection, we see that it is pivotal that we categorize disease correctly if our disease rates are to be meaningful. Mortality is easy to categorize—we can all agree on a case definition for "dead." If we want syndrome-specific death rates, however, we must correctly classify mortalities into the appropriate cause of death. We can do this based on clinical signs, or we may include necropsy and laboratory support. Whatever basis we use, the goal is to correctly classify mortalities by cause of death.

Standard criteria for classification of mortalities into various categories are helpful in minimizing

Table 3.6. Incidence density rate calculations

When the population is not static, that is, when animals may leave or enter the population during the period of interest, calculation of a cumulative incidence rate is not possible. Because animals are entering and leaving the population, the time at risk must be explicitly accounted for. We do this by counting animal-weeks (or months or years) at risk. In the example given in Table 3.6, we count calf-weeks at risk—one calf followed for 1 week provides 1 calf-week of risk; 100 calves followed for 2 weeks provide 200 calf-weeks of risk. In the week a calf is born, we count half a week of risk, assuming that some calves are born at the beginning of the week, some in the middle, and some at the end: week 1 calf-weeks at risk = calves beginning week + (new calves born − morbidities)/2 = 0 + (20 − 1)/2 = 9.5; week 2 calf-weeks at risk = 19 + (20 − 2)/2 = 28; and week 3 calf-weeks at risk = 37 + (20 − 4)/2 = 45.

If there are differences in risk over time (e.g., most morbidity occurs in the first 3 weeks after birth), then we need a different IDR for calves less than 3 weeks old and for calves over 3 weeks old. Therefore, in week 4, we must begin to subtract the calves that are over 3 weeks of age from the population at risk: week 4 calf-weeks at risk = 53 + (10 − 5)/2 − (20 − 1)/2 = 46; calves beginning week + (new calves born − morbidities)/2 − (calves born week 1 − morbidities week 1/2) = 46; week 5 calf-weeks at risk = 58 + (10 − 4)/2 − [20 − 1+(20 − 2)/2] = 33; calves beginning week + (new calves born − morbidities)/2 − (calves born week 1 − morbidities week 1 + calves born week 2 − morbidities week 2/2) = 33. The IDR is the number of morbidities divided by the calf-weeks at risk.

In this example, you can see that although the weekly morbidity count is highest in week 4, the incidence of calf morbidity is highest in week 6, so calves at risk are more likely to get sick in week 6, a fact that is only identified by the IDR calculations.

Week	Calves at Risk, Beginning Week	New Calves Born	Calves at Risk End Week	Calves < 3 Weeks of Age	Weekly Morbidity Calves < 3 Weeks of Age	Calf-Weeks at Risk	Incidence Density Rate Weekly Incidence Rate (cases per calf-week of risk)
1	0	20	19	20	1	9.5	0.105
2	19	20	37	40	2	28.	0.071
3	37	20	53	60	4	45.	0.089
4	53	10	58	60	5	46.	0.109
5	58	10	64	50	4	33.	0.121
6	64	10	71	40	3	22.5	0.133

bias and inconsistency in classification. Although no system that we may devise will correctly classify all cases, we must avoid misclassifying mortality causes such as acidosis in a calf (elevated respiratory rate and depression) as respiratory disease. Significant misclassification of mortality causes will obscure our ability to detect risk factors associated with a particular disease. Overspecification of cause of death into many different categories may also obscure important findings. We want to accurately sort mortality into meaningful categories for the purpose of management decision making. If diseases A, B, and C all have similar risk factors, then keeping separate data on them may not be useful from a management standpoint. For example, most cases of neonatal diarrhea in beef calves have similar risk factors regardless of the etiologic agent involved. Spending the money and time to differentiate them may not be financially wise or useful. Diagnostic laboratory support can be helpful in classifying mortality causes, but it may not be economical or practical to implement in all cases.

The problem is more serious for morbidity. Although everyone agrees on what mortality is, not everyone agrees on what morbidity is. That is, not everyone will agree on which calves are sick irrespective of specific morbidity categories. Within a ranch, different individuals may not have the same criteria for classifying calves as morbid or well. Even the same individual on the ranch may not classify calves as morbid or well in the same way every day. If disease rates are to be compared between ranches, then differences in morbidity classification between them further complicate the issue. When we attempt to classify morbidity into syndrome-specific categories, we further increase the problem, as outlined above. This leads us to the need for validating disease diagnoses. We need to correctly assign calves not only to "morbid" and "well" categories but also to specific morbidity categories, and to do so consistently from day to day and from person to person.

If morbidity records are to be useful in a production setting, they must represent what is happening in the herd and do so consistently from day to day and year to year. This can be facilitated by having a definition of what constitutes a morbidity for each syndrome of importance to the ranch. For example, we might set up a neonatal diarrhea case definition that involves calf age, assessment of depression, assessment of dehydration, presence of diarrhea, and temperature. If a calf fulfills certain criteria, it is a "case."

Obvious questions arise with this definition. Not all calves with neonatal diarrhea have diarrhea when the animal is first recognized as sick; the calf could be too dehydrated to have diarrhea. Nor do the calves always have an elevated temperature; in fact, some will have a subnormal temperature. This serves to remind us that the case definition must be carefully thought out and validated if it is to be useful.

A case definition intends to standardize the criteria for a case so that rates are comparable between groups. Take, for example, a ranch with two ranch hands, each responsible for calving and caring for half the herd. Ranch hand A calves out group A and Ranch hand B calves out group B. If the neonatal diarrhea rates are different between group A and B, is that a result of actual differences between the groups, or of differences in what Ranch hand A and B call a "sick calf"? A proper case definition can help resolve much of this type of discrepancy and make rates between the groups comparable.

The assessment of signs that make up a case definition is subjective in nature; for example, there are no objective criteria for what constitutes depression. Therefore, significant effort should go into validating case definitions. A case definition is validated when it can be shown to identify what it is intended to indicate; that is, it correctly identifies neonatal diarrhea cases, for example, and is consistently applied over time and between individuals. This is accomplished by examining the accuracy with which the case definition criteria identify cases compared to some other established method, perhaps including laboratory support. Some validation and standardization may be achieved by comparing the producer identification of cases with veterinary identification. In practice, case definitions are rarely rigorously validated, but some level of effort is needed if the resulting data are to be useful.

An extension of the case definition concept thought to be more objective is that of a clinical scoring system. The intent of a clinical scoring system is to more objectively assess the clinical picture of an animal in indicating that it is a case. Typically, each of several signs is given a numeric score, and the sum of the scores determines the category of the animal. The most extensive use in beef production of clinical scoring systems has probably been in feedlot respiratory morbidity trials. Although a clinical scoring system does appear to be more objective, by assigning a number to each sign and assigning the calf to a morbidity category based on the numeric score, it is important to remember that the underlying individual scores are subjective assessments. The final case assignment is no better than the validity of the underlying subjective assessments.

Comparing Disease among Herds or Risk Groups

Assuming we have a good case definition among groups, we compare disease rates through the use of measures of association such as odds ratios (ORs) and relative risk (RR). Each of these measure the increased (or decreased) likelihood of disease associated with some exposure. They require data collected in yes/no categories such as morbid or not morbid, exposed or not exposed.

The odds ratio (OR) can be calculated for any data that are classified for disease presence/absence and exposure presence/absence. When the disease is relatively rare (<5% of the population), the OR accurately estimates the risk associated with exposure. When disease is common, the OR overestimates the risk associated with exposure. The OR is calculated as the odds of being diseased if exposed divided by the odds of being diseased if not exposed.

RR is the best measure of the risk of disease associated with exposure, but it cannot be calculated in all instances, depending on how the data were collected. RR is calculated as the rate of disease in the exposed group divided by the rate of disease in the nonexposed group. If the rate of disease in each group is not known (such as in a case–control study), the RR cannot be calculated.

MEASURES OF ASSOCIATION: RELATIVE RISK AND ODDS RATIO

Relative risk (RR) is calculated as the rate of disease in exposed animals divided by the rate of disease in nonexposed animals. It quantifies how much more (or less) likely an exposed animal is to have disease when compared to a nonexposed animal. From the table below it is equal to

$$\frac{A/A+B}{C/C+D}$$

The odds ratio (OR) derives from the RR. If disease is rare, then A is small relative to B and C is small relative to D, so

$$\text{RR} \cong (A/B)/(C/D)$$

and

$$\text{OR} = \frac{(A \times D)}{(B \times C)}.$$

The OR will overestimate the risk compared to RR and, if disease is not rare (A is not small relative to B and C is not small relative to D), the OR will overestimate the risk significantly. If the data we have are from a case–control investigation of a disease outbreak, we do not know the rates of disease in the population (we fixed them by selecting a set number of cases and controls), so we cannot calculate a real RR, and we are left with the OR as an estimate of the RR.

	Disease Yes	Disease No	
Exposed Yes	A	B	A+B
Exposed No	C	D	C+D
	A+C	B+D	A+B+C+D

For the table below, we have calves divided into the cells based on whether they had neonatal diarrhea and whether they suffered dystocia. The question is: Are calves that suffer dystocia more likely to subsequently suffer diarrhea? The RR and OR estimate how much more or less likely diarrhea will occur in calves that suffer dystocia. Conversely, RR and OR tell us how much dystocia increases the risk for diarrhea.

$$\text{RR} = (50/150)/(90/850) = 0.33/0.106 = 3.14$$

We interpret this equation to mean that calves that experience dystocia are three times more likely to have a case of neonatal diarrhea when compared to calves that do not experience dystocia.

$$\text{OR} = (50 \times 760)/90 \times 100 = 38000/9000$$
$$= 4.22.$$

The interpretation is the same as for the RR, but here we estimate that dystocia quadruples the risk of neonatal diarrhea.

	Neonatal Diarrhea Yes	Neonatal Diarrhea No	
Dystocia Yes	50	100	150
Dystocia No	90	760	850
	140	860	1000

From a practical standpoint, the OR and RR are interpreted similarly. An OR or RR of 5 indicates that exposed animals are five times more likely to have (or get) disease than are unexposed animals. Ratios may range in value from zero to positive infinity. Ratios greater than one indicate increased risk; ratios less than one indicate decreased risk associated with the exposure.

Diagnostic Testing

Sensitivity, Specificity, Positive Predictive Values, Negative Predictive Values

We broadly apply the concept of "test" to anything we use to categorize animals as healthy or sick. A physical exam or a morbidity scoring system could be considered a "test," as could a clinical blood chemistry, polymerase chain reaction, or enzyme-linked immunosorbent assay. That is, a "test" does not need to be performed in a laboratory setting to be considered a test and to have a recognizable sensitivity and specificity.

Sensitivity is sometimes used to express a test's ability to detect a small amount of something, and specificity is sometimes used to express a test's low level of cross reactivity. These are the general usage meanings of the terms. In epidemiology, sensitivity and specificity have technical meanings that seem to depart from common usage. Sensitivity is the test's ability to correctly detect or identify positive animals, so that only positive animals are needed to calculate the test sensitivity. In a sense, sensitivity is a measure of how good, or "sensitive," the test is at detecting the presence of an attribute. It says nothing about how the test performs when the attribute is not present.

Specificity is the test's ability to correctly detect or identify negative animals. In a sense, it is a measure of how specific the test is to the particular attribute of interest or how often it is misled by other attributes. It says nothing about how the test performs when the attribute is present.

Sensitivity and specificity are ways we express the test's ability to correctly reveal the status of the animal. In brief, sensitivity is the probability that a positive animal will test positive, and specificity is the probability that a negative animal will test negative.

In clinical practice, you are not presented with a "positive" animal—if you were, you would not need to run the "test." You are presented with an animal of unknown status. When you run the test, you are then presented with a positive or a negative result. Predictive values attempt to quantify the probability that the test result is correct. They are dependent on the prevalence of disease in the population; that is, positive predictive values (PPVs) increase with an increasing prevalence of disease in the population and with increasing specificity. The actual PPV depends on the prevalence of the attribute in the population. Negative predictive values (NPVs) increase with a

SENSITIVITY AND SPECIFICITY

Sensitivity is the proportion of disease-positive animals that are test positive; only positive animals are required in order to calculate test sensitivity, which is the probability that an animal with disease will have a positive test result. The table below illustrates a test with 90% sensitivity. The test was performed on 100 known positive animals and identified 90 of the 100 as positive: 90 test positives/100 true positives =90% sensitivity.

	Disease Positive	Disease Negative	
Test Positive	90 (TP)	0	90
Test Negative	10 (FN)	0	10
	100	0	100

TP = true positives: disease-positive animals identified by the test as positive.
FN = false negatives: disease-positive animals identified by the test as negative.

Specificity is the proportion of disease-negative animals that are test negative; only negative animals are required to calculate specificity, which is the probability that an animal without disease will have a negative test result. The table below illustrates a test with 90% specificity. The test was performed on 100 known negative animals and identified 90 of the 100 as negative: 90 test negatives/100 true negatives = 90% specificity.

	Disease Positive	Disease Negative	
Test Positive	0	10 (FP)	90
Test Negative	0	90 (TN)	10
	0	100	100

TN = true negatives: disease-negative animals identified by the test as negative.
FP = false positives: disease-negative animals identified by the test as positive.

POSITIVE AND NEGATIVE PREDICTIVE VALUES

Positive Predictive Value (PPV)

PPV is the probability that an animal is disease positive given a positive test result; that is, the probability that a positive test result is correct. PPV is dependant on the prevalence of disease in the population and on the specificity of the test. Therefore, for a test that is 90% sensitive and 90% specific, applied to a population with prevalence of disease = 10%, the PPV is as follows.

	Disease Positive	Disease Negative	
Test Positive	90 (TP)	90 (FP)	180 (TP + FP)
Test Negative	10 (FN)	810 (TN)	820 (TN + FN)
	100 (TP + FN)	900 (TN + FP)	1000

TP = true positives: disease-positive animals identified by the test as positive.
FN = false negatives: disease-positive animals identified by the test as negative.
TN = true negatives: disease-negative animals identified by the test as negative.
FP = false positives: disease-negative animals identified by the test as positive.

Thus,

$$\text{PPV} = \text{TP}/\text{TP} + \text{FP} = 90/180 = 50\%$$

so when you get a positive test under these circumstances, it is only correct 50% of the time; half of the test positive cows are really negative.

Negative Predictive Value (NPV)

NPV is the probability that an animal is disease negative given a negative test result; that is, the probability that a negative test result is correct. NPV is dependent on the prevalence of disease in the population and on the sensitivity of the test. Therefore for a test that is 90% sensitive and 90% specific, applied to a population with prevalence of disease = 10%, the above table illustrates the NPV.

Thus,

$$\text{NPV} = \text{TN}/\text{TN} + \text{FN} = 810/820 = 98.7\%$$

so when you get a negative test under these circumstances it is correct 98.7% of the time; only 1.3% of the test-negative cows are really positive.

decreasing prevalence of disease in the population and with increasing sensitivity. We generally do not actually calculate predictive values when we run tests, but it is wise to realize the effect of prevalence on test performance. Even a test with excellent sensitivity and specificity will have poor predictive values in situations in which the prevalence is low. The point is to not indiscriminately use tests on populations with low probability of disease but to attempt to identify higher-prevalence populations to improve test performance. For diagnostic laboratory tests, this may be best done by giving a good physical exam and taking a good history to identify those animals with a higher likelihood of disease, and then to apply the test to them.

Sample Size Requirements

In testing a population for a disease, we often just need to take a sample from it, rather than test the entire herd. There are significant economic and convenience advantages to sampling a population. However, this raises the issues of "how" we obtain our sample and "how many" animals we need to include.

There are numerous ways to obtain a sample from a population. These include nonprobability sampling methods such as judgment sampling, where representative samples are selected by the investigator; convenience sampling, where the sample is selected because it is easy to obtain; and purposive sampling on the basis of known exposure or disease status. Each of these methods may be useful and indicated in some circumstances. None of these methods, however, will reliably result in an unbiased estimate of the population mean or proportion; therefore, a given sample's expected performance cannot be predicted and no reliable inference can be made.

To legitimately represent the population for decision making, we need to obtain a random sample; that is, the "how" of sampling. A random sample is one where every animal in the population has equal probability of being selected for sampling; it is an unbiased sample of the animals. A random sample gives us an unbiased estimate of the parameters we are interested in. This is just what we need from a sample if we are to make correct decisions from it. (By unbiased, we mean that if we sampled the population over and over again, the mean of all those sample means would equal the true mean of the population. That is, it would not be displaced away from the true value.)

We can obtain a random sample by using a method to identify which cows are to be tested that is independent of the cows or our manipulation. A table of random numbers can be used to select ID numbers for sampling. Drawing chips out of a hat is a simple way to select cows for sampling (so long as a selector does not peek in the hat for the "right" color of chip). The selector must have a chip for every cow, and if ten samples are wanted, he or she must put in ten blue chips and the rest must be white. Flipping a coin is an effective way to assign cows to two or three different categories, but it becomes difficult when more categories are involved. Modern spreadsheet programs can generate random numbers for selection of cows as well. If you have a list of cow ID numbers, you can enter them in the spreadsheet, generate a column of random numbers next to them, and then sort the IDs and random numbers in numerical order of the random numbers. The cow ID numbers for sampling will be sorted in random order.

SYSTEMATIC SAMPLING OF HERDS

If we have 110 cows and need 12 samples from them, we divide the number of cows (N) by the needed sample size (n) to get the sampling interval (k).

$$\left({}^{N}\!/_{n}\right) = k = \left({}^{110}\!/_{12}\right) = 9.17.$$

We round down to make sure we get enough samples, and we sample every ninth cow. We do need to randomly select the first cow to be sampled from the first nine cows through the chute. Randomly selecting a number between one and nine can work for a starting point.

In reality, a "true" random sample is sometimes difficult to get in a production setting. A realistic alternative is a systematic sample. In a systematic sample, we sample cows at regular intervals, such as every tenth cow, as they come through a chute. For example, sampling every tenth cow will result in sampling 10% of the herd. We do need to randomly select the first cow to be sampled from the first ten cows through the chute. Randomly selecting a number between one and ten can work for a starting point. Systematic sampling is probably the best, that is, the most practical and effective, way to get samples. You do need to know in advance how many animals are in the group to make the appropriate calculations.

Stratified random sampling is another method we may use. In this case, we may apply the same principles discussed above, but we sample randomly from strata defined by characteristics that are likely to affect the attribute being estimated. Sampling might be stratified by age, sex, breed, region, county, and so forth. In this case, we need to calculate sample size estimates for each stratum we are sampling.

Once we know how to get our sample, the next step is determining how many animals we need to sample. The answer depends on what information we need; the sample size for each will be different, so we must identify specifically what we need to know first. There are generally three questions to be answered at this point.

1. Is a particular disease or attribute present in a population? All we want to do is detect disease if it is present. This is the question we need to answer if we desire to categorize herds according to status, whether positive or negative.
2. What proportion of the population is affected by a particular attribute? Here we want to know what the prevalence of a disease in the population is. This might be important in assessing the potential economic return of a control program.
3. What is the mean value of an attribute in a population? This might commonly be a measure of performance; for example, the mean calf weight per day of age (WDA) for the population.

Sampling to Detect Disease

We'll start with the first question above, identifying whether a particular disease or attribute is present in a population. If we want the highest level of confidence possible, then we must test all the animals. If, however, we are satisfied with some lower level of certainty (say, 90% or 95%), then we can take a

SAMPLING TO DETECT DISEASE

Suppose we have a herd with a Bovine Leukemia Virus (BLV) prevalence of 10%. The probability that we would detect BLV in the herd by taking one random sample is 0.1, or 10%. Conversely, the probability of failing to detect an animal with BLV is 0.9, or 90% (1−prevalence).

The probability of failing to detect an animal with BLV after taking two samples is (0.9 × 0.9 = 0.81 = 81%, and after taking three samples it is (0.9 × 0.9 × 0.9) = 0.73 = 73% (or a 27% chance of detecting a positive).

The probability of failing to detect an animal with BLV after any number of samples is $(1 - \text{prevalence})^n$, where n is the number of samples taken.

Therefore, if we increase n until the probability of failing to detect an animal with BLV is less than 10%, we would be about 90% certain of detecting a BLV positive in a herd with 10% prevalence. That would be $(1 - 0.1)^{22} = 0.098$, or 22 samples, which leaves us with a 10% chance of failing to detect an animal with BLV if the prevalence is 10%. (We would be 90% sure that prevalence is less than 10%.) If we wanted only a 5% chance of failing to detect an animal with BLV given a 10% prevalence, we'd need to sample 29 cows.

That is the easy and intuitive answer to the questions of how many animals we need to test, and it may be useful for demonstration, but unfortunately, it is not entirely accurate. In reality, once we select one animal, we do not replace it to be selected again, so the probability of a positive changes as we sample. The equation to calculate an appropriate sample size is

$$n \cong 1 - \left(1 - \beta^{1/d}\right)\left(N - \frac{d-1}{2}\right),$$

where n is the number of animals to be sampled, β is the probability there are no positives in the sample, d is the number of positives in the population, and N is the number of animals in the population.

This equation allows for adjustment for herd size. An adjustment for herd size is intuitively obvious because, for example, if you had a herd of 20 cows with a BLV prevalence of 10%, you could not sample 22 of them to be 90% certain of detecting a positive.

sample of the herd. The number of samples we need to take is based on the expected prevalence in the population and on the level of certainty we want. As you might expect, the sample size needed to detect a disease goes down as the prevalence in the herd goes up. All we need is to take enough samples to have a sufficient probability of selecting a positive animal if the herd is positive at or above some expected prevalence (see Table 3.7).

Sampling to Estimate a Proportion

The second question we might need to answer is what proportion of the population is affected by a particular attribute. Suppose we have a herd in which we want to determine the prevalence of Bovine Leukemia Virus. We can take n samples and calculate the prevalence by

$$p = \frac{n}{N}.$$

The question then is, How accurate is our estimation of the prevalence? If the true prevalence is 25% and we take ten samples, how many positives will we get? If we took repeated samples of ten from the herd, the most common number of positives should be two or three, but how commonly would we get only one or zero? How commonly would we get six or seven or more positives? The answer to that question describes the sampling distribution for BLV prevalence in the herd. That distribution can be described by a mean and standard error. What we have to decide is how wide we want the distribution to be and how certain we want to be about the answer we get. Obviously, the more samples we take, the more accurate an estimate we will get of the population prevalence. If we take more samples, we will be less likely to get extreme estimates of the prevalence that are far off from the true population proportion (see Table 3.8).

Table 3.7. Disease detection sample size table: Sample size to detect a herd attribute.

Herd Size	Prevalence				
90% Confidence Interval	1%	5%	10%	25%	50%
10	10	10	9	6	3
20	20	18	13	7	3
30	30	23	16	7	3
40	40	27	17	7	3
50	50	30	18	7	3
60	59	32	18	8	3
70	68	33	19	8	3
80	76	34	19	8	3
90	83	35	19	8	3
100	90	36	20	8	3
125	105	38	20	8	3
150	117	39	20	8	3
175	128	40	21	8	3
200	136	40	21	8	3
225	144	41	21	8	3
250	150	41	21	8	3
275	155	41	21	8	3
300	160	42	21	8	3
400	174	42	21	8	3
500	184	43	21	8	3
600	190	43	21	8	3
700	195	43	22	8	3
800	199	44	22	8	3
900	202	44	22	8	3
1000	205	44	22	8	3
100,000	229	45	22	8	3
95% Confidence Interval					
10	10	10	10	6	4
20	20	19	15	8	4
30	30	26	18	9	4
40	40	31	20	9	4
50	50	34	22	9	4
60	60	37	23	10	4
70	69	40	23	10	4
80	78	41	24	10	4
90	87	43	24	10	4
100	95	44	25	10	4
125	114	47	25	10	4
150	129	48	26	10	4
175	143	50	26	10	4
200	155	51	27	10	4
225	165	51	27	10	4
250	174	52	27	10	4
275	182	53	27	10	4
300	189	53	27	10	4
400	210	54	27	10	4
500	224	55	28	10	4
600	235	56	28	10	4
700	243	56	28	10	4
800	249	56	28	10	4
900	254	57	28	10	4
1000	258	57	28	10	4
100,000	298	58	28	10	4

SAMPLING TO ESTIMATE A PROPORTION

The sample size required to estimate a mathematical mean value in two herds is estimated by the equation:

$$n_i \cong \frac{Z^2 p(1-p)}{d^2}$$

to give the sample size in an infinite size population, and adjustment for the size of the herd is made by the equation

$$n \cong 1/\frac{1}{n_i} + \frac{1}{N},$$

where Z is the confidence interval coefficient, p is the expected prevalence in the herd we are sampling, d is the confidence bound, n_i is the infinite sample size, n is the herd-adjusted sample size, and N is the size of the herd.

Suppose we have a herd in which we want to determine the prevalence of BLV. We can take n samples and calculate the prevalence by

$$p = \frac{n}{N}.$$

The confidence interval coefficient, Z in the above equations, is how certain we are that our sample estimate lies within the confidence interval we have defined: 1.68 defines a 90% confidence interval, and 1.96 defines a 95% confidence interval. The more certain we want to be, the more samples we need to take.

The expected prevalence is what we would estimate prevalence to be. We get the highest sample size at a prevalence of 50%. The confidence bound represents how wide we want our confidence interval to be (e.g., +/−5%, +/−10%).

Table 3.8. Proportion estimation sample size table: Sample size to estimate a proportion.

	Estimated Prevalence and Confidence Width (+/−)					
Herd Size	5% and 3%	10% and 5%	25% and 5%	25% and 10%	50% and 5%	50% and 10%
90% Confidence Interval						
10	9	9	10	8	10	9
20	18	17	18	15	19	16
30	25	23	26	19	27	21
40	32	29	34	23	35	26
50	37	34	40	26	42	29
60	43	38	47	28	49	32
70	48	41	53	30	56	35
80	52	45	58	32	62	37
90	56	48	63	33	68	40
100	60	50	68	35	74	41
125	68	56	79	37	87	45
150	75	61	88	39	98	48
175	80	64	96	41	108	50
200	85	67	103	42	117	52
225	90	70	109	43	125	54
250	93	72	115	44	133	55
275	97	74	120	44	139	56
300	100	76	124	45	145	57
400	109	81	138	47	165	60
500	115	84	149	48	180	62
600	119	87	156	49	192	63
700	123	89	163	49	201	64
800	126	90	167	50	209	65

Table 3.8. Continued

Herd Size	Estimated Prevalence and Confidence Width (+/−)					
	5% and 3%	10% and 5%	25% and 5%	25% and 10%	50% and 5%	50% and 10%
90% Confidence Interval						
900	128	91	171	50	215	65
1000	130	92	175	50	220	66
Infinite	149	102	212	53	282	71
95% Confidence Interval						
10	10	9	10	9	10	9
20	18	17	19	16	19	17
30	26	25	27	21	28	23
40	33	31	35	26	36	28
50	40	37	43	30	44	33
60	46	42	50	33	52	37
70	52	46	56	36	59	40
80	57	51	63	38	66	44
90	62	55	69	40	73	46
100	67	58	74	42	79	49
125	77	66	87	46	94	54
150	86	72	99	49	108	59
175	94	77	109	51	120	62
200	101	82	118	53	132	65
225	107	86	126	55	142	67
250	112	89	134	56	151	69
275	117	92	141	57	160	71
300	121	95	147	58	168	73
400	135	103	167	61	196	77
500	144	108	183	63	217	81
600	152	112	195	64	234	83
700	157	115	204	65	248	84
800	162	118	212	66	260	86
900	165	120	218	67	269	87
1000	169	121	224	67	278	88
Infinite	203	138	288	72	384	96

Sampling to Estimate a Mean

The third question that we might need to answer is, What is the mean value of an attribute in a population, say, calf WDA? This example is similar to the last one in that it involves some distribution issues related to how certain we want to be about our estimated mean. There are some differences because we are sampling a continuous distribution (calf WDA values might range from 1.5 to 3.0 lb), but we do not need to deal with them specifically. We do need to know what normal is and what an expected standard deviation is. Then we need to decide how sure we want to be about our estimate. We do this by selecting a confidence interval (90% or 95% usually) and by selecting how wide we want that confidence interval to be as a percentage of the standard deviation (see Table 3.9).

Outbreak Investigation

Outbreak investigation is a powerful tool to identify those factors in a herd that are contributing to increased disease rates or decreased production. It can serve to critically and objectively evaluate the effect in a particular herd of known risk factors. In addition,

SAMPLE SIZE TO ESTIMATE A MEAN

The equation for sample size is

$$n_i \cong \frac{Z^2}{d^2},$$

and the herd size adjustment

$$n \cong 1/\frac{1}{n_i} + \frac{1}{N},$$

where Z is the reliability coefficient (how certain are we that our sample estimate lies within the confidence interval we have defined): 1.68 defines a 90% confidence interval, and 1.96 defines a 95% confidence interval. The more certain we want to be, the more samples we need to take. The variable d is the confidence bound expressed as a percentage of the standard deviation (desired confidence width/expected standard deviation), n_i is the infinite sample size, n is the herd-adjusted sample size, and N is the size of the herd.

For example, say we want to estimate the mean weight per day of age for 200 calves. If we expect the standard deviation in weight per day of age to be 0.5 lb and we want an estimate with a 95% confidence interval for +/− 0.25 lb, then the reliability coefficient (Z) is 1.96 and the confidence bound (d) is 0.5, or 0.25/0.5. From table 3.9 we would need 14 samples. If we wanted our confidence bound to be +/−0.1 lb, we would need 65 samples. The right number of samples can tell us whether the herd average weight per day of age is normal or low. An inadequate sample size does not give us that information and might mislead us, and too large a sample size wastes money and time.

Table 3.9. Mean estimation sample size table: Sample size to estimate a mean.

	Standard Deviation Units							
Herd Size	0.1	0.2	0.3	0.4	0.5	0.6	0.7	0.8
90% Confidence Interval								
10	10	9	8	6	5	4	4	3
20	19	16	12	9	7	6	4	4
30	27	21	15	11	8	6	5	4
40	35	26	18	12	9	7	5	4
50	42	29	19	13	9	7	5	4
60	49	32	21	14	10	7	5	4
70	56	35	22	14	10	7	5	4
80	62	37	23	14	10	7	5	4
90	68	40	23	15	10	7	5	4
100	74	41	24	15	10	7	5	4
125	87	45	25	15	10	7	6	4
150	98	48	26	16	10	7	6	4
175	108	50	27	16	11	8	6	4
200	117	52	27	16	11	8	6	4
225	125	54	28	16	11	8	6	4
250	133	55	28	16	11	8	6	4
275	139	56	28	17	11	8	6	4
300	145	57	28	17	11	8	6	4
400	165	60	29	17	11	8	6	4
500	180	62	30	17	11	8	6	4
600	192	63	30	17	11	8	6	4
700	201	64	30	17	11	8	6	4
800	209	65	30	17	11	8	6	4

Table 3.9. Continued

Herd Size	Standard Deviation Units 0.1	0.2	0.3	0.4	0.5	0.6	0.7	0.8
90% Confidence Interval								
900	215	65	30	17	11	8	6	4
1000	220	66	30	17	11	8	6	4
Infinite	282	71	31	18	11	8	6	4
95% Confidence Interval								
10	10	9	8	7	6	5	4	4
20	19	17	14	11	9	7	6	5
30	28	23	18	13	10	8	6	5
40	36	28	21	15	11	8	7	5
50	44	33	23	16	12	9	7	5
60	52	37	25	17	12	9	7	5
70	59	40	27	18	13	9	7	6
80	66	44	28	18	13	9	7	6
90	73	46	29	19	13	10	7	6
100	79	49	30	19	13	10	7	6
125	94	54	32	20	14	10	7	6
150	108	59	33	21	14	10	7	6
175	120	62	34	21	14	10	8	6
200	132	65	35	21	14	10	8	6
225	142	67	36	22	14	10	8	6
250	151	69	36	22	14	10	8	6
275	160	71	37	22	15	10	8	6
300	168	73	37	22	15	10	8	6
400	196	77	39	23	15	10	8	6
500	217	81	39	23	15	10	8	6
600	234	83	40	23	15	10	8	6
700	248	84	40	23	15	11	8	6
800	260	86	41	23	15	11	8	6
900	269	87	41	23	15	11	8	6
1000	278	88	41	23	15	11	8	6
Infinite	384	96	43	24	15	11	8	6

an outbreak investigation can serve to identify factors or management practices that are increasing disease even if we do not know why. This is, of course, dependent on having good disease, management, and production records. Enzootic bovine abortion is a good example of this. Management factors to control the disease were identified through epidemiological studies before the causative agent was known, and indeed, it is still not known.

Outbreak investigation can be applied into two general areas: investigation of disease outbreaks, which can be defined as the occurrence of disease in a population at a level in excess of that expected, and investigation of impaired productivity, which is performance at a level below that expected or below target goals.

For the purpose of this discussion, we will divide outbreak investigation into five stages: planning, investigation, analysis, intervention, and monitoring.

Although they provide a good framework for discussion and implementation, these areas are not exclusive of each other, nor are they meant to be carried out in a rigid manner.

Planning

The first thing we need to do is develop a plan of approach to the investigation. What is the disease manifestation or production impairment, and how will we determine the cause? We need to decide what information we need to collect, and how we will collect it. This is probably the most important step if the investigation is to get off on the right foot. The goal is to identify "key determinants" responsible for increased disease in a herd. Key determinants are those factors that are under management control and that we can alter to affect disease rates.

This approach to disease looks beyond the infectious agent theory of disease: no single factor is considered the "cause," and "infection" and "disease" are not synonymous. Management factors are considered to be more important in dealing with disease and production than are individual infectious agents. Returning the herd to normal is the ultimate goal, but it is more accurately the goal of the intervention plan.

Much of what we do in the planning phase of an investigation and the construction of a path model is to evaluate and review risk factors that increase the risk of disease or decrease the level of performance in the herd. Therefore, we need to conduct a comprehensive review of management factors that contribute to risk.

Risk factors require careful interpretation and can be strongly influenced by confounding or by the mix of management factors among herds. An association between a factor and a disease does not necessarily equal causation. Our knowledge of factors that contribute to risk is based on three things: conjecture, informal observation, and scientific study.

In conjecture (or the "common sense" or "deductive" approach, we deduce specific risks for a disease based on our pathophysiologic model of it. Because our model is flawed, our deductions are often flawed as well. This method has a fairly poor record of identifying risk factors.

Informal observation, or experience, comes into play after a veterinarian has been practicing for a time, when he or she begins to associate certain risks with diseases. This is a better method than conjecture, but it is still subject to bias. Risk factor/disease interactions are complex and are often confounded by multiple other factors. It is probably impossible in most cases to sort out the confounding and bias in disease and production on an informal basis. This difficulty is well evidenced by the large number of competing and contradictory "clinical impressions" of veterinarians. Nonetheless, informal observation is a better method than conjecture, and often it is all we have to go on.

Scientific studies, when properly done, are able to assess the relative effect of a factor and to control for confounding. The effect of the factor is isolated through the use of design elements in the study and of statistical methods in the analysis. When properly done, this is the best method to assess risk factors. Well-done, field-based clinical trials are the most valuable studies for assisting in production decisions. Not all studies, however, are well done, and some should be relegated to the informal observation category. Well-done field trials should include the following factors: assessment of a clinically relevant outcome (morbidity, mortality, performance), random assignment of calves to treatment and control groups, blinded assessment of clinical outcomes (especially important for morbidity outcomes), and appropriate statistical analysis (correct application of statistical tests and control of herd clustering effects). Unfortunately, in many cases, well-done field trial data are lacking (Perino and Hunsaker, 1997).

From our assessment of the risk factors associated with disease, we can construct path models of the disease or production issue. Path models serve as a useful client education tool in helping producers understand both the complexity of the problem and that there may not be any quick, easy answers (see Figure 3.4) In a schematic form, path models serve to summarize the current understanding of disease causality, demonstrate the interrelationships among risk factors, and provide focus for planning an investigation.

Investigation

The second of the five steps of outbreak investigation listed above is investigation. Once we develop a plan for the investigation, we need to conduct the investigation and collect our data. We need to collect disease data on individual animals as well as on the herd as a whole. We need to look at the following patterns.

- Temporal patterns, or when cases of disease occurred in time: These patterns are important in characterizing the outbreak as well as associating it with management or environmental changes.
- Spatial patterns, or where cases occurred (and where they did not): These patterns are very important for finding initial clues about where to look harder for key determinants.

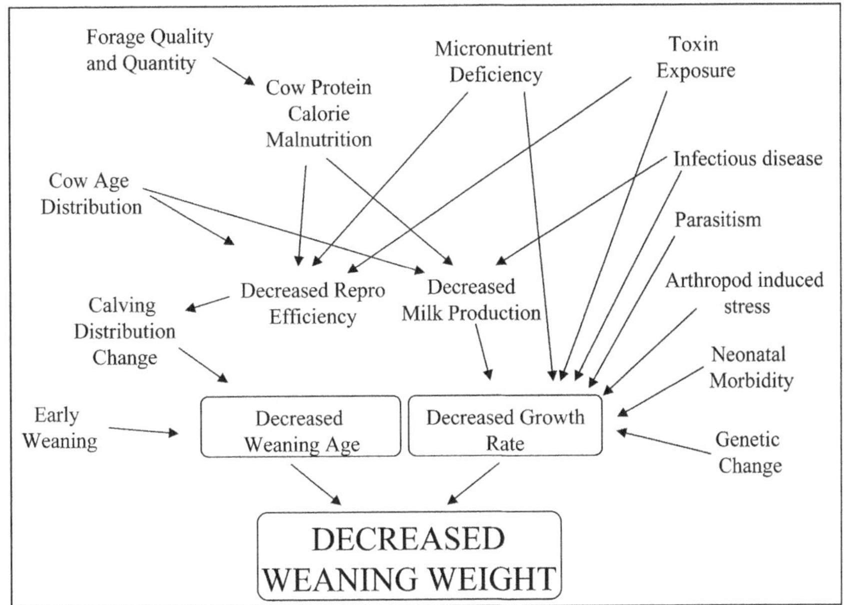

Figure 3.4. Path model for weaning weight depression.
Sanderson and Christmas, 1997

- Risk groups, or which categories of animals are most affected: For instance, first-calf heifers, old cows, newborn calves, and three-week-old calves.

The first rule of data collection is to collect actual data and not just client impressions. Sometimes impressions are all you have to go on historically, but they are subject to large bias and can be quite misleading. Always collect as much current and historical data as possible:

- List all cases
- Record the dates of disease
- Indicate group memberships such as age, breed, pen, string, diet, treatments, management procedures, and so forth
- Record the number of animals in each group.

The same rules apply to the collection of performance data: We need to have as much actual and detailed data as possible. Sometimes we have only group weaning weights and rough numbers of cows, but we need to collect as much actual data as possible rather than relying on the clients' impressions.

Target goals are expectations of production levels. In herds with which a veterinarian routinely works, he or she may have established goals for comparison. If not, there are expected levels of production for well-managed herds that can serve as a starting point. Historical performance is very important in establishing the existence of an outbreak: If we do not know what performance or disease rates were in the past, we cannot be sure whether they have changed now. The temporal onset of production depression or disease rate elevation can serve to focus the time parameters of the investigation. Unfortunately, many cow/calf producers do not keep performance records at a level that is useful for investigation.

Management data need to be collected in a similar way. Specific, actual management procedures should be collected, and collection should be targeted to the known and hypothesized risks based on the literature and intermediate results of the investigation. Initially, we need to list management factors to be evaluated. In some instances, it may be more efficient to prepare a questionnaire to provide a standard approach to data collection and to ensure that nothing is missed. The goal is to identify management factors that have changed and that could potentially be responsible for the outbreak.

Although we cannot do much about the weather, it is very important to collect data on environmental patterns because of the large effect they can have on disease rates and performance. Daily temperatures and fluctuations, precipitation, wind, and dust can be temporally associated with disease rates.

Once we have collected our data, we must stop and verify, or establish, the existence of an outbreak. That is, are the disease rates truly elevated or the performance truly depressed? Much time can be wasted on the detection of nonexistent epidemics resulting from client and veterinarian miscommunication or misconception. Apparent changes in disease or production can relate to a rising number of cases caused by increasing herd size or altered herd composition. Depressed weaning weights may be caused by a change in calving time, weaning time, or bull genetics.

Attack rates can be used to help verify an outbreak when they are compared with historical levels of disease. Attack rates are a form of CIRs and are calculated as

$$\text{Attack rate} = \frac{\text{Total number of cases during a defined time period}}{\text{Total number of animals at risk during that time}}.$$

Another useful measure is a true incidence rate of morbidity. True incidence rates take time into account, and so are dynamic measures of morbidity, giving us a day-by-day or week-by-week look at morbidity. They give us a real-time look at disease rather than a retrospective look.

$$\text{Morbidity rate} = \frac{\text{Number of new cases}}{\text{Animal time at risk}}.$$

Once we have convinced ourselves that an outbreak exists, we need to establish a diagnosis and define what a case is. An erroneous diagnosis may mislead the investigation by focusing on inappropriate risks and initiating inappropriate control strategies, and a single lab result may be inadequate to establish a diagnosis. We need to take into account how representative the samples are, how good the sample quality is, and what the test attributes are.

To be able to evaluate the effects of factors on the incidence of disease, a case definition is extremely important. The case definition can be an animal, group, or farm (a case can be an individual, a group of animals, or a farm), or it can be a clinical syndrome (age, sex, clinical signs, diet, lab results).

It is very important to set the case definition early in the investigation. Most techniques rely on comparison of cases and noncases with respect to a characteristic to evaluate risk. Extensive misclassification (placing true cases in the noncase category and vice versa) can be disastrous when trying to correctly identify risks. Some misclassification is unavoidable, but we need to minimize it.

Post mortem examinations can be very useful in case definitions. Any dead or terminal animals should be necropsied. (If economically plausible, the sacrifice of an animal in the early stage of disease for necropsy may be most helpful.) A small number of necropsies can have misleading results if they are not representative of the disease at hand. Before beginning to necropsy animals, it is useful to define the preliminary epidemiological characteristics of the syndrome at hand and to prepare a list of diagnostic possibilities you want to rule in or rule out. Sample according to the pathology you encounter, as well. In most instances in which you may be doing a full investigation, post mortem sampling will not provide a definitive diagnosis but, rather, will limit the diagnostic possibilities and provide clues for further investigation.

Analysis

After we have planned and carried out the investigation and collected our data, we need to analyze this information to see what answers or clues it may provide us. The first thing to do is to establish the temporal pattern of disease. We need a chronological occurrence of the cases, which is best graphed as a histogram, or a graphical representation of a frequency distribution. Once this is done, it can be quite useful to juxtapose this histogram with management and environmental data collected in the investigation. Association of increased disease rates with environmental or management change in a visual form is a powerful tool in investigation and client education.

We also need to establish the animal pattern of the outbreak. This includes specific information about the individual and class of animals involved. The pattern is established by comparing cases and noncases according to breed, sex, age, feed, period of gestation, and exposure to whatever other factors might be pertinent. We can then attempt to quantify the risk these factors have for disease by using odds ratios. Odds ratios are a comparison of the disease rates in exposed and nonexposed animals. They can help us see whether a particular factor is increasing disease rates.

It is also important to establish any spatial pattern to disease; that is, disease occurrence relative to place and animal movement patterns. Establishing a spatial pattern is best accomplished using a map of the farm, on which we can identify areas of morbidity or mortality and dates of animal movements. Mapping can be very useful in identifying clusters of disease associated with a particular area.

Intervention

Our intervention during an outbreak investigation needs to be based on the "key determinants" identified in the investigation and analysis steps. These "key determinants" are the management factors we can change that will affect disease rates in the herd. The power of this process is in its ability to identify practices that are affecting production or disease rates, even if we do not know why. We may be able to identify a particular ration, pasture, or management practice that elevates disease (or depresses production) and change our exposure to that factor even if we do not know why, or how, it caused the problem. Other factors are more commonly known and associated with disease rates, such as colostral management or proper replacement heifer development, and they may not require as formal an investigation. All of our management recommendations must take into account the economic reality of the farm: Recommendations must be profitable and take cash flow into consideration to be realistic and improve the financial status of the farm. All of this should be communicated to the client in a concise letter outlining the problem, the investigation conducted, and the recommended changes. This professional method of communicating with the client supplies a clear record of the service provided, recommendations made, and expectations for improvement.

Monitoring

After we have implemented our recommendations to control the outbreak, it is imperative that we monitor disease rates or performance data to evaluate the effectiveness of our management recommendations. First, we need to assess the effect of our recommendations and their adequacy in dealing with the problem. Second, we need to establish for clients the benefits accrued by following our recommendations. In many instances, this may require the establishment of new records in addition to continuing current record-keeping practices to assess improvements. The importance of this step in helping clients to recognize the value of the veterinarian in health management cannot be overemphasized. If we fail to show clients the profit accrued by following our recommendations, over time they will attribute it solely to themselves or to the problem just "going away."

Prospective Studies in Investigations

In many instances, there are inadequate records available to identify factors that are active in modulating disease rates. Many times we are called to do investigations when the problem is identified long after it occurred. In these "late investigations," the "smoking gun" of causation has already cooled (and rusted in many instances). Under these circumstances, we often come up short in identifying the "key determinants."

In prospective investigations, we need to focus on which questions must be answered to solve the problem. Once this is done, we can set up a records system to collect the necessary data to evaluate unknown key factors. This may include formal epidemiological studies such as cross-sectional, case–control, or cohort studies. Clients need to thoroughly understand the time and effort involved in this type of investigation before it is embarked on. Poor communication at this time almost guarantees an unhappy outcome.

Outbreak investigation is a powerful tool for use in clinical herd medicine to minimize disease rates and optimize production. When properly applied, it can serve to highlight the value of a veterinarian to the well-being of a farm enterprise.

BIOSTATISTICS

Purpose and Goals of Statistics

Statistics has been much maligned as a tool for those who want to lie with numbers, and indeed, statistics can be misused to obscure real relationships or generate spurious ones under the veneer of technicality. Properly used, however, statistical procedures are a valuable tool to sort out the complex data of production systems and support rational decision making. A thorough coverage of biostatistics is far beyond the scope of this chapter, and other texts devoted to this subject are available. Instead, this chapter will provide a brief introduction to biostatistics and an understanding of its usefulness in decision making for production systems.

We commonly need to make decisions about the difference between two or more production groups; for example, performance between different breeds or sire groups, or between the results of different supplements or implants on those groups. How do we know if there are real differences? Say we examine calves born in the first half of the calving season and those born in the last half of the calving season for differences in WDA. The mean WDA for the two groups will always be different, but how different does it need to be for us to conclude that one group grew faster than the other?

Statistics help us to sort out whether the difference is normal population variation or is caused by a real difference between the groups. As the question to be answered becomes more complex, statistical procedures become more important for differentiating between normal variation and variation resulting from extraneous factors, not from the variable of interest. In production management, we are often forced to make decisions based on limited data, and statistical tests may not provide the level of probability required in scientific journals. They can, however, help us assess both the quality of our data and the probability that real differences exist that would warrant management action.

Statistical Variable Types

Qualitative categorical variables are those variables where the subjects fit into one of two or more categories. There are many ways of dealing with this type of data. This chapter will cover calculation of the relative risk and odds ratio and of a chi-square test on these types of data. These variables may be divided into three kinds: dichotomous (or binary), nominal, or ordinal.

A dichotomous (or binary) variable is a yes/no proposition, such as morbidity status or pregnancy status. A dichotomous variable (such as pregnancy status) that is recorded as 0 for not pregnant and 1 for pregnant can be averaged to give the proportion of animals pregnant (the average of nine 1's and one 0 is 0.90, or 90% pregnant). We can also calculate a standard deviation for this type of data.

A nominal variable is usually a named category with no order, associated with different categories such as breed, color, or gender. One category is not clearly "better" than another.

An ordinal variable is an ordered category. There is an order to the different categories, but there is not a defined interval. For example, we might grade incoming calves as poor, fair, and good. Good is "better" than fair, but it is not definite how much better. We could also categorize pregnant cows into those that were bred in the first 21 days of the breeding season, those bred in the second 21 days, and so forth, and have ordered categories of expected breeding time. This would provide some additional information compared to having just pregnancy status, but less information than having the specific duration of pregnancy estimates.

The mean value of a nominal or ordinal variable is meaningless. Were we to label bulls, steers, and heifers as 1, 2, and 3, the average of the numbers would not be helpful. What would 2.4 mean? We summarize nominal and ordinal data by reporting the number or proportion of animals in each category, so we might have 10% bulls, 40% steers, and 50% heifers. In addition, there is no measure of variability for nominal or ordinal variables other than the distribution of animals within categories.

As opposed to qualitative categorical variables, quantitative variables are those that are measured on a continuous scale. They have a unit of measure and may take on any value. Examples include weight, age, duration of pregnancy, and postpartum interval. There are many ways of dealing with this type of data as well. In this chapter we cover calculation of a simple *t*-test to compare two means.

Summarizing Data

We commonly summarize data in many areas of life with the use of a mean or of the arithmetic average of a series of numbers. The mean WDA of a group of calves, however, does not capture all the information we need to make decisions. Not every calf in the population has the same WDA; there is variability in the population. We commonly quantify this variability by the standard deviation.

The variability in the herd as a whole is related to a number of variables. These include the genetic diversity in the herd in such issues as growth potential and milk production, as well as cow age and nutritional or environmental differences among groups. There is also variability related to how we measure the attribute we are interested in; for example, if we estimate calf weights with a heart girth tape, we will likely have a less precise estimate than if we use an actual weight scale.

Probability Distributions

A probability distribution is the frequency of occurrence of the values of a variable. It relates how probable a certain value is, given a certain mean and standard deviation. Two common probability distributions used in production medicine are the normal distribution and the binomial distribution.

The normal distribution (the bell-shaped curve) has nothing to do with how normal the subjects are (as opposed to abnormal). Rather, it merely describes a particular frequency distribution of a continuous variable. It is described by its mean and standard deviation and is symmetric around the mean of the distribution, with approximately 95% of its observations falling within two standard deviations on either side of the mean (see Figure 3.5). A major assumption of many statistical tests is that the data are normally distributed. If the distribution of the production data departs significantly from this normal distribution, the statistical test results may not be valid.

MEAN AND STANDARD DEVIATION

Continuous Data

The arithmetic mean of a sample is a measure of its center. For a continuous variable, it is obtained by adding all the observations in the sample together and dividing by the number of observations. For the ten WDA observations below, therefore, the mean is

$$\frac{\sum x}{n} = 2.5 + 1.7 + 1.1 + 1.6 + 1.7 + 2.1 + 1.8 + 2.7 + 1.9 + 3.0/10 = 2.0.$$

Calf	Weight per Day of Age
101	2.5
102	1.7
103	1.1
104	1.6
105	1.7
106	2.1
107	1.8
108	2.7
109	1.9
110	3.0

The standard deviation is a measure of dispersion; that is, how far the individual measurements vary from the mean. It is obtained for these ten observations by the equation

$$\sqrt{\frac{\sum (x - \bar{x})^2}{n - 1}} = \sqrt{(2.5 - 2.0)^2 + (1.7 - 2.0)^2 \ldots (3.0 - 2.0)^2/10 - 1} = 0.6.$$

Dichotomous (0/1) Data

For a dichotomous (0/1) variable, we calculate the mean in the same way as for a continuous variable. This value is the probability of "success," or 0.90 if 90% of the cows in a sample are pregnant. This probability of success is designated p. The standard deviation of this dichotomous variable is based on the probability of success p and the number of samples n. For a 90% pregnancy rate in 100 cows, the standard deviation is

$$\sqrt{\frac{p - (1 - p)}{n}} = \sqrt{\frac{0.9 - (1 - 0.9)}{100}} = 0.089.$$

The binomial distribution, as the name implies, can only take on one of two values commonly recorded as 0 (failure) or 1 (success). It is described by two parameters: n, the sample size, and p, the probability of a "success," usually recorded as a 1.

Pregnancy proportion is a binomial variable—cows are either pregnant or open—and we designate pregnant as a "success" and open as a "failure." When the sample size is moderately large ($n \times p$ and $n \times [1 - p]$ are both >5), the binomial distribution is approximately equal to the normal distribution.

As we increase the number of animals we sample, we increase the precision of our estimate of the mean value of those animals. For example, if we sample 50 animals, our estimate of the herd mean WDA is more precise than if we sample only 10 animals. So if we have more animals in our sample, we increase our ability to distinguish between two groups (See Figure 3.6). Our ability to distinguish between groups, then, is a function of two factors: the diversity within the herd genetically and environmentally, and the number of animals in our sample.

COMPARING MEANS

The t-test is the simplest method of comparing the mean of two continuous variables. For example, suppose we have 30 calves from each of two bulls, and we want to know whether the average daily gains from birth to weaning are different for the two bulls. First of all, to answer that question fairly, we need to have randomly divided the cows between the two bulls so that the cows with higher milk production or greater growth genetics are not more commonly bred by one bull rather than by another. If the cow groups are comparable, and the nutritional resources available to each are equivalent as well, we can compare the bulls by comparing the mean average daily calf gains with a t-test. The t-test will compare the

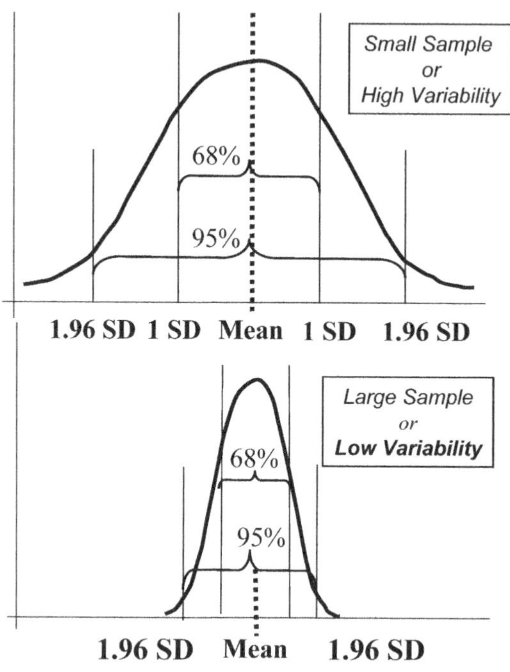

Figure 3.5. Normal and distribution effects of sample size.

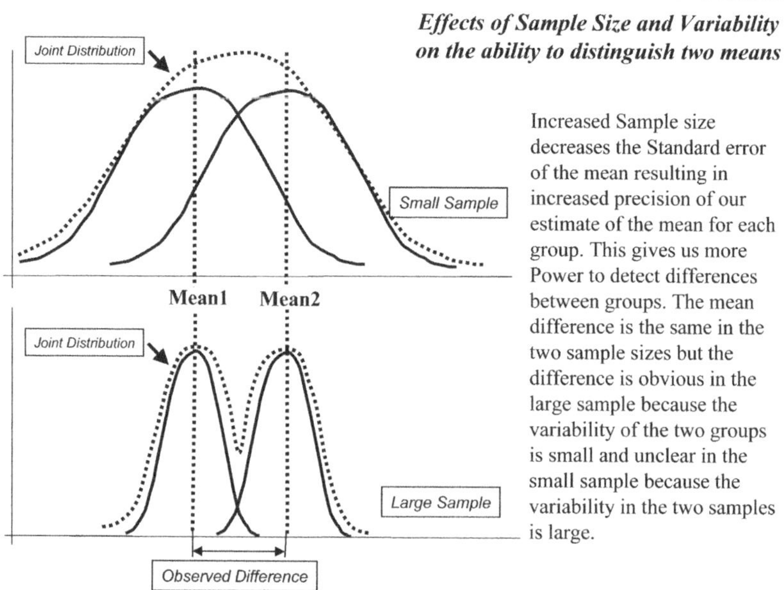

Effects of Sample Size and Variability on the ability to distinguish two means
Joint Distribution
Small Sample
Mean1 Mean2
Joint Distribution
Large Sample
Observed Difference
Increased Sample size decreases the Standard error of the mean resulting in increased precision of our estimate of the mean for each group. This gives us more Power to detect differences between groups. The mean difference is the same in the two sample sizes but the difference is obvious in the large sample because the variability of the two groups is small and unclear in the small sample because the variability in the two samples is large.

Figure 3.6. Effects of sample size and variability on the ability to distinguish two means.

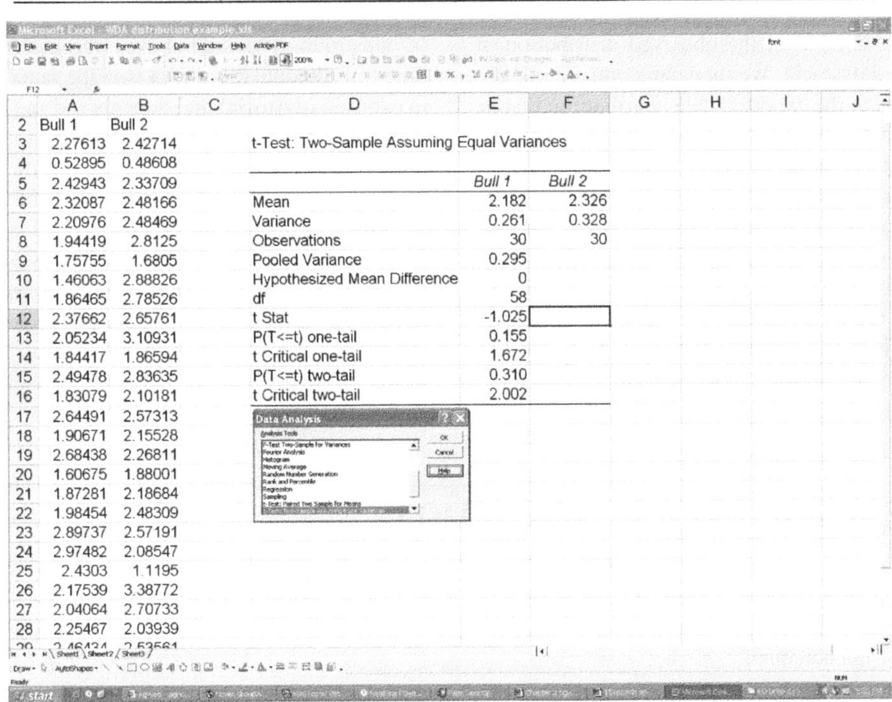

	A	B
2	Bull 1	Bull 2
3	2.27613	2.42714
4	0.52895	0.48608
5	2.42943	2.33709
6	2.32087	2.48166
7	2.20976	2.48469
8	1.94419	2.8125
9	1.75755	1.6805
10	1.46063	2.88826
11	1.86465	2.78526
12	2.37662	2.65761
13	2.05234	3.10931
14	1.84417	1.86594
15	2.49478	2.83635
16	1.83079	2.10181
17	2.64491	2.57313
18	1.90671	2.15528
19	2.68438	2.26811
20	1.60675	1.88001
21	1.87281	2.18684
22	1.98454	2.48309
23	2.89737	2.57191
24	2.97482	2.08547
25	2.4303	1.1195
26	2.17539	3.38772
27	2.04064	2.70733
28	2.25467	2.03939

t-Test: Two-Sample Assuming Equal Variances

	Bull 1	*Bull 2*
Mean	2.182	2.326
Variance	0.261	0.328
Observations	30	30
Pooled Variance	0.295	
Hypothesized Mean Difference	0	
df	58	
t Stat	-1.025	
P(T<=t) one-tail	0.155	
t Critical one-tail	1.672	
P(T<=t) two-tail	0.310	
t Critical two-tail	2.002	

Figure 3.7. Microsoft Excel calculation of two sample *t*-tests for the difference in calf average daily gain.

difference in the two means in relation to the amount of variability in the data (standard error).

A full treatment of the *t*-test is beyond the scope of this book, but a simple implementation of it is included in spreadsheet programs such as Microsoft Excel. The data need to be entered in columns and "Data Analysis" chosen from the "Tools" menu. The on-screen menus will guide a user through selecting the columns of data and generating the results (see Figure 3.7). For our example, average daily gain difference between the two bulls in Figure 3.7, we get a *P* value of .30, and we would not conclude that there was a difference between the two groups. What the *P* value really means is that if there truly is no difference between the two groups in average daily gain, we would expect to observe a difference this big or bigger about 30% of the time, so the observed difference between the two bulls would be fairly common even if there is no true difference between them. We usually use a *P* value of .05 or .10 as the point at which we conclude there really is a difference. If we had calculated a *P* value of .05, we would have interpreted it as follows: If there were no difference between the bulls, we would expect to see this big a difference only rarely (5% of the time), and so we conclude there is probably a difference between the two bulls.

Comparing Proportions

The chi-square test is a common way to compare proportions between two groups. It is based on a comparison between the observed proportions in each group and the expected proportions if there were no relationship between the groups and the exposure. Suppose, for example, we want to compare the proportion of calves that get sick in the first 21 days of life between those calves that experienced dystocia and those with a normal birth. As in comparing means, the groups need to be comparable except for the attribute of interest. Potential confounders in this comparison would include failure of passive transfer and dam age. If we really want to assess the effect of dystocia on morbidity with a simple chi-square test, then the two groups must have equal numbers of first-calf heifers and calves with failure of passive transfer. (If the groups are not truly comparable, we can use more sophisticated statistical methods to control for confounding.)

For the chi-square test, we can categorize our data into a 2 × 2 table of the observed distribution of disease and dystocia. We then can compare our observed data to the expected data distribution; that is, the distribution we would expect if there were no relationship between dystocia and disease. The result is a chi-square statistic that we compare with a chi-square table to arrive at a P value. For our example, we find $P < .01$ and conclude that there is a relationship between dystocia and disease. We may also set up an Excel spreadsheet to calculate the chi-square and P value when we type in our data (see Figure 3.8). We can also estimate the direction and magnitude of the relationship using relative risk or odds ratio, discussed above. Because our example is a full population, we know the rate of disease in exposed (dystocia) and nonexposed (no dystocia) animals. Therefore, we can calculate the relative risk as 2.5. In this example, calves that experienced dystocia were 2.5 times more likely to be diseased.

At the end of the day, biostatistics is a tool that we use to try and improve our ability to make good decisions. They can be misused (accidentally or intentionally) and may then mislead us and even increase our confidence in wrong decisions. Used

CHI-SQUARE TEST FOR PROPORTIONS

The chi-square test examines the difference between the observed proportions and the expected proportions if there were no relationship between exposure and outcome.

For a group of 100 calves, the observed outcome for dystocia and neonatal disease is summarized in this table:

		Disease		Totals
		Yes	No	
Dystocia	Yes	15	18	33
	No	12	55	67
Totals		27	70	100

Expected Outcome

The expected outcome is obtained by multiplying the row total by the column total and dividing by the table total for each cell, as shown here.

		Disease		Totals
		Yes	No	
Dystocia	Yes	33×27/ 100 = 8.9	33×63/ 100 = 24.1	33
	No	67×27/ 100 = 18.1	67×63/ 100 = 48.9	67
Totals		27	63	100

The chi-square statistic is obtained by summing the squared difference between the observed and expected cell frequencies divided by the expected frequency across all the cells:

$$\chi^2 = \Sigma \frac{(\text{Observed} - \text{Expected})^2}{\text{Expected}};$$

$$\chi^2 = \frac{(15 - 8.9)^2}{15} + \frac{(18 - 24.1)^2}{18} + \frac{(12 - 18.1)^2}{12} + \frac{(55 - 48.9)^2}{55} = 8.5.$$

The larger the number, the more likely the difference is significant; that is, a large chi-square value indicates that the observed values are very different from the expected values. For a 2 × 2 table, we look up this number in a chi-square table with one degree of freedom. (All chi-square statistics coming from a 2 × 2 table have one degree of freedom.)

Chi Square Critical Value Table				
	P value			
Degrees of Freedom	0.2	0.1	0.05	0.01
1	1.65	2.71	3.84	6.64

Our observed distribution of disease is different from the expected distribution, with $P < .01$ (chi square = 8.5 > 6.64)

The relative risk is

$$\frac{\text{Rate of disease in exposed}}{\text{Rate of disease in nonexposed}} = \frac{15/33}{12/67} = 2.5.$$

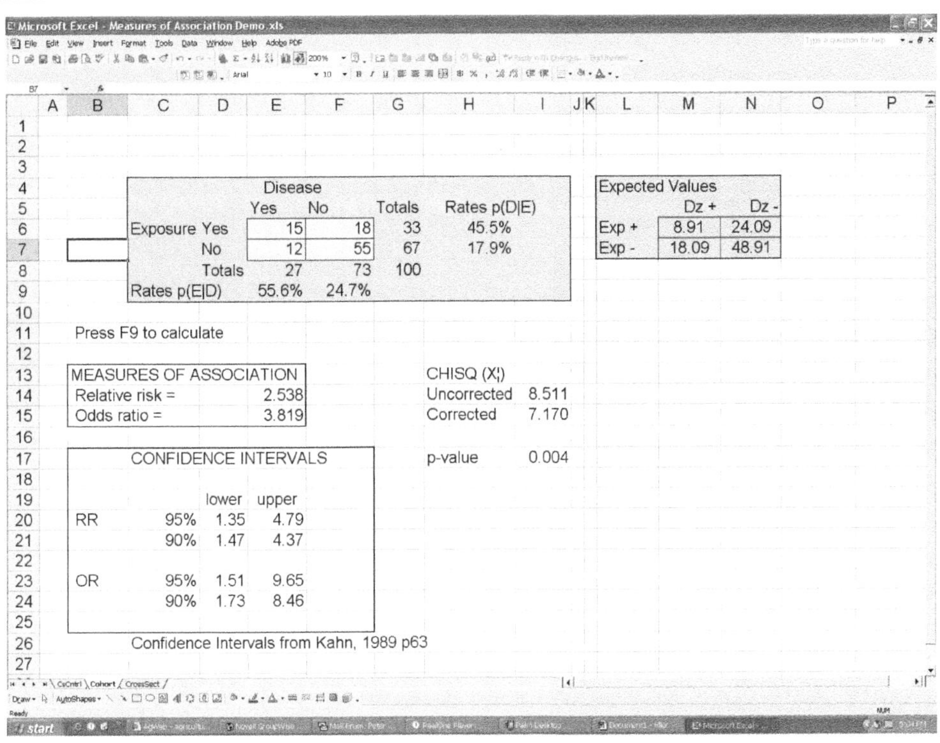

Figure 3.8. Microsoft Excel calculation of chi-square test for difference in proportion of calves sick following dystocia.

correctly, biostatistics can dramatically improve our ability to identify real differences and effects.

EPIDEMIOLOGY SOURCES

Slenning, B.D. 2001. Quantitative tools for production-oriented veterinarians. In *Herd Health: Food Animal Production Medicine*, edited by O.M. Radostits. Philadelphia, PA: WB Saunders.

Martin, S.W., A.H. Meek, P. Willeburg. 1987. *Veterinary Epidemiology: Principles and Methods*. Ames, IA: Iowa State University Press.

Dohoo, I., S. Martin, H. Stryhn. 2003. *Veterinary Epidemiologic Research*. Charlottetown: PEI Canada AVC Inc.

Smith, R.D. 1995. *Veterinary Clinical Epidemiology: A Problem-Oriented Approach*, 2nd ed. Boca Raton, FL: CRC.

REFERENCES

Chenoweth, P.J., M.W. Sanderson, 2001. Herd health and production management in beef cattle breeding herds. In: *Herd Health Food Annual Production Medicine*, 3d ed., edited by O.M. Radostits. Philadelphia, PA: WB Saunders.

Perino, L.J., B.D. Hunsaker. 1997. A review of bovine respiratory disease vaccine field efficacy. *Bovine Practitioner* 31(1):59–66.

Sanderson, M.W., R.C. Christmas, 1997. Investigation of depressed weaning weights in beef cow-calf herds. Compared. Contin. Educ. Pract. Vet. 19(3): 395–399.

Sanderson, M.W., J.M. Gay. 1996. Veterinary involvement in management practices of beef cow-calf producers. *Journal of the American Veterinary Medical Association* 208:488–491.

Spire, M.F. 1991. Cow/calf production records: Justification, gathering, and interpretation. *Proceedings of the American Association of Bovine Practitioners* 1:93–95. Indianapolis, IN: AABP.

4
Herd Health Management

Peter J. Chenoweth

INTRODUCTION

A cost-effective herd health program provides a solid foundation for economic viability—if not survival—in cow/calf operations. Here, herd health is often considered within the context of protection from diseases caused by infectious organisms. This causal relationship concept is, however, outdated, as we now recognize that a dynamic equilibrium usually exists between organism and host—an equilibrium that can be upset by a number of factors, leading to the expression of overt disease or lost production (Leman 1988). The interactions between pathogen and host response may be represented as below (Figure 4.1).

The term "immune system management" provides an inclusive concept of herd health (Stokka and Falkner 1998), whereby reproductive, nutritional, behavioral, environmental, and health management considerations overlap.

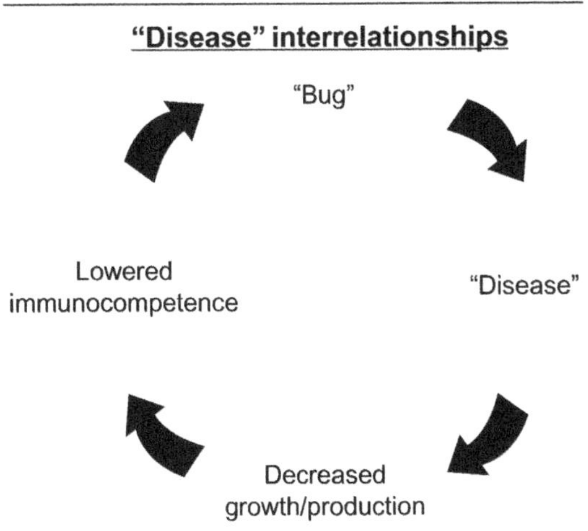

Figure 4.1. "Disease" interrelationships.

With most diseases in a herd, there are usually considerably fewer clinical cases than there are subclinical ones (Gay 1999), with the ratio typically ranging from 1:5 to 1:20. This phenomenon may be illustrated as the "disease iceberg" (Figure 4.2). Many infectious agents are endemic within herds, with most animals becoming exposed and infected at some stage. Young animals often acquire a "silent" subclinical infection, with the subsequent development of protective immunity. This should not, however, imply that production may not be adversely affected, even though individual animals do not become obviously "sick" (Gay 1999). Rather, when signs of overt sickness occur, they are usually indicative of a more widespread problem. This also means that effective remedy of the problem generally requires more than just the treatment of the clinically sick individual or individuals.

Thus, a good herd health program is one that manages risk of disease and lowered productivity at a number of levels, including considerations of biosecurity, management, and nutrition, as well as judicious use of biologicals and pharmaceuticals.

The role of the herd health provider in developing programs for clients has a large educational component (Spire 1988). Producers in the United States rely heavily on veterinarians for relevant information on herd health (National Animal Health Monitoring System [NAHMS] 1994a). However, despite the privileged position of veterinarians as advisors on herd health, many producers do not have health programs that supply even minimal protection against common pathogens. For example, in a survey of 169 producers in northwest Kansas, Spire (1988) indicated the following vaccine usage in cows and heifers, respectively: none (31.3% and 28.4%), *Campylobacter* (55.6% and 44.4%), leptospirosis (62.1% and 53.2%), *Clostridia* (5.3% and 23.6%), infectious bovine rhinotracheitis (IBR)

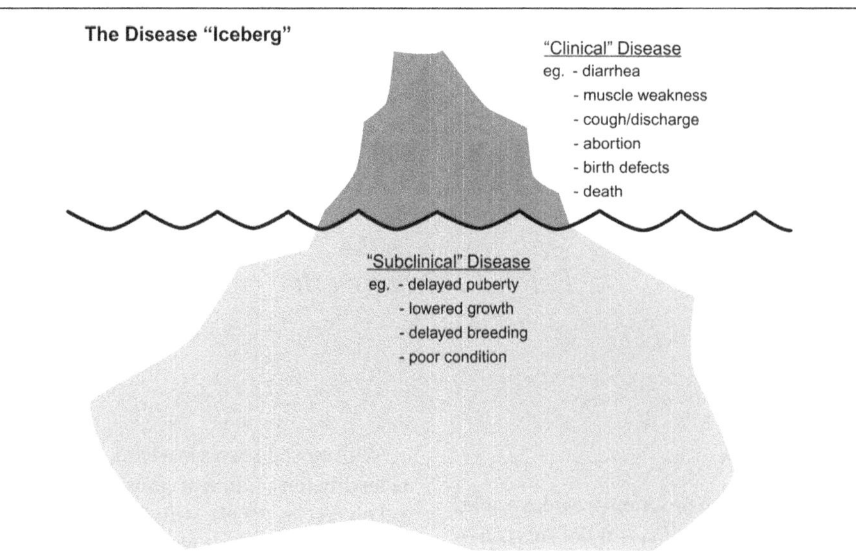

Figure 4.2. "The disease iceberg."

(11.2%, 19.3%), bovine viral diarrhea (BVD) (11.2 and 13.6%), and rotavirus-coronavirus (10.6% and 5.9%).

A survey from a wider geographical sampling of cattle breeding operations in the United States (NAHMS 1998a) indicated even less vaccination activity. Here, the most common vaccination was for clostridial diseases in calves. Most cow vaccinations were for reproductive disease, such as leptospirosis (28.5% of operations) and *Campylobacter* (20.1%). Other vaccinations included bovine respiratory disease (BRD) and *Clostridia*, both approximately 14%; *Escherichia coli,* 4.7%; rotavirus and coronavirus, 4%; and *Trichomonas fetus,* 1.1%. Only 11% of heifers were vaccinated against *Campylobacter fetus* before breeding, and 0.2% against *Trichomonas fetus.*

A greater level of vaccination activity emerged from a survey of 53 cow/calf producers in the Rocky Mountain West (Sanderson and Gay 1996). Here, over 60% of producers vaccinated cows for *Clostridia*, 87% for leptospirosis, and 85% for *Campylobacter fetus*. However, although this survey was conducted among progressive ranchers, the results indicated that management practices still did not adequately reflect relative risk, especially when purchasing cattle of unknown origin. This was confirmed by data showing that only a third of U.S. operations quarantined new additions to the herd (NAHMS 1997b). Over 15 years ago, U.S. veterinarians identified herd health management services as the area with greatest potential for revenue growth (Wise 1987). Today it would appear that this is still an area of largely untapped potential.

MARKETING HERD HEALTH

The sheer number of potential disease threats, as well as the biological and therapeutic products that are available for their prevention or treatment, can be overwhelming. Veterinarians can play an invaluable role in simplifying and translating this information into effective, practical, cost-effective herd health programs for producers. In the past, such advice has been given generously and freely, with veterinarians relying on associated services and product sales to generate income. This system is changing as producers themselves perform more and more services that were previously regarded as falling solely within the veterinary domain, and as veterinarians are finding it increasingly difficult to compete with other sources for product sales. Today's veterinarians need to better market their greatest asset—health–based knowledge—within the context of economic advantage to producers.

COSTS OF ANIMAL DISEASE

Considerations of an appropriate herd health program should include relative costs of disease and prevention. These are usually an estimate only, as many disease entities can go unrecognized. In a survey of

86 selected Colorado beef herds (Salman et al. 1990, 1991), total annual disease cost (including deaths, culls, reduced production, veterinary costs, and miscellaneous expenses) was estimated at $20.56 per cow. Of this cost, approximately 10% ($2.04 per cow) was spent on veterinary services for disease treatment (Salman et al. 1991). The most expensive disease category on a total cost basis was the miscellaneous category, as it covered many diverse cases. Sudden death was the next most expensive category because of the high individual cost. Enteric disease was the least expensive class, despite being common, because the average cost per case was low. The cost of veterinary services for treating diseased animals represented a relatively small portion (approximately 5%) of the total costs for each disease category.

For disease prevention measures, mean annual costs varied from $6.39/cow to $11.29/cow (Salman et al. 1990, 1991). Approximately 60% of the total costs of preventive vaccines and drugs were for miscellaneous diseases. The prevention of reproductive diseases accounted for almost the entire cost of preventive veterinary services; of which only about $1/cow was spent directly on veterinarians and their services. Here, the three main veterinary procedures were pregnancy examination, brucellosis vaccination of replacement heifers, and breeding soundness examinations of bulls.

In 57 California beef cow/calf herds studied in 1988 and 1989, the mean cost associated with disease was $33.90/cow-year, with $0.78 and $1.37/cow-year spent for veterinary services and drugs, respectively (Hird et al. 1991). For disease prevention measures, the mean expenditure for veterinary services was $1.67/cow-year, most of which was spent on reproductive problems.

In a survey of 60 cow/calf herds in Tennessee in 1987 and 1988, 4% of the total cost of disease was attributable to veterinary services and 2.3% to the purchase of drugs for the treatment of sick animals (New 1991). The cost of preventive veterinary services accounted for 8.8% of the total cost of all preventive veterinary services. Approximately 68% of the total cost of preventive actions was spent on drugs, with drugs to prevent intestinal and external parasites being the most costly ($5.99 and $1.80/cow-year, respectively). Only 28% of the producers used veterinarians to implement animal health management actions, only 15% used veterinarians to monitor the pregnancy status of cows and heifers, and only 3% used veterinarians as consultants.

In contrast to actual expenditures, estimates have been provided of the financial effect of different veterinary interventions on U.S. operations (Toombs 1996). Based on calf prices of 80 cents/lb (an estimated calf value at weaning of $400), a combination of decreasing calf morbidity (from 50% to 10%) and mortality (from 5% to 1%) led to savings of an estimated $20/cow (or $2,000 for a 100-cow herd). A program that increased weaning weights (from 450 lb to 500 lb), increased pregnancy rates (from 90% to 95%), and decreased calf mortality (from 5% to 2%) caused an estimated improved return of approximately $45,000 over 5 years for a 250-head cow herd.

HERD HEALTH CONCEPTS

Whenever possible, herd health should involve simple concepts and relatively easy implementation. The major difference between herd health and individual animal medicine is that in the former, the herd, or group, is the unit of reference. However, just as with individuals, there are indicators, or clinical signs, of group health and sickness, as described in chapter 3.

A herd is healthy when its collective resistance is greater than the disease challenge. Effective herd health programs act to raise resistance, reduce the risk of disease challenge, or both. A challenge may occur when a pathogen is introduced to a susceptible population (or animal), or when an overwhelming infection occurs in a normally resistant population (or animal). The primary herd health goal is to raise herd resistance (or protection) levels, to reduce susceptibility (or challenge) levels, and to maintain a healthy safety buffer between these two aspects. This can be achieved via a number of methods (Figure 4.3).

Different options exist to keep a herd healthy and disease free. To devise the best option for each operation requires a good understanding of the animal populations involved, and their environment, as well as management priorities and constraints. The risk level of the herd varies according to its potential to be exposed to particular pathogens, with the terms "open," "closed," and "modified open" being used to categorize operations in terms of degree of risk (see Table 4.1) (Spire 1988).

Herd health concepts may also be depicted graphically, as in Figures 4.4 and 4.5. Here, average disease resistance of the herd will vary according to a normal population distribution curve (Figure 4.4).

Each herd, or group, contains animals of varying disease susceptibility to a given challenge. The number of "sick" animals can be influenced by the proportion of animals susceptible to the given challenge, by a change in challenge level, or by an interaction

Table 4.1. Herd potential risk categories.

Category	Closed Herd	Modified Open Herd	Open Herd
Cattle movement and introductions	Minimal animal introductions (often breeding bulls only); no animal reentry; no fenceline contact with other herds	Limited introduction of new animals (e.g. for herd expansion or replacement heifers); reentry allowed from, for example, livestock shows; fenceline contact with other herds	Routine purchase, reentry, and movement of animals; direct commingling of introduced animals (including those stressed or of unknown background) with herd animals
Biosecurity precautions	Introduced animals have known health background; new arrivals are isolated	Introduced animals are of known compatible health background; new arrivals are isolated	Introduced animals are of unknown health background
Risk	Low	Moderate	High

Adapted from Spire 1988.

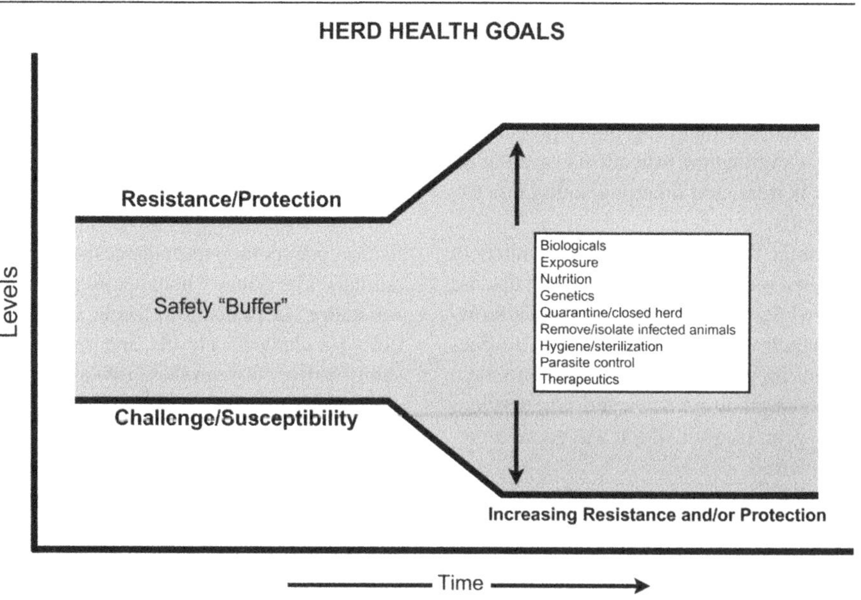

Figure 4.3. Herd health goals.

between these factors. A number of factors can increase herd susceptibility to disease, and thus change this profile, as seen in Figure 4.5.

Cow/calf producers often need some persuasion to adopt specific herd-health measures, especially when there appears to be no obvious direct benefit. It can be useful to provide economic estimates of the value of different herd health components. For example, estimates of the returns obtainable with a number of practices are shown below, as is their adoption rate in U.S. operations (Table 4.2). It is apparent that many producers are not availing themselves of the benefits of a number of cost-effective practices. It should, however, be noted that few cost–benefit studies are available for livestock vaccination programs.

Access to herds and the development of owner confidence can be achieved with marketable skills—veterinary procedures or techniques that are proven

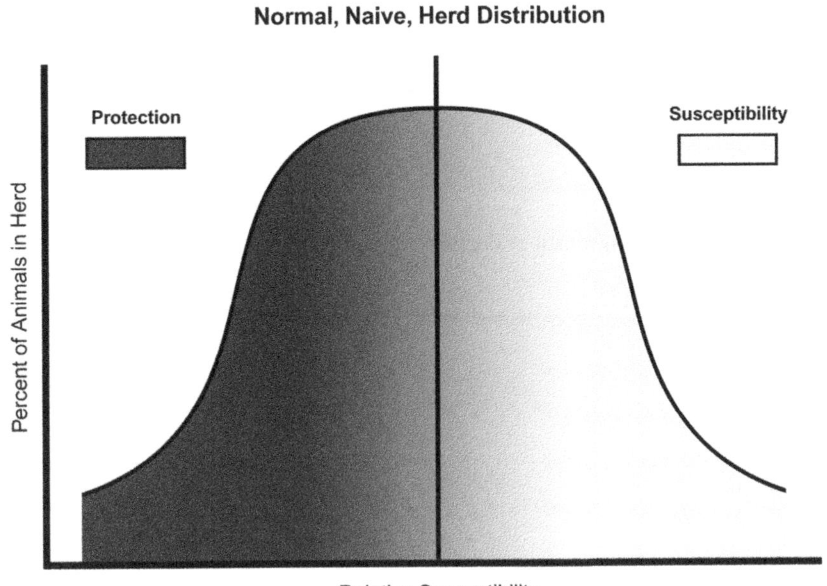

Figure 4.4. Average herd health profile.

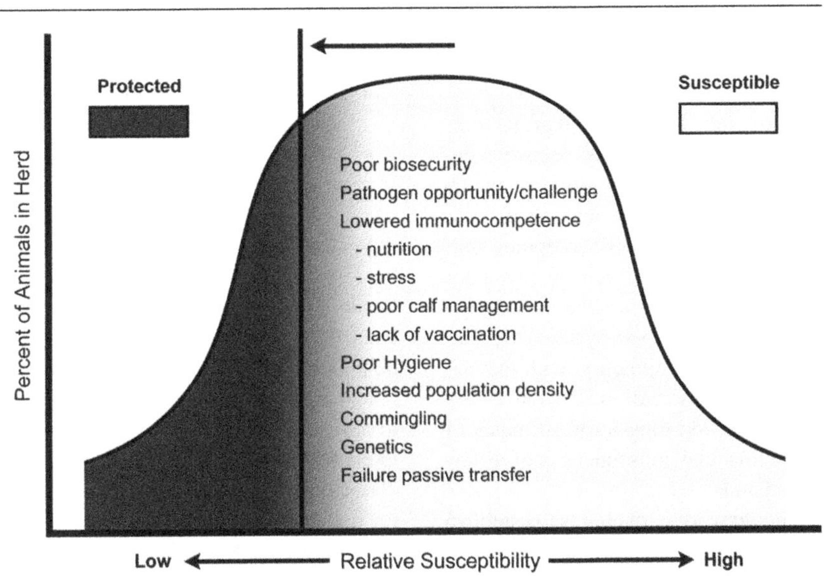

Figure 4.5. Susceptible herd profile.

cost-effective (Chenoweth 1989). A number of these skills have been identified, ranging from individual animal diagnosis and treatment to palpation skills, bull evaluation, financial management, and risk assessment. Within the context of herd health, appropriate marketable skills could also include diagnostic and economic analyses, herd health and biosecurity planning and monitoring, nutrition counseling, data handling/analysis, and personnel training.

For a particular herd, a logical starting point is to identify current production losses by comparing the performance of the herd with relevant benchmarks,

Table 4.2. Estimated economic returns and adoption rates of different management practices.

Practice	Return/Cow[a]	% U.S. Operations[b]
Pregnancy test/cull	$30	34.5
Heifers		15.9
Cows		17.7
Health program (vaccination)	$5–$20	
Bull breeding soundness evaluation/semen evaluation	$3–$50	39.9
Before sale		57.3
Annual		17.3
Restricted breeding season	$50	46.4
Heifer early breeding	$10	12.7
Heifer—easy calving sires	$10–$60	
Performance tested sires	$30–$90	
Growth implants—calves	$7–$25	18.4
Estrus synchronization	$3–$12	11.9
Cows		4.3
Heifers		3.0
Artificial insemination	$7–$30	13.3
Use of standardized performance analysis		4.2
Use of body composition score		23.3

Adapted from [a]Toombs 1996 and [b]NAHMS 1997a, 1997b, 1998a.

as discussed in chapter 3. If performance is below par, then an estimate of the economic loss involved should be attempted.

The next step is to conduct cost–benefit projections of different herd health options. Final decisions depend on a number of factors, only some of which might be economic. The decision-making process can be improved by using "what-if" computer simulation software such as DECI (Decision Evaluator for the Cattle Industry; http://www.marc.usda.gov/genome/genome.html), developed by the U.S. Department of Agriculture in conjunction with the National Cattlemen's Beef Association (Roybal 1999). Such programs can provide long-term estimates of the production and financial consequences of different managerial decisions.

Veterinarians should work with clients to establish a list of disease threats to the herd, rank them according to their potential to cause disaster, and estimate the cost–benefits of elimination or control of those that are a real hazard.

Appropriate steps in the disease threat decision making process include the following list (adapted from Richey 1991):

- Recognizing the disease challenge (diagnosis and surveillance)
- Knowing when and why a particular challenge will occur (pathogenesis)
- Estimating cost–benefits of elimination or control (risk assessment)
- Lowering (or eradicating) the challenge (management)
- Raising resistance both to the specific disease and in general (immune system management).

Recognizing potential disease challenges for a beef herd in any region can be problematic. For a particular locale, local knowledge and advice from producers, extension personnel, and diagnostic laboratories can help identify the most likely problems. Once a list of potential pathogens is established, the next steps are to assign a relative risk to each class of cattle most likely to be affected and to then determine when these classes are most at risk. For example, a list of the major pathogens associated with reproductive wastage in beef cows in North America, with relative risk for each gestational period, may be tabulated (Table 4.3). Such a list of infectious disease entities and their relative risk can be used to determine the cost–benefits of different approaches to herd protection or, at least, risk reduction.

HERD HEALTH CALENDAR

The major disease risks for a given herd, as well as appropriate preventive measures, should be established

Table 4.3. Relative risk of beef cow reproductive pathogens per gestational period.

Pathogen	Gestational Period			
	Early	Mid	Late	Periparturient
Aspergillus fumigatus			L	
Bluetongue virus	L		M	L
Bovine viral diarrhea virus	V	H	L	L
Campylobacter ferus spp venerealis	H	M		
Chlamydia psitacii			L	
Haemophilus somnus	L		H	M
Infectious bovine rhinotracheitis virus	L	M	H	L
Leptospira	L		H	V
Listeria monocytogenes			L	
Mixed bacteria			L	L
Parainfluenza virus 3 (PI_3 virus)				L
Ureaplasma diversum	V	V	V	V

Adapted from Spire 1988.
H = high, M = moderate, L = low, V = variable/unknown.

in consultation with herd owners. For a given disease, the options may be to ignore it, minimize risk managerially, or use biologicals. When opting for the latter, veterinarians should inform the client of the relative efficacy of different products, their formulation and duration of effect, handling and storage information, contraindications, and cost-effectiveness. A good vaccine should decrease the number of susceptible animals, increase the level of collective herd immunity, and have minimal deleterious effects (Spire 1988).

The best times for herd intervention should be established. These will often coincide with other major managerial tasks or events for the cow/calf operation. A good strategy is to devise a herd-health calendar in which the health events are coordinated with major operational events. An example of such a calendar for a spring calving herd is shown in Table 4.4.

The normal activities, or events, that occur annually in a beef breeding herd may also be illustrated diagrammatically (Richey 1992). For example, Figure 4.6 shows the cow "year" for a generic herd as a major circle, with a smaller circle representing the breeding season (in this case, 60 days) and another circle showing the corresponding calving period. Other major management events may be added. To construct such a herd health program for a client, several key dates representing normal herd management events need to be inserted. These should include the start of calving season, end of calving season, start of breeding season, end of breeding season, time of pregnancy test, and time of weaning.

Veterinarians should work closely with individual producers to develop such plans, as each operation is different, having its own calendar, risk factors, economics, and priorities. The outcome should be an individually tailored, documented, herd health program that is subject to regular review and updating (Gay 1999). Herd health procedures should be integrated with major, routine managerial events as much as possible to minimize herd disruption and labor costs. Coordination of health and managerial events may be achieved as shown later.

Major herd health intervention opportunities are synchronized with managerial events as much as is feasible. Possible intervention opportunities are represented in Figure 4.6 by numbers within small, dark-edged rectangles. Intervention windows of opportunity vary with different enterprises, and some may be combined (e.g., precalving and prebreeding). Similar cycles, with inclusion of their own specific health intervention windows, may be constructed for each class of cattle, such as heifers, bulls, and calves.

Intervention Opportunities

Precalving

Precalving vaccination of the dam helps to ensure optimal colostral quality and passive transfer of immunity to the calf. In turn, good passive transfer lays the foundation for a fully functional immune system

Table 4.4. Herd health calendar for a spring calving herd.

Season/Group	Cow	Heifer	Calf	Bull
Late winter–early spring	**Calving:** Dystocia check. **Prebreeding:** Check BCS, udder, teeth. Vax–vib, trich Rising nutrition. Deworm, defluke. Control external parasites. Cull problem animals.	**Prebreeding:** Check target weight, BCS, RTS, pelvic measure. Vax–vib, trich, Lepto (5), Clostridia, BRD. Deworm, defluke. **First calving:** Separate and monitor Control external parasites.	**At birth:** Dystocia check. Check newborn: Birth weight, navel, colostrum, ID. **Working:** Brand, castrate, dehorn, implant Vax (calves should be at least 8 weeks old): Clostridia, BRD, deworm.	**Prebreeding:** BSE. Libido/serving capacity test Vax–Lepto (5), vibriosis, Clostridia, BRD Deworm, defluke Control external parasites Bull selection or purchase: Check EPDs, do BSE, quarantine, check vib/trich
Late spring–early summer	**Breeding:** Control external parasites.	**Breeding:** Breed early. Control external parasites.		**Breeding:** Monitor. Control external parasites.
Late summer–early fall (autumn)	**Gestation:** Pregnancy test. Cull open/problem cows. Parasite control. Supplementation.	**Gestation:** As for cows.	**Pre-weaning:** Calf boosters—vax. Clostridia, BRD. Implant, parasite control.	**Post-breeding:** Check soundness, nutrition, parasite control.
Late fall–early winter	**Precalving:** Vax–lepto (5), BRD, Clostridia, calf scours. Vitamins A and D. Lice control. Check BCS, supplementation. Check calving supplies and details.	**Precalving:** As for cows.	**Weaning:** Second vax.	Monitor herd bulls.

Vax = vaccination, BCS = body condition score, RTS = reproductive tract score, BRD = bovine respiratory disease, vib = vibriosis (campylobacteriosis), trich = trichomoniasis, lepto = leptospirosis, (5) = five serotypes, BSE = breeding soundness evaluation, EPD = estimated progeny difference.

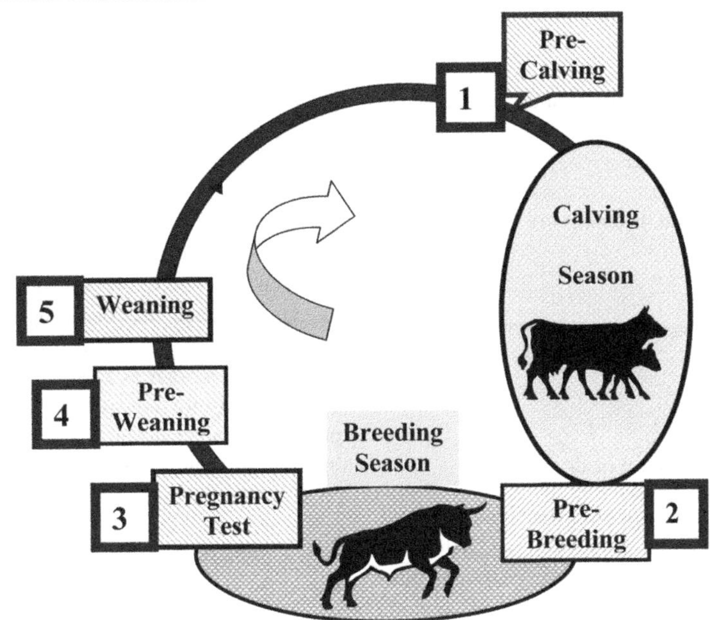

Figure 4.6. Major management "events" and intervention opportunities. (Adapted from Richey 1992.)

from calfhood to maturity. The ideal time for precalving immunizations is approximately 1 month (or at least 3 weeks) before commencement of calving. Antigens that are often considered to be important (although regional differences will affect relative importance) include those of the major clostridial diseases (particularly *Clostridium perfringens*), leptospirosis, calf scour pathogens (*E. coli*, rotavirus, and coronavirus), respiratory disease (IBR, BVD, parainfluenza virus 3 or PI_3, bovine respiratory syncytial virus, or BRSV, plus *Haemophilus somnus* and *Manheimia haemolytica*), as well as anaplasmosis.

Calving

Here, as mentioned above, effective passive transfer of immunity is the prime concern. This transfer can be compromised by stressful environmental conditions, so appropriate protection is indicated. The only respiratory vaccines that the newborn calf can respond to are those that are temperature-sensitive, modified live, such as a modified living intranasal IBR/PI_3. More complete discussion of calving and neonate management is provided in chapter 11.

Prebreeding

Prebreeding vaccinations should be completed within 2–4 weeks of the start of breeding. The major diseases to consider here are campylobacteriosis, trichomonosis, leptospirosis, IBR, and BVD.

Pregnancy Examination

Ideally, pregnancy examination should be conducted as soon as possible after breeding, allowing for detection of those animals that got pregnant last. For practical purposes, this is probably 45 days or more after the end of breeding. For some operations, it may be most convenient to schedule this examination to coincide with weaning.

Pregnancy examination represents an important management event during which cull females are identified. These include some or all animals that are classified as nonpregnant and those that qualify on the grounds of age, soundness, or disease. Veterinary procedures such as removal of cancer eye (ocular squamous cell carcinoma) may be performed or scheduled at this stage.

For pregnant animals, important information should be recorded such as stage of pregnancy, age, parity, body condition, and lactation status (wet or dry). Relevant information for breeding season evaluation of the beef herd includes

- Breeding herd inventory
- Vaccination history
- Sires used (ID, age, breed, group composition)

- Breeding pastures
- Bull-to-female ratios
- Breeding season dates
- Pregnancy rates (overall and per 21-day breeding period)
- Expected calving dates
- Calving distribution
- Herd averages (and distribution)
- Female factors
  - Breed type
  - Age
  - Parity
  - Body weight
  - Frame score
  - Body condition
  - Lactation status (wet or dry)

For convenience, cows are often vaccinated for reproductive pathogens such as leptospirosis, IBR, and BVD at the time of pregnancy exam, even though this is probably not the best time for effective protection from these diseases. Producers may think that they are saving money by vaccinating only cows identified as pregnant; however, risks associated with this approach include errors in records, animal identification, and pregnancy diagnosis, as well as lack of consistency in immune system management of the herd.

Preweaning

Preweaning represents an important strategic intervention, as it can help prepare calves for the stress of weaning and can reduce the possibility that such stress will compromise the efficacy of biologicals. Therefore, preweaning has been the focus of many preconditioning programs, partly because of the erroneous assumption that it represents the first time that calves will effectively respond to BRD vaccines. A number of procedures have been recommended at this time, including castration of bull calves, dehorning, and treatment for appropriate internal and external parasites. Common vaccinations at this juncture include the clostridials (seven-way), BRD (IBR, BVD, PI_3, BRSV plus *H. somnus*, and *M. haemolytica*, and *P. multocida*), and leptospirosis. For BRD vaccinations, only products licensed for suckling calves should be given. A broad-spectrum anthelmintic may also be given at this time to ensure that calves cause minimal pasture infestation when released into the weaning area.

Weaning

Good weaning management includes preparing calves by allowing them prior access to the weaning facility, feed, and water. Here, the emphasis should be on stress reduction. Surgical procedures (castration, dehorning) ideally should have been performed before weaning. If not, this is the latest time at which these procedures should be performed. Good weaning management also entails weaning into a clean area containing nutritious, palatable feed and water. A palatable, well-designed starter ration is important at this time. At weaning, a second vaccination should be given for products requiring a booster. Modified live viral products may be applied at this stage, as the opportunity for the calf to transmit live organisms back to the mother is removed. Further clostridial vaccination is not indicated at weaning if calves were previously vaccinated at the time of original processing and preweaning. Implanting may be performed on animals not intended for reproduction. Parasite and grub control also may be performed at this point, with regional calendar recommendations being followed for the latter. For areas in which brucellosis is under regulatory control, heifer vaccination with either Strain 19 or RB51 should be given within the age window stipulated.

Recommendations have been made for preweaning and weaning practices for beef calves, including a number of recommended programs that indicate the type and degree of processing that calves have received before intensive feeding. Terms that have been associated with such processes include "preconditioning," "backgrounding," and "preshipment." They imply preparatory procedures that enable the calf to better withstand stressors such as those associated with marketing, transport, and adaptation to the new environment. Preconditioning programs, for example, vary with the origins and backgrounds of the calves and of the programs for which they are intended. Examples of such programs and their benefits are provided on the Web sites listed at the end of the chapter. Economic analyses of preconditioning have generally shown cost–benefits in terms of marketable product.

Management procedures that can be used at or around weaning to reduce stress (and subsequent increased susceptibility to disease) are described in chapter 7.

Replacement Females

Replacement females should be subjected to the same vaccinations as those given to other females before commencement of the breeding season. Readers are directed to chapter 8 for a more thorough discussion of this topic, as well as to the discussion by Sanderson and Gnad (2002).

Bulls

In general, the bull team should be subjected to the same vaccinations as is the cow herd, with several major exceptions.

First, bulls should not be vaccinated for brucellosis. Second, the trichomoniasis vaccine currently approved for use in the United States is not approved for bull use. Caution is advised in using modified live virus IBR vaccines in bulls, as the virus may recrudesce and be shed in semen. Also, semen shipment to other countries may be jeopardized if virus is detected in the semen. Otherwise, recommended bull vaccinations include leptospirosis and campylobacteriosis, in accord with recommendations for the cow herd. Greater detail on the subject of bulls is provided in chapter 9.

Purchased bulls should be segregated from the resident herd to acclimatize them to the farm environment, as well as to allow them to receive appropriate vaccinations and tests, which should include tests for BVD and trichomoniasis (for bulls more than 2 years old).

ABORTION OUTBREAKS

One of the greatest challenges facing the cow/calf veterinarian is that of establishing the cause of abortions. Although abortions represent a significant source of economic loss to beef cattle operations, the cause is often undetermined for several reasons.

First, there is a normal level of pregnancy loss between the time of pregnancy diagnosis and subsequent calving. This is usually on the order of 2%–3%. (A level of 5% or greater in beef herds indicates a problem.) Early (<4 months gestation) abortions often go unnoticed. Overall, infectious causes are believed to account for little more than half of abortions in beef cattle. Definitive diagnosis of the cause of abortion by diagnostic laboratories commonly occurs in a minority (approximately 33%) of cases, even when appropriate samples are submitted. This may partly reflect the fact that there has been a marked decrease in the occurrence of those diseases (e.g., IBR) associated with abortion storms, as well as in the number of calves with congenital defects that might lead to abortion (Nietfeld 2003). Last, a large number of factors and organisms have been implicated in bovine abortion, with new ones (e.g., *Neospora caninum*) being periodically identified. A current list of causes of abortion in cattle is shown in Table 4.5.

The ability to accurately diagnose the cause of abortion is linked with competent, knowledgeable clinical diagnostics, timely submission of appropriate history and samples (ideally the entire fetus and placenta and a sample of the dam's blood) to the diagnostic laboratory, and quality and amount of background information that is obtained.

This information can be used to ascertain the importance of different risk factors. Thus, it is necessary to establish an accurate and detailed history of the animal, herd, and environment. Local knowledge is also useful in helping to determine which risk factors are most likely to be causal.

A positive agent diagnosis of abortion should be based not only on serologic or culture evidence of a specific agent but also on gross or microscopic lesions characteristic of the agent (Wikse 2002). Serological findings alone are often of limited use, and although pathologic findings alone can provide important clues, definitive diagnosis often requires additional evidence. The application of epidemiological principles often helps to identify key determinants (risk factors alterable by management) that have allowed the problem to occur. These might include poor biosecurity, environmental stress, nutritional deficiency, or other factors. Diagnosis and management of abortions require attention to both the primary agent or implicated factor and those environmental and managerial determinants that have allowed them to create an abortion problem in the herd.

CULLING PROGRAMS

Cull cows represent 15%–20% of gross income for cow/calf operations in the United States (National Cattlemen's Beef Association 1999). Cull animals represent an important component of farm economics, with approximately 11% of cows being sold annually for slaughter in the United States (NAHMS 1998a). The identification of cull cows as a profit center provides the opportunity to develop programs that optimize cull cow management and returns.

Although each operation has its own unique biological or economic criteria for culling animals (Spire and Hotz 1995), four major reasons for culling cows were identified in the USDA/NAHMS Beef '97 survey (NAHMS 1997b): age/teeth (39.8%), pregnancy status (24.3%), economic reasons (18.5%), and poor production (5.7%). Other reasons for culling may include animal-related factors such as physical or temperament problems, as well as economic or environmental reasons.

Culling patterns differed with size of operation. For example, the use of pregnancy status as a basis for culling (average 25.6% of operations) increased

Table 4.5. Causes of abortion.

Infectious	Nutritional
Epidemic	Deficiencies
Brucellosis (Br abortus)	Protein
Leptospira spp	Vitamin A
Campylobacter fetus	Iodine
Infectious bovine rhinotracheitis	Vitamin E/Selenium
Bovine viral diarrhea	Excesses
Bovine respiratory syncytial virus	Protein
Tritrichomonas foetus	Urea
Neospora caninum	Copper
Sporadic	Iodine
Hemophilus (H. somnus)	Selenium
Listeria monocytogenes	Toxic
Actinomyces pyogenes	Bacterial
Bluetongue	Gram-negative endotoxins
Ureaplasma diversum	Plants
Mycotic	Pine needles
Escherichia coli	Broomweed
Salmonella spp	Locoweed
Genetic	Narrowleaf sumpweed
Inbreeding	Plant estrogens
"Problem" sires	Clover
Genetic defects (e.g., Robertsonian)	Mycotoxins
Aged/damaged gametes	Aflatoxin
Hormonal	Ergotamines
Estrogens (silage, poultry litter)	Zearalenone
Progesterone	Chemicals
	Nitrate
	Organophosphates

Adapted from Wikse 2002, Nietfeld 2003.

from 16.2% in operations with fewer than 50 females to 69.4% in those with 300 or more females (NAHMS 1997b). Regional differences also occur. In North Dakota, for example, the largest cause of female culling was open/fertility problems (36%), followed by old age (17%), and then selling females as replacement breeding stock (15%) (Hughes 2003).

Elective culling is a powerful tool for improving herd performance by replacing poor-producing animals with those that are more productive within the on-farm environment. Good culling decisions are essential for profitability in today's economic climate. Considerable progress can be made in improving herd fertility by regularly identifying and removing infertile animals. Despite this benefit, many cows are apparently culled for the wrong reasons. For example, only 14.1% of operations culled cows for economic reasons, and 11.7% culled for poor production (NAHMS 1997b), although larger operations (>300 head) were more represented in these categories than were smaller operations. In addition, approximately 43% of U.S. cull cows were pregnant (NAHMS 1997b).

Other problem areas for cull animals were identified in the United States 1999 *National Market Cow and Bull Quality Audit*, conducted by Colorado State University (National Cattlemen's Beef Association 1999). In this report, it was established that too many producers were harvesting cows and bulls too late, when they had lost too much body condition and were disabled. It should be noted that recent (December 2003) regulatory decisions in the United States now mandate that nonambulatory animals not be sent for slaughter. In addition, the 1999 report concluded that too many cattle (and carcasses) were condemned. It recommended that producers improve the value and welfare of cull animals by managing to minimize problems, monitoring health and condition, and marketing in a timely manner. Here, analysis of high-profit IRM herds in the United States indicates that marketing strategies of cull cows are an important component of profitability (Hughes 2003).

Programs have been developed to improve the value and marketability of cull cows. Work in South Dakota, for example, established that feeding cull cows could increase their grading by one or two categories (Hughes 2003), which, in turn, increases the price received. Seasonal differences also occur in cull cow prices in the United States, with prices increasing from a low point in November (when many cull animals are being marketed) to a high point in March/April. Thus, a combination of feeding fall cull cows and delayed marketing to capitalize on improved prices in the spring can represent a viable market strategy for cull cows (Hughes 2003). One option, dependent on feed and cattle markets, is to overwinter and breed the cull cows to sell as a pregnant cow the following year.

Culling of open (nonpregnant) females is one area in which producers can benefit from professional advice and input. If we assume that females are correctly classified as nonpregnant, this may be the result of a variety of reasons, many of which may not be related to female inherent fertility. Such reasons may include bull fertility, nutrition, and infertility disease. However, errors in pregnancy diagnosis can also occur, leading to erroneous culling decisions. Errors may occur because of faulty technique, as well as because of inherent difficulties in detecting early pregnancies. In addition, open females are not necessarily infertile (Neville et al. 1987). It is not always good policy to automatically sell cows with late calves. In the United States, such cows generated a profit when the beef price cycle was high and lost money when it was low (Hughes 2003).

Strategic culling of females represents a strategy to improve cash flow and reduce costs at times of economic or environmental stress. Here, a cull female priority list decision tree can be used to cull animals in groups that represent their increasing importance to the future economic viability of the cow/calf operation:

1. Females that have had dystocia, eye cancer, or other severe problems
2. Females that have problems rearing calves (e.g., bottle teats)
3. Nonpregnant, dry (nonlactating), aged females
4. Nonpregnant, dry (nonlactating), younger females
5. Nonpregnant, eligible heifers (those that have achieved target weights at breeding)
6. Nonpregnant, wet (lactating), older females
7. Nonpregnant, wet (lactating), younger females.

In summary, cull cows represent a potentially valuable source of income that should be managed accordingly. Quality, food safety, and welfare issues (as discussed in chapter 14) are important, as is the need to maximize returns from these animals. Veterinarians should play a pivotal role in ensuring that these considerations are integrated in the best interests of both the owner and the animals involved.

DISEASE CONTROL IN BOVINE SEMEN USED FOR ARTIFICIAL INSEMINATION

Many pathogenic microorganisms can be semen-borne, and these can survive processing, freezing, and storage at least as well as sperm can (Bartlett 1976). Because artificial insemination allows widespread use of semen from individual bulls, the potential for disease transmission is great. Diseases that are capable of being transmitted in bull semen include the following (adapted from Eaglesome and Garcia 1992, Eaglesome et al. 1992):

- Bovine brucellosis
- Blue tongue
- BVD
- Chlamydia
- Foot and mouth disease
- *H. somnus*
- IBR
- Leptospirosis
- Mycoplasmosis
- Trichomoniasis
- Bovine tuberculosis
- Ureaplasmosis
- Vibriosis (Campylobacteriosis)
- Potential contaminants/pathogens
  - *Pseudomonas aeruginosa pyogenes*
  - Staphylococci
  - Streptococci
  - *E. coli*
  - Proteus spp
  - Mycotic agents.

Methods used to reduce the risk of disease transmission in frozen bovine semen include testing of both animals and semen for specific diseases, implementation of biosecurity practices at artificial insemination centers, and addition of antibiotics to extended semen. In the United States, health recommendations for bulls in artificial insemination centers and for semen processing are made by Certified Semen Services, a subsidiary of the National Association of Animal Breeders (http://www.naab-css.org). In Australia, recommended tests and techniques for bull certification are documented in *Australian Standard Diagnostic Techniques for Animal Diseases* as published by the Commonwealth Scientific and Industrial Research Organisation (1993).

VACCINOLOGY

The term "vaccination" refers to the act of applying a vaccine to an animal. This, however, by no means guarantees protection or immunization. Vaccine failure can occur via a number of reasons:

- Inability of any vaccine to confer 100% immunity
- Improper storage, handling, or injection
- Poor timing (e.g., interference with colostral antibodies in young animals, or overdelay between initial and booster vaccinations)
- Poor immunocompetence at the time of vaccination (e.g., disease or stress may compromise response)
- Overwhelming infection
- Poor vaccine efficacy
- Emergence of new strains or resistance
- Interactions with other biologicals
- Different risk factors for different infectious agents.

Managerial steps to minimize vaccine failure include

- Ensuring vaccines are stored, constituted, and handled according to direction
- Avoiding heat stress (>85°C) at the time of vaccination.
- Avoiding calving stress (−4 to +7 days) as much as possible
- Avoiding behavioral stress as much as possible
- Waiting until calves are 5 days old or older for injectable vaccines; intranasal and oral vaccines may be given earlier
- Checking micronutrients that could lower immunocompetence (e.g., selenium, copper).

The term "immunization" is preferred to "vaccination," as the former term implies an appropriate animal response. The objectives of animal immunization include (Schultz 1993):

- A good humoral, cellular and local immune response similar to that with natural infection
- Protection against clinical disease and reinfection
- Protection over several years (preferably the lifetime)
- Minimal immediate side reactions
- Simple administration, in a form acceptable to both the producer and practitioner
- Clear cost-effectiveness when compared with the risk and cost of natural disease.

Not all vaccines meet all of these requirements. For example, some cannot be given to certain groups of animals (e.g., pregnant females), some require booster vaccinations (e.g., killed vaccines), many do not provide lifelong immunity, and some are not cost-effective.

Several different vaccine types are available to cattle, including modified live and killed/inactivated. These differ in important aspects, as shown below in Table 4.6.

A subgroup of modified live vaccines is the group that is modified live and temperature sensitive. These vaccines will only replicate in surface cells, such

Table 4.6. Types of vaccines employed in beef cattle operations.

Modified Live Vaccines (Replicating)	Killed–Inactivated Vaccines (Nonreplicating)
Provide longer and more complete immunity	Provide short-lived systemic immunity
Should produce cellular and secretory immunity	Provide poor cellular and secretory immunity
Do not require multiple vaccinations to ensure immunologic memory	Often require revaccination to ensure immunologic memory
Require less lifetime revaccination	Require multiple vaccinations for active immunity
Rarely cause hypersensitivity reactions	Often cause hypersensitivity reactions
May be virulent in certain individuals	Cannot cause disease, even in immunocompromised animals
May cause abortions in pregnant animals	

as those in the upper respiratory tract, which are at a lower temperature than those within the body. They stimulate local immunity in the area of replication. Examples include intranasal vaccines (e.g., IBR, PI_3).

When using vaccines in cow/calf herds, important decisions must be made concerning the type of product to be used and the timing of its administration for the best effect. Timing considerations include the effects of maternal (colostral) antibody on active immunization, interval between booster vaccinations, age or phase at which animals are most at risk for a particular disease, and immunocompetence of the group.

RESOURCES

WEB SITES

GENERAL

http://gpvec.unl.edu/sites.htm
http://www.vetmed.wsu.edu/courses-jmgay/FDIUCowCalfHH.htm
http://cowcalfcorner.okstate.edu/archive.htm
http://www.aphis.usda.gov/vs/ceah/cahm/
http://www.oznet.ksu.edu/dp_ansi/bcattle.htm
http://www.aabp.org/

PRECONDITIONING CALVES

http://farwest.tamu.edu/animsci/rd-calves.htm

SHIPPING AND RECEIVING CALVES

http://edis.ifas.ufl.edu/VM081
http://edis.ifas.ufl.edu/VM04

REFERENCES

Bartlett D.E., L.L. Larsen, W.G. Parker, T.H. Howard. 1976. Specific pathogen free (SPF) frozen bovine semen: A goal? *Proceedings of the Sixth NAAB Technical Conference on Artificial Insemination and Reproduction*. Columbia, MO: National Association of Artificial Breeders, pp. 11–22.

Chenoweth, P.J. 1989. Marketable skills: Beef cattle services that produce economic gain. *Large Animal Veterinarian* 44: 26–28.

Commonwealth Scientific and Industrial Research Organisation. 1993. Australian Standard Diagnostic Techniques for Animal Diseases. Commonwealth Scientific and Industrial Research Organisation.

Eaglesome, M.D., M.M. Garcia. 1992. Microbial agents associated with bovine genital tract infections and semen. Part I. *Brucella abortus*, Leptospires, *Campylobacter fetus* and *Tritrichomonas foetus*. *Veterinary Bulletin* 62:743–775.

Eaglesome, M.D., M.M. Garcia, R.B. Stewart. 1992. Microbial agents associated with bovine genital tract infections and semen. Part II. *Haemophilus somnus*, Mycoplasma spp and Ureaplasma spp, Chlamydia; pathogens and semen contaminants; treatment of bull semen with antimicrobial agents. *Veterinary Bulletin* 62:887–910.

Gay, J.M. 1999. *Basic concepts for cow-calf herd health programs*. College of Veterinary Medicine, Washington State University, Version 2.2. Available at: http://www.vetmed.wsu.edu/courses-jmgay/FDIUCowCalfHH.htm. Accessed January 15, 2004.

Hird, D.W., B.J. Weigler, M.D. Salman, et al. 1991. Expenditures for veterinary services and other costs of disease and disease prevention in 57 California beef herds in the National Animal Health Monitoring System (1988–1989). *Journal of the American Veterinary Medical Association* 198:554–558.

Hughes, H. 2003. The economics of culling cows. *Moorman's Feed Facts*. Quincy, IL: Moorman's Feeds, ADM Alliance Nutrition Inc.

Leman, A. 1988. Diagnosis and treatment of food animal educational diseases. *Journal of the American Veterinary Medical Association* 193:1066–1068.

National Animal Health Monitoring System. 1994a. *CHAPA Part II and Part III: Beef cow-calf reproductive and nutritional management practices* and *Beef cow-calf health and health management*. USDA:APHIS:VS, CEAH. N135.0194, pp. 1–46. Available at: http://www.aphis.usda.gov/vs/ceah/cahm. Accessed July 18, 2004.

National Animal Health Monitoring System. 1997a. Beef '97 Part I: *Reference of 1997 beef cow-calf management practices*. USDA:APHIS:VS, CEAH. N233.697, pp. 1–55. Available at: http://www.aphis.usda.gov/vs/ceah/cahm. Accessed July 18, 2004.

National Animal Health Monitoring System. 1998a. Beef '97 Part III: *Reference of 1997 beef cow-calf production management and disease control*. USDA:APHIS:VS, CEAH N247.198, pp. 1–42. Available at: http://www.aphis.usda.gov/vs/ceah/cahm. Accessed July 18, 2004.

National Animal Health Monitoring System. 1997b. Beef '97 Part II: *Reference of 1997 beef cow-calf health and health management practices*. USDA:APHIS:VS, CEAH. N238-797, pp. 1–38. Available at: http://www.aphis.usda.gov/vs/ceah/cahm. Accessed July 18, 2004.

National Cattlemen's Beef Association. 1999. *Market Cow and Calf Quality Audit*. Englewood, CO: National Cattlemen's Beef Association.

Neville, W.E., K.L. Richardson, D.J. Williams, P.R. Utley. 1987. Cow breeding and calf growth performance as affected by pregnancy status the previous year. *Journal of Animal Science* 65:872–876.

New, J.C. 1991. Costs of veterinary services and vaccines/drugs used for the prevention and treatment of diseases in 60 Tennessee cow-calf operations (1987–1988). *Journal of the American Veterinary Medical Association* 198:1334–1340.

Nietfeld, J.C. 2003. The abortion work-up: When to sample, what to sample and what to expect. *Proceedings of the Bovine Conference on Investigating Pregnancy Wastage in Cattle Herds*. pp. 2–9. Manhattan: College of Veterinary Medicine, Kansas State University.

Richey, E.J. 1991. Facilitating communication, Part II: Herd health. *Large Animal Veterinarian* 46(3):8–16.

Richey, E.J. 1992. *Constructing diagrams to represent the management system of a beef herd*. Florida Cooperative Extension Service/IFAS Bulletin 278.

Roybal, J. 1999. A model of efficiency. *Beef* Spring:60–61.

Salman, M.D., M.E. King, T.E. Wittum, et al. 1990. The National Animal Health Monitoring System in Colorado beef herds: Disease rates and their associated costs. *Preventive Veterinary Medicine* 8:203–214.

Salman, M.D., M.E. King, K.G. Odde, et al. 1991. Costs of veterinary services and vaccines/drugs used for prevention and treatment of diseases in 86 Colorado cow-calf operations participating in the National Animal Health Monitoring System (1986–1988). *Journal of the American Veterinary Medical Association* 198:1739–1744.

Sanderson, M.W., J.M. Gay. 1996. Veterinary involvement in management practices of beef cow-calf producers. *Journal of the American Veterinary Medical Association* 208:488–491.

Sanderson, M.W., D.P. Gnad. 2002. Biosecurity for reproductive diseases. *Veterinary Clinics of North America: Food Animal Practice* 18:79–98.

Schultz, R.D. 1993. Certain factors to consider when designing a bovine vaccination program. *The Bovine Proceedings* 26:19–26.

Spire, M.F. 1988. Immunization of the beef breeding herd. *Compendium on Continuing Education for the Practicing Veterinarian* 10:1111–1118.

Spire, M.F., J.P. Hotz. 1995. Establishing culling criteria in beef cow-calf operations. *Veterinary Medicine* 90:693–700.

Stokka, G.L., T.R. Falkner. 1998. *Preventive herd health program*. EP-50. Manhattan: Kansas State University, K-State Agricultural Experiment Station and Cooperative Extension Service, pp. 1–4.

Toombs, R.E. 1996. Setting fee schedules: How much are veterinarians worth to beef cattle operations? *Compendium on Continuing Education for the Practicing Veterinarian* 18:S85–S93.

Wikse, S.E. 2002. Approach to investigation of abortions in beef and dairy cattle. *Proceedings of the Australian Association of Cattle Veterinarians Annual General Meeting*, pp. 33–36. Queensland: Australian Association of Cattle Veterinarians.

Wise, J.K. 1987. *The U.S. Market for Food Animal Veterinary Medical Services*, pp. 1–185. Schaumburg, IL: American Veterinary Medical Association.

5
Biosecurity for Beef Cow/Calf Production

Michael W. Sanderson and David R. Smith

INTRODUCTION

Biosecurity is the sum of actions taken to prevent the introduction of disease-causing agents. Biocontainment consists of the actions taken to control the spread of disease-causing agents already existing within a herd. A biosecurity plan is concerned with preventing the introduction of pathogens or toxins that have the potential to damage the health or productivity of a herd of cattle or the safety and quality of a food product (Dargatz et al. 2002).

Biosecurity has become an increasingly important issue for U.S. beef cattle operations because of trends in the industry that boost the risk of disease, such as the increasing movement of cattle, a growing reliance on purchased rather than home-raised replacements, a more global marketplace, and increasing pressure on producers and veterinarians to minimize antimicrobial use. Production medicine programs have not typically emphasized biosecurity; however, biosecurity has more recently become an important component of an integrated production management program. A recent review of biosecurity in cattle operations is available (Dargatz 2002).

Ultimately, it is livestock producers themselves who are responsible for the biosecurity of their herds. However, others may be responsible for implementing and enforcing biosecurity plans at other (e.g., geopolitical) levels. For example, the U.S. Department of Agriculture determines which actions are necessary for protecting the borders of the United States from the introduction of foreign animal diseases such as foot-and-mouth disease or rinderpest.

At this level, the practicing veterinarian and the livestock producer are responsible for obeying regulations concerning animal movement, and they serve as the first line of defense in recognizing the occurrence of a potential foreign animal disease. Producers have the responsibility to take biosecurity actions to protect their own herds from exposure to these agents and to promptly report suspicious disease occurrences. Once a foreign disease agent has entered the country, the U.S. Department of Agriculture implements control and eradication procedures. The strength of each producer's biosecurity plan largely determines how safe his or her own cattle are from exposure.

The U.S. Department of Agriculture also has responsibility for developing programs to control the spread of certain diseases that exist within U.S. borders. Again, practicing veterinarians and livestock producers are responsible for adhering to the program rules. These may include testing and quarantine procedures to prevent transmission of the agent from one herd to another, as well as identification and removal of positive individuals or herds. For other diseases, sole responsibility for biosecurity lies with individual producers.

Biosecurity for cow-calf producers is challenging because of the extensive nature of cow-calf production and the complex and diverse environment to which cattle are exposed (Figure 5.1). Cattle are ruminant animals designed to consume forage and convert it to meat and milk. Typically, in cow-calf production systems, the cattle harvest forages in extensive pastures—an environment in which cattle may be exposed to disease agents originating from neighboring herds and from a wide variety of wildlife. Available evidence indicates that cow-calf herds are commonly exposed to outside herds and wildlife and that producers do not always properly adjust management practices to protect against the increased risk of these exposures (Sanderson et al. 2000).

The producer, using the advice of the veterinarian, decides which actions to take to prevent exposing his or her herd to disease agents. Biosecurity strategies should be developed by rationally assessing the

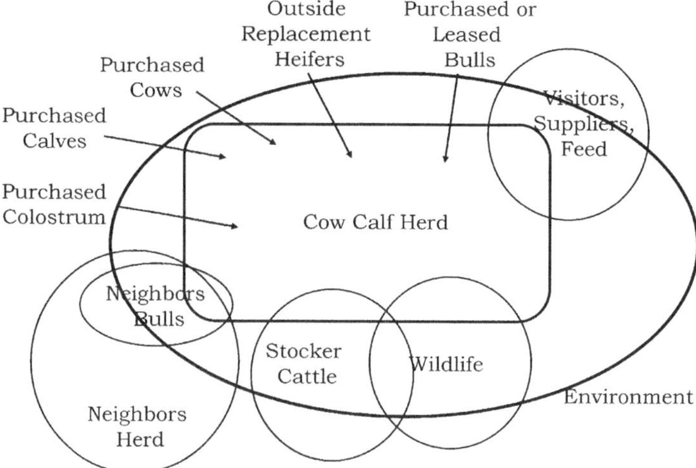

Figure 5.1. Cow/calf production systems provide numerous interactions that may result in the introduction of disease agents.

risk and cost of introducing a disease, as well as the costs and effectiveness of the risk mitigation plan. The value producers place on a biosecurity plan will differ based on a number of factors, including the value of the cattle, whether the producer sells breeding stock, his or her reputation, and the producer's adversity to risk.

A seed stock producer may realize marketing value out of a biosecurity plan by providing assurance of low disease risk to his or her customers. Therefore, a risk analysis model is a rational approach for developing and applying biosecurity programs. Biosecurity programs to control disease transmission may be integrated into other risk analysis programs (e.g., Hazard Analysis and Critical Control Point–like programs) that aim to control food quality and safety and minimize antimicrobial use in cattle (Hogue et al. 1998, Griffin chapter 13 of this book).

The risk analysis approach to biosecurity involves processes of risk assessment, risk management, and risk documentation. Risk assessment is the process of identifying diseases of biosecurity concern, estimating the probability of the disease being introduced to the herd, and determining the consequences of its introduction. The consequences of introducing disease may obviously be economic, but they may also include less quantifiable effects such as loss of reputation and social stigma.

Risk management involves determining which biosecurity actions can be taken to reduce the probability that the disease will occur and what it would cost to take those actions. Documentation involves using health and production records to make risk assessment and risk management decisions, as well as communicating the risk analysis process to all interested parties. The use of records is an integral part of biosecurity. Without records, it is difficult to identify needs, document actions, or benchmark progress. Two-way communication is a critical component of a biosecurity plan. Communication between the veterinarian and the producer ensures that the intent of an action plan is understood and followed, and it also ensures that there is a mechanism for receiving feedback about the usefulness of particular biosecurity practices.

IMPLEMENTING BIOSECURITY

In one form or another, all biosecurity and biocontainment plans apply the principles of infectious disease control. These principles are to maximize the resistance of the host animal against the pathogen, prevent the entry of disease agents, and prevent effective contacts that result in transmission of the pathogen (Figure 5.2). Many biosecurity plans also employ some form of diagnostic surveillance to detect the introduction of pathogens. In most cases, biosecurity is not achieved unless multiple actions addressing more than one of these principles are applied.

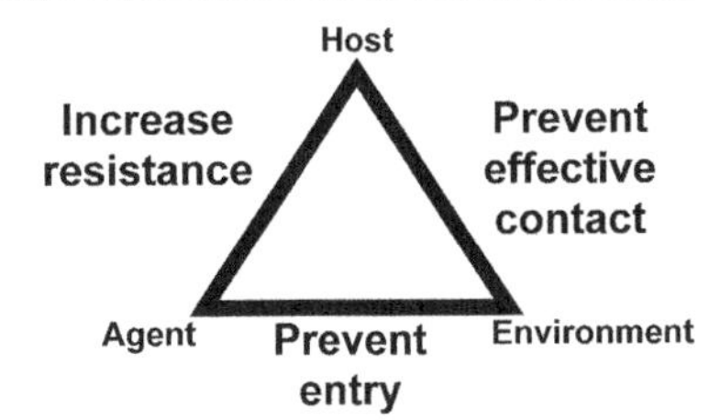

Figure 5.2. Biosecurity and biocontainment are achieved by applying the principles of infectious disease control.

Maximizing Resistance to Disease

Maintenance of immunity in the resident herd through the use of vaccination is a familiar way to manage risk of disease transmission and has been the most common way veterinarians and producers have attempted to mitigate biosecurity and biocontainment risks. Unfortunately, the effectiveness of most vaccines is limited, and vaccination should not be looked on as the only, or even a primary, means of decreasing the probability of infection or disease. Even under optimal conditions, not all cattle will respond to vaccination, nor will all that respond to vaccination be protected from infection. In most cases, vaccines do not prevent infection but work to decrease clinical disease and reduce shedding of pathogens.

Unfortunately, no effective vaccine exists for many diseases. Vaccination programs, however, can be useful adjuncts to other management practices. Producers may vaccinate the resident herd, to increase immunity to pathogens that may be introduced by imported cattle, or they may require vaccination of imported cattle before entry into the herd, to decrease infection and shedding of disease agents.

A herd may be protected from an outbreak of disease even though not all of the animals are immune from the disease. This phenomenon is known as herd immunity. Transmission of the pathogen through the herd is halted because a proportion of the animals in the herd have immunity and are therefore protected from clinical disease and are less likely to shed infectious amounts of pathogen following exposure. The nonimmune animals within the herd benefit because they have less opportunity for exposure and because transmission is unlikely to be maintained. The proportion of immune animals in the herd that is necessary to provide herd immunity varies with the pathogen and with other management factors such as population density. Vaccination can elevate the level of herd immunity even though not all animals mount a protective response to the vaccine. Vaccination may also decrease the effect of a disease introduction by decreasing clinical disease and shedding, thus decreasing exposure.

Preventing the Introduction of Disease Agents

Reservoirs of disease agents are the sources of pathogen exposure. Sources of contagious pathogens to cattle include other cattle, other species of animals (including man), and inanimate objects. The most important reservoirs of disease to cattle are other cattle. Imported cattle that arrive infected with pathogens are a major biosecurity concern. A special concern is that some healthy-appearing cattle may be carriers of disease, meaning they are infected and capable of transmitting the agent, though they do not show signs of illness. Biosecurity actions often include methods to prevent the exposure of the native herd to imported cattle harboring disease agents. These methods include limiting import sources to low-risk herds, monitoring imported cattle for clinical signs of disease, and using diagnostic testing to detect carriers.

Using Herd-of-Origin Records to Prevent Entry of Disease

Herd production and health records can be a valuable tool to identify low-risk source herds. Herds with documented high reproductive rates are less likely to harbor a reproductive pathogen; for example, herds that have documented records of aged cow health and disease are less likely to harbor Johne's disease if there has been no clinical syndrome recognized that is consistent with a diagnosis of Johne's. In many cases, it may not be possible to determine whether an individual animal is a carrier of disease; however, health and production records from the herd of origin may provide sufficient evidence that infection in the individual is unlikely. At present, cow-calf records kept by seed stock producers may not provide all the information a buyer would need, particularly that related to health issues.

Using Diagnostic Tests to Prevent the Entry of Disease

It is important that the veterinarian understand diagnostic tests and how they might be used to meet the biosecurity objective. Testing imported cattle might decrease the risk of introducing disease into the herd. Testing is no panacea, however. Tests must be carefully evaluated to ensure they achieve the desired goal of decreasing risk of disease entry at an acceptable cost.

Testing can be a valuable tool in a biosecurity program when appropriately applied. The diagnostic strategy will differ greatly depending on the diagnostic objective. Determining which cows to test, how many cows to test, and when to perform the test depends on the diagnostic objective. The various reasons for using diagnostic tests in a biosecurity setting may be to determine whether or not an individual is infected, to determine whether or not the infection exists within a population, or to determine the prevalence of infection in the population. (For a more thorough discussion of sampling and sample size, see chapter 3). In many cases, it may be more valuable to limit imports to those from herds of known status (those that have been tested and found free) than to test individual imports. Records of testing from the herd of origin that establish freedom from disease make it unlikely that the individual animal to be imported is diseased.

Regardless of the diagnostic objective, livestock producers and veterinarians are most interested in whether or not to believe a test result. This is an issue of test interpretation. However, to understand test interpretation, it is necessary to first understand how test performance is evaluated.

Test Performance

Diagnostic tests serve as indicators of an animal's disease (or infection) status, and the results are often reported as positive or negative. All diagnostic tests are imperfect, and the performance of a test is characterized by how likely false-positive or false-negative errors are to occur. The performance of a test is described in terms of its sensitivity and specificity. Sensitivity is the proportion of truly diseased (or infected) individuals that are expected to test positive (i.e., have a true positive test result). Specificity is the proportion of nondiseased (or noninfected) individuals that are expected to test negative (i.e., have a true negative test result). As sensitivity decreases, the proportion of false-negative results increases. As specificity decreases, the proportion of false-positive results increases (Martin 1977).

Test Interpretation

Producers and veterinarians are most interested in whether the test result truly reflects the infection status of the animal (Spangler 1992). If the test result is positive, a veterinarian must estimate the probability that it is a true-positive rather than a false-positive test result. If the test result is negative, a veterinarian must estimate the probability that it is a true-negative and not a false-negative test result. These estimates are called predictive values or posttest probabilities. The posttest probability is a function of the test's performance and the probability that the animal is infected. The probability that the animal is infected is estimated by judgment as well as by estimating the prevalence of infected individuals expected in a representative population.

When the prevalence of infection is expected to be low and the specificity of the test is imperfect, then there may be more false-positive results than true-positive results. In this case, the positive predictive value would be low and the positive result might be noninformative or completely misleading. If the cost of a false-positive result is high, for example, leading to the unnecessary exclusion of valuable genetics, then testing in circumstances in which the posttest probability of a positive test result is low may be unacceptable. In contrast, if a significant level of false positives is acceptable to avoid false negatives, then this may be an acceptable testing strategy.

In circumstances in which the prevalence of infection is high and the test sensitivity is imperfect, then the proportion of negative results that are false negative may be high. Knowing that the negative predictive value is low in these circumstances shows that diagnostic testing will not reliably keep out infected animals, because there are too many opportunities for infected animals to be accepted with a false-negative test result. (For a more thorough discussion of diagnostic testing, see chapter 3.)

Because tests are imperfect, their usefulness in certain circumstances is limited, and there are circumstances in which the desired diagnostic objective cannot be achieved (Smith 2002). Consider, for example, the use of the Johne's disease enzyme-linked immunosorbent assay as a prepurchase test such that cattle testing negative would be imported into the herd. The goal, of course, is to prevent infected cattle from entering the herd. However, published data indicate that the test is only about 25% sensitive in young, nonsymptomatic cattle (Dargatz et al. 2001). Sensitivity increases as the disease progresses, but for the typical young imported cow or heifer, 25% is a good estimate. The enzyme-linked immunosorbent assay is a reasonably specific test, correctly identifying about 97% of negative animals. Available evidence, however, indicates that prevalence of Johne's disease in the general beef cow population is approximately 1% (Dargatz et al.). When this test is applied to a group of animals with a 1% prevalence of disease, the posttest probability of a positive test result is 8%, meaning that 92% of the positive test results are false positives. The posttest probability of a negative test result is 99.2%, which is an excellent value until one remembers that there is already a 99% probability that any given animal was negative before the test was run (i.e., if 1% of cows were positive, then 99% were negative). Under these circumstances, a positive test result is 12 times more likely to be wrong than right, and a negative test result provides little more new information than was known before the money was spent for testing, and it also provides a false sense of security that Johne's disease has been excluded from the herd.

As the previous example illustrates, it is not always possible to identify cattle that serve as reservoirs of infection, because they are unapparent carriers of disease and because diagnostic testing may fail to indicate their infection status. In this case, it may be more valuable to have knowledge about the source herd than to base decisions on clinical signs and test results of individuals. Gaining knowledge of the source herd may include evaluating herd health and production records, questioning the owner or veterinarian about the health history of the herd, and conducting thorough diagnostic screening of the source herd (Martin et al. 1992).

Other species of animals may serve as a source of infection to cattle. These vectors of disease may be infected themselves or may merely be carrying the agent mechanically from place to place. Examples include *Salmonella* from wild turkeys or geese, *Leptospira* from rodents, and *Neospora* from dogs or wild canids. Contact between wildlife and cattle should be controlled as much as possible. Feed sources should be stored to prevent contamination by animal urine and feces as much as possible. Domestic pets such as cats and dogs should also be prevented from contaminating feedstuffs.

In addition, pathogens may be carried onto the farm on inanimate objects such as vehicles, clothing, feed, or water drainage from other operations. Preventing pathogen exposure via contaminated objects calls for traffic control and sanitation. For example, it may be possible to prevent vehicles from other livestock operations (e.g., delivery and veterinary trucks) from driving into areas where cattle may be exposed.

When reservoirs of infection cannot be totally eliminated, then actions must be taken to prevent effective contacts between the cattle herd and the source of disease. A particularly notable example of a disease that may enter through feed is bovine spongiform encephalopathy (BSE). BSE is not a contagious disease (i.e., it is not transmitted from cow to cow) but is introduced to a herd through feed containing meat and bone meal contaminated with the agent from the carcass of a BSE-infected ruminant. This highlights the importance of adherence to the mammalian-to-ruminant feed ban in effect in much of the world to prevent introduction of BSE into the herd.

Preventing Effective Contact with Disease Agents

An effective contact is contact made by an uninfected animal with a source of pathogens sufficient to result in infection and disease. Effective contacts can be prevented by using various methods to physically separate animals or herds including isolation, segregation, or quarantine; minimizing the dose-load of exposure through sanitation or use of prophylactic medication; or minimizing the duration of contact time.

Physical separation involves putting sufficient distance between animals or groups of animals to prevent transmission of the agent. Proper physical separation involves preventing more than nose-to-nose contact. For disease agents that are shed in feces or urine such as *Salmonella*, *Leptospira*, or bovine viral diarrhea virus (BVDV), contact with contaminated drainage must be prevented. For diseases such as bovine leukosis virus, which may have a significant insect vector, separation must be far enough away to prevent travel of biting flies.

Isolation is the physical separation of individuals to prevent disease transmission. Strict isolation is rarely practiced in cow/calf operations, but it is exemplified in the dairy industry by the use of calf hutches to isolate newborn calves from other sources of pathogens and to keep them from contacting each other to transmit disease. Segregation is the long-term physical separation of groups of cattle within herds. An example of segregation is the protection of the pregnant cow herd from potential sources of BVDV by physically separating them from neighboring herds or commingled feeder cattle.

Quarantine is the physical separation of infected or potentially infected cattle until disease-free status can be established. Quarantine can be a valuable biosecurity tool if properly understood and applied. Quarantine is most effective for preventing transmission of diseases that are characterized by short incubation periods and obvious clinical signs of infection, such as bovine respiratory disease. Quarantine is useful for preventing exposure of the native herd to disease from new arrivals. Regulatory officials often use quarantine to prevent movement of cattle from herds infected with dangerous, contagious diseases until the infection is eliminated. Quarantine may not be effective against diseases that have an unapparent carrier state or a prolonged incubation period that cannot be reliably detected during the quarantine period by observing clinical signs or by diagnostic testing. Because of this, introduction of diseases such as Johne's disease, brucellosis, leptospirosis, neosporosis, salmonellosis, and leukosis may not be prevented by quarantine alone.

Thomson (1997) outlines six levels of exposure to disease agents, with increasing risk of disease exposure moving from level 1 to level 6:

1. Closed herd—"specific pathogen free" (SPF) herd
2. Closed herd—no entry or reentry of animals
3. No entry of new animals, reentry allowed
4. Entry of new animals (known medical records) and quarantine
5. Entry of new animals (known medical records) and no quarantine
6. Entry of new animals (no medical records), and no quarantine.

Levels 1 and 2 provide the least exposure and risk for cow/calf herds. At Level 1, an SPF herd maintains freedom from specified pathogens by utilizing strict biosecurity and active surveillance. A herd might fit into level 2 without establishing freedom from any particular disease. There may be, however, little advantage in excluding animals from entry that are carrying pathogens already endemic in the resident herd. Therefore, knowledge of the current herd status is important in establishing rational biosecurity plans.

Level 3 herds allow for exit of resident animals to cattle shows, leasing of cattle to other herds, or offsite replacement heifer development and subsequent reentry into the resident herd. This results in varying levels of exposure to pathogens from other herds, depending on sexual contact, population density, and environmental conditions. Knowledge of the biosecurity practices, health, and type of contact with other herds is important in evaluating the need for quarantine and testing on reentry. This level would exclude entry of new genetics through purchase of bulls.

The differences between the categories are found in whether medical or production records from the source herd are known and whether the imported animals are quarantined for a period after arrival. We could add an additional category between levels 5 and 6 for herds that allow entry of new animals with no medical or production records but do practice quarantine of imports.

Only level 1 would result in complete physical separation of the resident herd from all outside herds. Maintenance of SPF status could likely be very difficult depending on the agent, although at least one beef herd has maintained its status for multiple agents (Lees 1991). Level 2 separates the resident herd from outside herds without attempting to establish SPF status. Levels 3–6 allow increasing exposure to outside herds and greater probability of disease introduction if other prevention procedures are not in place. Most cow/calf herds fall into levels 4–6.

Wildlife may be a source of cattle exposure to *Leptospira, Salmonella,* and *Neospora.* Preventing effective contacts between cattle and wildlife is a particularly difficult task given the extensive nature of the cow/calf environment. Physically separating the cattle herd from wildlife is often not possible or economically feasible; however, it may be possible to take practical steps to minimize the dose-load or duration of exposure to wildlife. For example, preventing wildlife access to stored feeds is prudent to minimize contamination by pathogen-containing feces or urine. In addition, control of the cow environment to prevent accumulation of standing water in corrals and pens will limit exposure to *Leptospira* or other waterborne pathogens. Finally, population control of wild animals that are in contact with the herd may be prudent. Canids, opossums, squirrels, rats, raccoons, and similar varmints, as well as birds, have spread disease to cattle.

A biosecurity plan must be designed for a specific operation based on the operation's specific risks to be effective (Table 5.1). Operations may be at elevated risk for introduction of a specific agent because of their location or management practices. On the basis of the assessment of risk for the operation, specific agents may be identified to target biosecurity efforts. Knowledge of the disease status for the herd is a critical first step in planning a biosecurity program. If a particular disease agent is already present on a ranch, time, effort, and money spent on its exclusion will be wasted until an effective eradication plan for the agent is in place. For agents not already endemic on the ranch, effective biosecurity practices must be based on the epidemiology of the disease agent. The interaction of the disease agent, the cattle, and the environment must be understood, including how the disease is transmitted, possible outside reservoirs, duration of shedding, and unapparent carrier states. Biosecurity is most simple for agents with simple epidemiology. For example, agents such as *Campylobacter fetus* ssp venerealis and *Trichomonas fetus* are only transmitted by venereal contact. We do not have to control

Table 5.1. Checklist for herd biosecurity implementation.

Herd management

1. Value of the cattle
2. Presence of specific disease agents in the herd
3. Presence of the specific disease agents in the local area
4. Prevalence of disease agents in areas from which cattle are purchased
5. Type of cattle purchased
   Stockers
   Breeding (virgin or nonvirgin)
6. Origin of cattle purchases
   Auction market
   Seed stock
   Seed stock with health and production records
7. Contact of cattle with neighboring herds
   Fenceline contact
   Intermingling
   Import practices of neighboring herds
8. Contact of cattle with other cattle at shows
9. Different classes of cattle worked through the same working facilities without cleaning/disinfection between

Feed and water

1. Is feed purchased from outside sources?
   Quality assurance plan in place on farm and at supplier?
   Assurance of no ruminant protein in feed?
2. Do wildlife or pets have access to stored feed or water sources?
3. Do surface water sources originate on neighboring farms (drainage or flowing water)?

Ranch traffic

1. Do vehicles with access to other farms enter cattle areas (veterinarian, feed supply, drug supply, rendering)?
2. Do owners or employees have exposure to other cattle outside the herd?
3. Is there a visitor protocol and log?

Record keeping and monitoring

1. Are health and production records kept on the herd?
2. Are health and production records analyzed routinely to detect emerging production depression or disease outbreaks?

outside reservoirs or drainage of feces and urine. We can limit imports to virgin animals or prevent sexual contact with animals of unknown status until test results can be obtained. There is, however, significant opportunity for an "effective contact" during the breeding season if a producer fails to control the entry of sexually mature animals such as the neighbors' bull—or the producer's own bull after he has visited the neighbors.

Agents of disease that are shed in feces or urine and transmitted by oral ingestion (e.g., BVDV, *Salmonella*, *Leptospira*) or through mucus membranes or even intact skin (e.g., *Leptospira*) require more effort for exclusion. Animals that are not sexually mature may be a source of infection for the herd, so all imported animals are a potential risk, as are contacts with neighboring cattle. Adequate quarantine for these agents will require that the resident herd not be exposed to any drainage from the quarantine pens.

Disease agents that have a significant nonbovine reservoir (e.g., *Leptospira*, *Neospora*, *Salmonella*) add additional complexity to biosecurity plans. For these agents, we need to not only control the import of cattle to the resident herd and prevent contact with neighboring herds but also prevent contact with another domestic or wildlife reservoir. In general this cannot be completely accomplished, so efforts to control the number of effective contacts are necessary through securing feed storage, ensuring proper water drainage, controlling the wildlife population, and generally managing to minimize contact.

Disease agents that may be transmitted by an insect vector (e.g., *Anaplasma*, bovine leukemia virus) require that imported animals be quarantined far enough from the resident herd to prevent transmission of disease by biting insects. The distance should be equal to or greater than the flying distance of the insect responsible for transmission.

IMPLEMENTING BIOCONTAINMENT

In circumstances in which the disease agent already exists within a herd, the disease control objective is biocontainment. The principles of infectious disease control used for biosecurity also apply to biocontainment; only the goals differ. Biocontainment actions are directed at controlling transmission of disease agents within the farm with the goal of either eliminating the agent from the herd or minimizing the disease effect of infection. The goal of biocontainment depends in large part on the nature of the agent. For example, BVDV infection can be eliminated from herds with biocontainment actions that protect cattle in early gestation from exposure to cattle transiently or persistently infected with BVDV, which might result in the subsequent persistent infection of the fetus. Although it may be impossible to eliminate the common agents that cause neonatal calf diarrhea from a farm, biocontainment actions to minimize dose-load of exposure and transmission still may effectively prevent outbreaks of disease.

CONCLUSION

The tools of biosecurity and biocontainment are not new. However, cattle production systems have largely evolved with little consideration for the transmission of infectious agents. More recently, world events, economics, and trends in the cattle industry have changed how cow/calf operators perceive the importance of disease control. There is wider recognition of the cost of accidentally or intentionally introduced livestock diseases. Cattle producers increasingly wish to protect their herds from infectious diseases, and the veterinarian must be prepared to provide sound strategies for biosecurity and biocontainment tailored to the circumstances.

REFERENCES

Dargatz, D.A. 2002. Biosecurity of cattle operations. *Veterinary Clinics of North America: Food Animal Practice* 18(1):205.

Dargatz, D.A., F.B. Garry, J.L. Traub-Dargatz. 2002. An introduction to biosecurity of cattle operations. *Veterinary Clinics of North America: Food Animal Practice* 18(1):1–5.

Dargatz, D.A., B.A. Byrum, L.K. Barber, et al. 2001. Evaluation of a commercial ELISA for diagnosis of paratuberculosis in cattle. *Journal of the American Veterinary Medical Association* 218:1163–1166.

Hogue, A.T., P.L. White, J.A. Heminover. 1998. Pathogen reduction and hazard analysis and critical control point (HACCP) systems for meat and poultry. *Veterinary Clinics of North America: Food Animal Practice* 14:151–163.

Lees, V.W. 1991. Developing a model specific pathogen free beef herd. *Proceedings of the 6th International Symposium on Veterinary Epidemiology and Economics*, pp. 364–366. Ottowa, Canada, August 12–16, 1991.

Martin, S.W. 1977. The evaluation of tests. *Canadian Journal of Comparative Medicine* 41:19–25.

Martin, S.W., M. Shoukri, M.A. Thorburn. 1992. Evaluating the health status of herds based on tests applied to individuals. *Preventive Veterinary Medicine* 14:33–43.

Sanderson, M.W., D.A. Dargatz, F.B. Garry. 2000. Biosecurity practices of beef cow-calf producers. *Journal of the American Veterinary Medical Association* 217(2):185–189.

Spangler, L. 1992. Using and interpreting diagnostic tests. *The Bovine Proceedings* 24:22–28.

Smith, D.R. 2002. Epidemiologic tools for biosecurity and biocontainment. *Veterinary Clinics of North America: Food Animal Practice* 18(1):157–175.

Thomson, J.U. 1997. Implementing biosecurity in beef and dairy herds. *Proceedings of the 30th Annual Convention of the American Association of Bovine Practitioners*. Montreal, Canada, September 18–21, 1997. 30:8–14.

6
Beef Cowherd Nutrition and Management

T. T. Marston

INTRODUCTION

Nutritional requirements of beef cowherds are dynamic because of differences in season, stages of production, genetics, and ranch resources. Because feed will often represent 60%–70% of the cowherd budget and controls the majority of a ranch's output, it is one of the most scrutinized areas of interest. The common goal of all cattle producers is to optimize reproduction and animal performance while minimizing costs. Variability in input values makes basic understanding of nutritional principles necessary for veterinarians, producers, and industry support personnel. This chapter will contain a review of ruminant nutrition, cattle requirements and entities, and finally, common feedstuff characteristics.

GENERAL RUMINANT FACTS

In the simplest of terms, a cow's digestive system is designed to crush and soak food particles, sustain microbial fermentation, reduce molecular structures through chemical reactions, absorb nutrients, and finally, to expel any leftover materials. This makes for interesting relationships between the characteristics that affect the physical breakdown, solubility, bacterial activity, volatile fatty acid production, and rate of passage of a diet, all of which will have a dramatic effect on animal performance.

Particle size can influence both digestibility and intake. Generally speaking, the smaller the particle size, the greater the surface area. Greater surface area provides a greater number of sites for ruminal bacteria to attack a feed particle. With more bacteria attacking the feed particle, the rumen is emptied faster, which in turn allows for an increase in intake to satisfy the full feeling that cattle try to maintain. Reducing particle size by mechanical methods is quite common in feedlot and cowherd rations. Grinding, rolling, pelleting, tub grinding, and chopping are common means of reducing particle size.

Benefits such as improved animal performance and reduced feed handling or transport costs have to offset the cost of the procedure. Processing feedstuffs is usually economically advantageous in feedlots but can often be unprofitable under normal ranching conditions. Under extreme conditions, when feedstuff availability is limited (such as during drought), or when animal performance needs to be enhanced to meet an impending challenge, the decision to reduce feedstuff particle size becomes more advantageous.

Reducing particle size to the point that fines (small pieces of feed that will separate themselves from the rest of the feed mix) are generated will cause cattle to sort or even refuse their diets. Therefore, in most cases, rations containing dusty, fine particles should be avoided. The exception to this rule is grain sorghum. If grain sorghum is used, it should be finely ground to increase digestibility. When tub-grinding, producers should try to minimize losses caused by small particles being lost in the wind. The majority of these small particles are leaf material, which is the most nutrient-dense part of the plant.

The relationship between ruminal microflora and beef cattle is one of the few true symbiotic relationships found in nature. The cattle provide an ideal, anaerobic environment for bacteria, protozoa, and fungi. In turn, the microorganisms grow (through multiplication) by breaking down plant material (primarily cellulose). They then generate compounds that will satisfy the host animal's nutrient requirements. The microorganisms themselves are washed from the rumen and are broken down into nutritional building blocks. As the ruminal bacteria grow, they produce volatile fatty acids, mainly acetate, propionate, and butyrate. These volatile fatty acids can be absorbed through the rumen wall and used by cattle as energy sources. The rumen bacteria can also build water-soluble vitamins. Therefore, under normal conditions, the complex of B vitamins does not have to be fed directly to ruminants.

The rumen environment allows bacteria to build amino acids. Bacteria can combine urea, carbon chains, and energy, which in turn can balance protein needs of the host cattle. Rumen-degradable proteins are used by rumen microflora, and supplementation will increase forage intake and digestibility by stimulating bacterial growth. Urea as a nitrogen source does have some limitations, even though it is considered to be 100% rumen degradable. Supplements should not contain more than 20% of equivalent crude protein from urea to have positive effects on forage intake and digestibility. Furthermore, cattle have the ability to recycle nitrogen via saliva and to transport it across the rumen wall. This allows producers to feed cattle protein supplements two or three times weekly but achieve the same nutritional effect gained by feeding daily.

The rumen is the most advanced and complex digestive system studied. Pregastric fermentation makes it possible for bacteria to digest cellulose and make energy available to the host ruminant, giving it a competitive advantage over the nonruminant. Both nutrient requirements of the animal and the ruminal microflora must be considered when predicting performance or building diets. Luckily, this complex system follows simple principles, making ruminant nutrition an exciting field of study.

FACTORS THAT INFLUENCE NUTRITIONAL REQUIREMENTS

Biological Priority

Beef cattle have biological priorities for nutrients. The breeding female's highest priority is for survival (maintenance). Following maintenance comes growth, milk production, and finally, reproduction. Therefore, reproduction has the lowest priority and will be sacrificed first under adverse conditions. The composition of growth will change with the proximity of mature weight. Cows changing from one body condition score (BCS) to another will gain or lose not only energy stores (fat) but also protein (muscle). Their skeleton (bone weight and size) will change very little in mass with concurring weight changes once their mature frame size is achieved. Meeting the nutritional requirements of beef cows at minimal cost is complicated by the many variables that influence cow requirements and the nutritional composition and density of feedstuffs.

Stages of Production

Nutritional requirements of beef cows vary throughout the year. For the most part, these changes are gradual from day to day, with exceptions at parturition and weaning. Dividing the calendar year into four stages of production can simply reflect the differing pregnancy and lactation statuses of the cow (Table 6.1).

Period 1

This period begins with parturition and is the most critical period in the beef cow–year. Both production and reproduction will be heavily influenced by the nutritional status of the cow. Not only must the cow nurse her new offspring, she must rebreed within 80 to 85 days to calve at the same time next year. As a consequence, her nutrient requirements are greatest during this period. In addition, inadequate nutrition (particularly energy) during this period will result in lower milk production and calf weaning weight, as well as compromised reproductive performance.

Requirements for protein and energy rise dramatically with calving. Depending on the genetic potential for milk, protein and energy requirements may increase by 20%–30% or more in just a few days. Typically, feed intake is reduced shortly before and after calving, but this phenomenon lasts a few days

Table 6.1. The 365-day beef cow year by stages of production.

Period	1	2	3	4
Days	85	123 (varies with weaning date)	85 (varies with weaning date)	75
Stage	Postcalving Early/heavy lactation Uterine involution	Pregnant and lactating	Midgestation Dry	Late-gestation (rapid fetal growth) Preparation for lactation

only. In fact, in less than a week after calving, the cow's capacity to eat is enhanced by as much as 20%–30%. Fortunately, this increased intake helps cattle meet their increased nutritional requirements.

It is not uncommon for cows in early lactation to lose weight. Studies have indicated that weight reductions greater than 17% of precalving weights will result in impaired breeding (Ewing et al. 1966). Such weight loss may at first seem quite dramatic, but one should realize that subsequent weight loss from expelling the fetus, fetal fluids, and placenta during birth could equal up to a 13% loss of the dam's total body weight.

Lactation causes a major demand for energy and protein. Factors that need to be accounted for when estimating nutrient requirements for milk production are daily milk yield, fat, and protein content. Because of the high energy content of fat, milk fat can be used to estimate the energy required for milk production. The average beef cow will produce milk that contains about 4% fat. Expressed as net energy for maintenance (NEm), the Mcal of NEm required per pound of milk is equal to 0.045 times the milk fat percentage plus 0.16 (National Research Council 1984, 2000). For example, a cow with a daily milk production level of 22 lb will need to have 7.88 Mcal of NEm added to her maintenance energy requirement to satisfy that level of production. In addition, the same cow will require about 0.85 lb of net protein daily to satisfy her needs (for milk averaging 3.4% protein). Research indicates that milk production levels can vary greatly, with reported daily maximum or peak yields from 8 to 44 lb. Expected peak milk yield production is very dependent on cow genotype and age. Literature indicates that 2- and 3-year-old cows will have about 26%–12% lower milk production levels than will mature cows (National Research Council 2000).

Period 2

The cow is pregnant and lactating during the second production period. Here, the requirements for pregnancy are quite small, so they make up an extremely small part of the animal's total requirements. Often producers will align period 2 to coincide with high-quality forage production. As a consequence, period 2 is not considered critical for immediate reproductive performance, but under certain and continuing circumstances, period 2 can be used to maintain a herd's long-term reproductive goals. One description of period 2 could be the time from breeding to weaning dates. Adjusting the age at which calves are weaned may greatly influence cow body weight, condition score, and pasture conditions. (A more detailed discussion of weaning age will follow.) Setting the weaning age will influence the length of both periods 2 and 3. In fact, period 2 can be completely eliminated with early weaning (see chapter 11). This may have a significant effect on annual feed costs, as well as on cowherd production and resource management.

Period 2 is important to calf performance and weaning weight. Calves' appetites will generally outgrow their dams' ability to produce milk during this time. Therefore, they will begin to consume available feedstuffs to supplement their milk diets. Many producers will creep feed calves to enhance calf performance. Non–creep fed calves will supplement their diets with forages that are also available to their mothers.

Period 3

Midgestation is the time when a cow is not nursing a calf and the requirements for the developing fetus are relatively low. The cow's nutritional needs are at their lowest levels, so period 3 is considered a time to economize on a cowherd's annual feed cost by using low-quality, inexpensive feedstuffs. It is also a time when body weight and condition score or flesh can be most easily manipulated. For spring-calving cows in the Midwest United States, period 3 is timed to use crop aftermath left on harvested fall-crop fields. This type of forage can be quite abundant and very inexpensive to graze. Maximizing the grazing of crop residues will usually substantially reduce annual cow cost.

BCS will be discussed in greater detail later. However, in lean years, period 3 can be an excellent time to increase the energy reserves stored in the body. The process of changing from period 2 to period 3 decreases protein and energy requirements by about 20%, because lactation ceases. However, little change is noted in dry-matter intake between the two periods. Therefore, nutrient density of the diet can be a major factor in determining cow performance. High-BCS cows can be placed on diets that will cause them to lose weight and thus take advantage of stored energy reserves. However, thin cows (often those completing their first or second lactation) can be supplemented or fed high-quality diets, with rapid, positive weight responses.

Period 4

This period, from 60 to 90 days before calving, is critical in a cow's reproductive cycle. During this time, approximately two-thirds of the fetal growth will occur, which can happen in excess of 1 lb per day. In addition, the cow is trying to lay on fat stores for the impending lactation and is beginning to synthesize colostrum (which is extremely high in energy

Table 6.2. NRC[a] Requirements for a 1100-pound beef cow producing 15 pounds of milk at peak lactation.

	Stage of Production			
	Period 1	Period 2	Period 3	Period 4
	Early Lactation	Late Lactation/ Early Gestation	MidGestation, No Lactation	Late Gestation
TDN (lbs/day)[b]	14.5	11.5	9.5	11.2
Net energy for maintenance (Mcal/day)	14.9	12.2	9.2	10.3
Protein (lbs/day)	2.3	1.9	1.4	1.6
Calcium (g/day)	33	27	17	25
Phosphorus (g/day)	25	22	17	20
Vitamin A (U/day)	39,000	36,000	25,000	27,000

[a]Nutrient Requirements of Beef Cattle (NRC 1984).
[b]Total digestible nutrients.

Table 6.3. Wind chill index chart.

	Temperature (°F)						
Wind speed (mph)	−10	0	10	20	30	40	50
Calm	−10	0	10	20	30	40	50
5	−16	−6	3	13	23	33	43
10	−21	−11	−1	8	18	28	38
15	−25	−15	−5	4	14	24	34
20	−30	−20	−10	0	9	19	29
25	38	−27	−17	−7	2	12	22
30	−46	−36	−27	−16	−6	3	13
35	−60	−50	−40	−30	−20	−10	0

content). The consequences of inadequate nutrition during period 4 include lighter birth weight calves (without a corresponding decline in dystocia), lower calf survival, lower milk production and calf growth, and delayed estrus, which means a later calf next year and subsequent reduced weaning weight (Ludwig et al. 1967).

Other factors can influence nutrient requirements including temperature, humidity and precipitation, forage availability (exercise), and genetics. Table 6.2 shows the nutritional requirements for a 1100-lb beef cow by period.

Weather

Cold weather can greatly increase energy requirements. Cattle perform most efficiently within a thermal neutral zone, where temperatures are neither too hot nor too cold. When the effective ambient temperature, an index of heating or cooling power of the environment, is outside the thermal neutral zone, animal performance is depressed. Effective ambient temperature is an index that combines temperature, wind, humidity, and solar radiation. The most common situation is when the ambient temperature (wind chill) is below the lower critical temperature, or the lowest point of the thermal neutral zone. Table 6.3 reviews the wind chill index for varying combinations of temperature and wind speeds. If cows are sheltered from the wind, the effective ambient temperature will rise toward the actual air temperature. In addition to actual weather conditions, the amount of insulation on the animal influences the lower critical temperature.

Table 6.4. Estimated lower critical temperatures of beef cattle by hair coat type.

Coat Description	Lower Critical Temperature (°F)
Wet or summer	59
Dry fall	45
Dry winter	32
Dry heavy winter	18

Table 6.5. Increased maintenance energy requirements for cattle per degree Fahrenheit below the lower critical temperature in percentage increase per degree below lower critical temperature.

	Cow Weight (lbs)			
Coat Type	1000	1100	1200	1300
Wet or summer	2.0	2.0	1.9	1.9
Dry fall	1.4	1.3	1.3	1.3
Dry winter	1.1	1.0	1.0	1.0
Dry heavy winter	0.7	0.7	0.6	0.6

Source: Ames 1987.

Table 6.4 shows the estimated lower critical temperatures for cattle with varying quality of hair coats.

The only adjustment in cow diets necessitated by weather changes is in maintenance energy. Protein, mineral, and vitamin requirements are not affected by weather stress. The general rule of thumb is to increase winter ration energy 1% for each degree Fahrenheit (or about 1.5% per degree Celsius) below the lower critical temperature. Table 6.5 shows the percentage increase in energy required for various cow weights and types of hair coats.

When making predictions for winter-feeding programs, weather databases can be extremely useful. Historical data can be used to estimate levels of ambient temperature so that proper adjustments can be made to more correctly achieve targeted animal performance goals. Because of differences in energy concentrations of feedstuffs, diet adjustments must take into account the fill and substitution factors associated with combinations of diet ingredients. Tables 6.6 and 6.7 are presented to aid understanding of these concepts. In particular cases of cold stress, one can easily tell that simply feeding a cow a larger amount of low-quality forage will not meet the animal's added energy demands, in which case the energy density of the diet must be increased by feeding the cow high-quality forage, grain, or a grain byproduct.

Body Weight

There is a direct relationship between body weight and nutritional requirements. Studies conducted mostly in the areas of protein and energy indicate that adding pounds of body weight will proportionally add to the animal's daily need for calories and protein. Table 6.8 shows the energy and protein requirements for cows of differing weights at moderate BCSs. Theoretically, the NEm requirements of beef cattle are estimated to be 0.077 Mcal/empty body weight (kg)$^{0.75}$ in a nonstressful environment. However, in general, each 100-lb change in body weight will result in a difference of 0.57 Mcal NEm daily.

By using nitrogen balance studies, the Institut National de la Recherche Agronomique (1989) determined that the metabolizable protein maintenance requirement for beef cattle is about 3.25 g/kg of shrunk body weight$^{0.75}$ (shrunk body weight is equal to the body weight after a 12–18-hour deprivation of feed; it helps to standardize and minimize the effect of gut fill). This is a much simpler system for estimating overall requirements when compared to one based on the demands that protein makes of various body functions, and it is similar in concept to the net energy system used to determine maintenance energy requirements. Metabolizable protein is assumed to be 64% of the diet's crude protein value. This conversion accounts for true protein content (80%) and an average protein digestibility of 80% (National Research Council 2000). There is a relationship between total digestible nutrients (TDN) and bacterial crude protein produced within the rumen. The bacterial crude protein production will equal about 13% of the TDN consumed. Therefore, under normal conditions, 3.8 g of metabolizable protein per kilogram of body weight to the three-quarter power should be the best estimate of the maintenance protein requirement. Figure 6.1 depicts the metabolizable protein requirements of a 1200-lb beef cow.

Many producers, nutritionists, and consultants are not comfortable with the metabolizable protein concept. Simply put, the metabolizable protein requirement is the amount of degradable intake protein (DIP) required by the rumen microflora, and the amount of undegradable intake protein (UIP) needed to supplement the microbial protein for a given animal

Table 6.6. Additions in dietary components (adding either extra hay **or** extra grain) to correct cold stress requirements of cattle with a dry winter coat.

Effective Temperature (°F)	Increase Percentage in Energy	Amount of Extra Hay Needed	Amount of Extra Grain Needed
50	0	0	0
30	0	0	0
10	20	3.5–4 lbs/cow	2–2.5 lbs/cow
−10	30	7–8 lbs/cow	4–5 lbs/cow

Table 6.7. Dry matter intake guide of mature beef cows.

	% Body Weight	
Roughage Type	Dry, Gestating Cow	Lactating Cow
Low-quality roughages (dry, mature grass; straw; etc.)		
Unsupplemented	1.5	2.0
Supplemented >30% crude protein	1.8	2.2
Supplemented <30% crude protein[a]	1.5	2.0
Average-quality roughages (Bermuda, native, maturing cool-season grasses, etc.)		
Unsupplemented	2.0	2.3
Supplemented >30% crude protein	2.2	2.5
Supplemented <30% crude protein[a]	2.0	2.3
High-quality roughages (Alfalfa hay, silage, green pastures)		
Unsupplemented	2.5	2.7
Supplemented >30% crude protein	2.5	2.7
Supplemented <30% crude protein[b]	2.5	2.7

[a]Above 4 lb of supplement, each pound of supplement will decrease forage consumption by about 0.6 lb.
[b]Pound for pound substitution of supplement for forage.
Source: Lusby et al. 1985, Hibberd and Thrift 1992.

Table 6.8. Energy and protein requirements for different cow body weights nonlactating, mature cows—middle trimester of pregnancy.

Cow weight, lbs	TDN (lbs/day)	Net Energy for Maintenance(Mcal/day)	Protein (lbs/day)
1000	8.8	7.57	1.3
1100	9.5	8.13	1.4
1200	10.1	8.68	1.4
1300	10.8	9.22	1.5
1400	11.4	9.75	1.6

Source: NRC 1984.

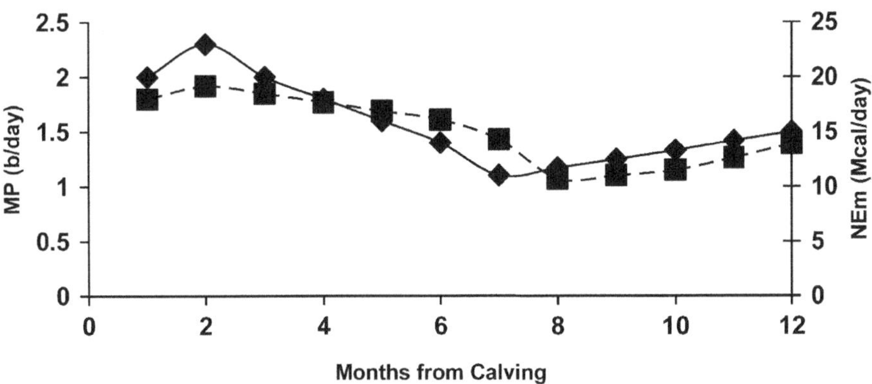

Figure 6.1. Energy and protein requirements of a 1200-lb beef cow with 23 lbs/day peak milk production, weaned at 7 months (Iowa Beef Center 2002).

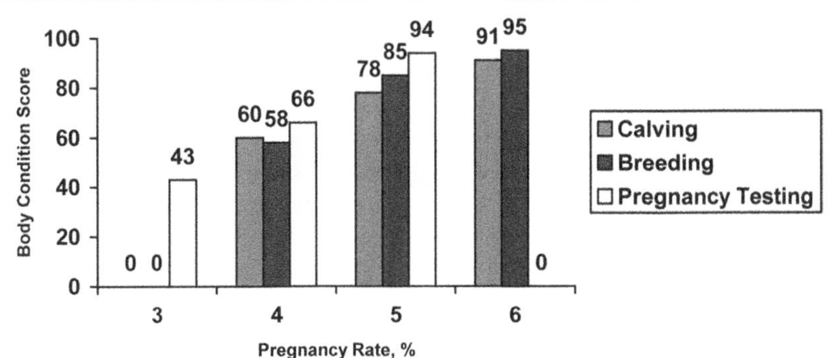

Figure 6.2. Relationship of body condition to pregnancy rate when scored at calving, breeding, or pregnancy testing.

response. Crude protein requirement is the amount of protein in the diet that will satisfy the metabolizable protein requirement. Because of inefficiencies in the system, the crude protein requirement will always be equal to or greater than the metabolizable protein requirement. Referring again to Table 6.8, a difference in 100 lb of body weight will directly correspond to about a 0.1-lb difference in crude protein requirement.

Cow BCS

Body condition scoring allows cow/calf producers to evaluate their nutritional programs every time they observe their cows. Body condition is a subjective measure of stored energy reserves or flesh and is an excellent method of describing cows. Visual inspection alone or in combination with palpation is used to assign the condition score, although the preferred procedure would be to combine both visual and palpation techniques. By evaluating cowherds at strategic times of the year, producers can coordinate their forage resources, supplemental feeding programs, and cowherd management. Reducing the amounts of supplemental hay and feed used is often a big factor in improving profitability. The relationship between BCS and reproductive efficiency has been studied such that the body composition score has become a powerful management tool (Figure 6.2 and Table 6.9).

Several body condition scoring systems have been developed. The most popular system in the United States uses a 9-point scale for beef cows (Wagner 1985, Momont and Pruitt 1998, Blasi et al. 1999). The BCS scale ranges from 1 to 9, with a score of 1 used to describe an extremely thin cow and 9 to describe an obese cow. Most cows will have a BCS

Table 6.9. Cow body condition and probability of cycling by the beginning of the breeding season.

Condition Score	Probability Based on Precalving Condition Score	Probability Based on Prebreeding Condition Score
2	—	.05
3	.09	.12
4	.19	.28
5	.35	.52
6	.55	.74
7	.74	.89
8	.86	—

Source: Momont and Pruitt 1998.
Dash indicates not available (no data were reported).

between 3 and 7. A BCS of 3 indicates a cow has a body composition containing about 9% fat, whereas BCSs of 5 and 7 will indicate fat stores of 18% and 27% of the carcass weight. The difference between successive condition scores is usually between 60 and 90 lb of body weight. The exact difference is dependent on the frame score of the cow. Tables 6.10 and 6.11 explain the different BCSs. Many producers will simplify the 9-point scale into three sections: thin, moderate, and heavy flesh.

Late-calving cows in thin condition have a difficult time becoming pregnant in a time-limited breeding season. This is mainly because of the female's inability to initiate the estrus cycle before the end of the breeding season. Cows that calve early may be one BCS point lower than late-calving cows at the beginning of the breeding season but still have the same conception rates. Late-calving, thin cows should be put on a diet that allows them to gain weight before the start of the breeding season if they are to maintain the herd's productivity. This can be extremely difficult and costly, with the added demands of early lactation and questionable weather conditions.

Many studies indicate that cows in BCS 5 at calving and breeding will have an excellent chance of cycling and conceiving early in the breeding season (see Table 6.12). One should remember that several factors influence reproductive performance. The ideal body condition for the breeding herd will vary with region, season, breed, and desired level of performance. Feeding programs are often based on a management group's average body condition. Cows with a high BCS, for example, 7, at calving can lose nearly 140 lb of body weight and still be in condition score 5 at the start of the breeding season. In contrast, a cow calving in BCS 3 may need to gain 140 lb by the start of the breeding season to conceive. If the time from calving to breeding is 80 days, that translates to cows gaining nearly 2 lb daily during early lactation. This will take a particularly energy-dense diet, which most cow/calf producers are unaccustomed to feeding.

Body fat can be used as an energy source by lactating cows. Heavy-milking cows will deplete greater amounts of body energy stores during lactation, especially during early lactation, as compared to average- and low-milk-producing cows on the same nutritional

Table 6.10. Key points for condition scoring beef cows.

	Body Condition Score								
Reference point	1	2	3	4	5	6	7	8	9
Physically weak	Yes	No	No	No	No	No	No	No	No
Muscle atrophy	Yes	Yes	Slight	No	No	No	No	No	No
Outline of spine visible	Yes	Yes	Yes	Slight	No	No	No	No	No
Outline of rib visible	All	All	All	3–5	1–2	0	0	0	0
Fat in brisket and flanks	No	No	No	No	No	Some	Full	Full	Extreme
Outline of hip, pin bones visible	Yes	Yes	Yes	Yes	Yes	Yes	Slight	No	No
Fat udder, patchy fat around tailhead	No	No	No	No	No	No	No	Slight	Yes
Backfat estimate, inches	0	0	.05	.11	.19	.29	.41	.54	≥ .68

Source: Momont and Pruitt 1998.

Table 6.11. Nine-point body condition scoring system.

1. Bone structure of shoulder, ribs, back, hooks, and pins is sharp to the touch and easily visible. Little evidence of fat deposits or muscling.
2. Little evidence of fat deposition but some muscling in the hindquarters. The spinous processes feel sharp to the touch and are easily seen with space between them.
3. Beginning of fat cover over the loin, back, and foreribs. The backbone is still highly visible. Processes of the spine can be identified individually by touch and may still be visible. Spaces between the processes are less pronounced.
4. Foreribs are not noticeable, but the twelfth and thirteenth ribs are still noticeable to the eye, particularly in cattle with a big spring of rib and width between the ribs. The transverse spinous processes can be identified only by palpation (with slight pressure) and feel rounded rather than sharp. Full, but straight muscling in the hindquarters.
5. The twelfth and thirteenth ribs are not visible to the eye unless the animal has been deprived of feed and water for 12–16 hours. The transverse spinous processes can only be felt with firm pressure and feel rounded, but they are not noticeable to the eye. Spaces between the processes are not visible and are only distinguishable with firm pressure. Areas on each side of the tail head are well filled but not mounded.
6. Ribs are fully covered and are not noticeable to the eye. Hindquarters are plump and full. Noticeable sponginess over the foreribs and on each side of the tail head. Firm pressure is now required to feel the transverse processes.
7. Ends of the spinous processes can only be felt with firm pressure. Spaces between processes can barely be distinguished. Abundant fat cover on either side of the tail head, with evident patchiness.
8. Animal takes on a smooth, blocky appearance. Bone structure disappears from sight. Fat cover is thick and spongy and patchiness is likely.
9. Bone structure is not seen or easily felt. The tail head is buried in fat. The animal's mobility may actually be impaired by excessive fat.

Source: Corah et al. 1991.

Table 6.12. Relationship of body condition and percentage of cows cycling 60 days after calving.

Condition at Calving	Weight Change Precalving	Weight Change Postcalving	% Cycling 60 Days Postcalving
Good	Lost	Lost	90+
Moderate	Gained	Lost	74
Moderate	Lost	Lost	48
Thin	Lost	Gained	46
Thin	Lost	Lost	25

Source: Whitman et al. 1975.

regimen. This can result in the thinnest cows weaning the heaviest calves. Some studies have shown that no relationship exists between calf weaning weights and dam BCSs. This may mean that calves of low-milking dams forage more to satisfy their appetites than do calves of high-milking dams, especially after they are 90–120 days of age. Only under severe nutritional restrictions, such as when cows lose two or more BCS points, has it been shown that weaning weights of calves also suffer (Table 6.13).

Milk Production

Milk production can be related to BCS, especially during limiting nutritional programs. Milk production has the greatest nutritional demand of all beef cow activities. Five pounds of daily milk production will require 1.7 Mcal NEg, 0.3 lb crude protein, 0.012 lb calcium, and 0.006 lb phosphorus. Daily milk production changes with the shift from parturition to weaning, and also with age. Young 2- and

Table 6.13. Effects of cow condition score change from March until May on calf performance.

	Condition Score Change, March to May		
	Maintained	Lost One Point[a]	Lost Two or More Points
205-day adjusted weaning weight, lb	607	606	586

Source: Momont and Pruitt 1998.
[a]The body condition score system has a scale of 1–9 points: 1 = emaciated; 9 = obese.

Table 6.14. Peak and average milk production for common beef breeds.

Breed	Peak Milk Production (lb/day)	Average Milk Production (lb/day)
Angus	20.7	14.9
Charolais	21.6	15.1
Gelbvieh	25.3	17.8
Hereford	18.7	12.5
Limousin	20.9	14.1
Simmental	24.1	16.8

3-year-old cows will increase milk production with succeeding lactations. During a single lactation period, beef cows will reach peak daily milk production at 60–90 days after calving. This is most unfortunate in terms of breeding, because peaking at this time places the greatest nutritional demands just before and during the breeding season. As a consequence, attempting to feed a thin cow more just before the breeding season to increase body condition may only result in greater milk production. Table 6.14 reviews average peak and total milk production for the most common beef breeds. Studies measuring milk production have indicated that milk production rises quickly from day 1 until peak milk yield (about 80 days) and then slowly declines until the calf is weaned at between 200 and 230 days.

Age

Cows reach their mature size when they are 4–5 years of age. Therefore, 2- and 3-year-old cows have additional nutrient demands because of their growth requirements. Growth is second in nutrient priority behind maintenance, making the management of young cows more difficult in terms of their reproductive function and milk production. Clients will complain that their young cows are the largest group of open cows. In most cases, the failure for these females to rebreed is the result of a failure of nutrition to meet the requirements of maintenance plus growth plus milk production.

A 2-year-old heifer producing approximately 10 lb of milk daily will need to consume about 12.9 lb of TDN to be able to grow at the modest rate of 0.5 lb per day. A mature 1200-lb cow producing 10 lb of daily milk requires 12.8 lb of TDN daily. Both cattle need to consume nearly equal amounts of total energy, but the mature female will have the advantage of capacity and will be able to eat nearly 26 lb of dry matter daily, whereas the young cow will only be able to consume about 22 lb of dry matter in the same period of time. This increases the importance of nutrient density within the diet. The mature cow will be adequately supplied by a diet containing 49% TDN, whereas the 2-year-old cow will need nearly 59% TDN (a 17% increase in energy concentration) in her diet to perform and develop in a desirable fashion.

Typically, producers should try to split their cowherds into management groups. Young cows need extra management to maintain their productive presence in a herd. Thin, mature cows may also need preferential treatment. Cowherds using low-quality forages during stressful times, such as in winter-feeding programs, are extremely susceptible to culling a high percentage of young cows from the herd. By splitting the cowherd into management groups, young and thin cattle can receive higher-quality diets that would be nutritionally and economically wasteful for the better-conditioned, mature cow group. Forage testing to allot feed resources by quality will add to the efficiency of the split-herd management system.

Physical Activity

Physical activity is part of the maintenance requirement. As a general rule, the NEm requirement increases about 0.9 Mcal/day in grazing cows versus drylot cattle. Grazing activity is directly related to forage availability. Cattle like to be full; therefore, they are constantly looking for something to eat. Pastures and grazing lands with sparse, low-quality vegetation will automatically increase the maintenance requirement of cattle when compared to lands covered with lush, high-quality forages.

PRACTICAL CONSIDERATIONS

Forage Intake

The ruminant digestive system evolved to take advantage of forage as the primary constituent of the diet. Several plant and animal factors work in harmony to control the forage intake.

Quality of Forage Available

Forage quality, defined as energy and protein content, has a major effect on intake. There is an inverse relationship between forage quality and structural components of the plant. As a plant matures, it increases its stem-to-leaf ratio and lignin content (Ugherughe 1986). Lignin gives the plant structure and the ability to stand upright. However, lignin is unfortunately not digestible by the ruminant. Forage testing procedures have been developed to measure the quality or digestibility of the plant. The Van Soest system is most commonly used and measures the acid detergent fiber content of the sample. Increasing acid detergent fiber of a sample is associated with decreasing digestibility.

Forage quality is extremely important when buying and feeding mechanically harvested forages. Many authors and feeding manuals have stressed the importance of harvesting forages at the proper stage of maturity. This is because of the inverse relationship between plant maturity and digestibility. The rate and extent of forage digestibility will determine the disappearance of forage from the rumen. Duble et al. (1972) concluded that daily gain of yearling mixed-breed heifers was highly correlated with in vitro dry matter digestibility estimates ($r^2 = .85$). Rate of removal from the rumen (digestibility and passage rate) will ultimately dictate the animal's performance (Paterson et al. 1994).

Quantity of Forage Available

Beef cattle are selective grazers. When given a choice of plants and plant parts, they will consume diets that are greater in crude protein and energy content than those from an unselected forage sample. Guerrero et al. (1984) found that, within classified levels of forage quality, animal performance increased with forage availability up to the point at which diet selection was not hindered by the amount of specific plants and plant parts.

Protein Content of the Diet

Beef cows are often reared on diets that are exclusively made from low-quality forages. These types of diets are usually low in protein content. Limiting dietary protein will decrease rumen fermentation, and thus will decrease forage intake and energy consumption and will constrain animal performance. In most cases, increasing DIP consumption will enhance forage intake or digestibility and will elevate animal performance. Good sources of DIP come from processing of oilseeds such as soybeans, cottonseed, linseed, and sunflowers. UIP supplementation may correct dietary protein deficiencies, but it will have minimal if any effect on forage intake and digestibility. Common sources for UIP originate from animals—typically products such as hydrolyzed feather meal and fish meal. Because there are numerous combinations of feedstuffs for diets that will satisfy production goals, it is extremely important to price the different feedstuffs when developing cattle diets. As forage crude protein increases, the production response to protein supplementation will probably change from an intake response to changes in forage digestibility or metabolic efficiency. It would seem reasonable that collecting samples of grazed forages would provide a good representation of grazing cattle's diets.

Environment

Weather conditions can change forage intake by affecting eating behavior and grazing patterns. Cold weather will generally cause cattle to have increased dietary intake. Early research has shown that when animals are placed in cold environments, they have an increased energy demand, and their intake increases in an attempt to match their increased energy expenditure (Stevenson 1954). Conversely, windy or wet conditions will cause cattle to decrease daily food intake.

However, producers should be concerned with hot weather as well. Much research has been done with feedlot and confinement-fed animals, but little information is available for grazing animals. Table 6.15 is provided to assist in making decisions about heat stress.

As cattle mature, their thermocomfort zone increases. Young cattle will be comfortable between 45°F and 80°F, whereas feedlot and mature cattle will be comfortable from 0°F or below in the winter to around 75°F in the summer. The comfort zone will

Table 6.15. Temperature humidity index.

	Relative Humidity					
Temperature (°F)	35%	45%	55%	65%	75%	85%
100	85	87	90	92	94	97
98	84	86	88	90	93	95
96	82	85	87	89	91	93
94	81	83	85	87	89	91
92	80	82	84	85	87	89
90	79	80	82	81	86	87
88	77	79	81	82	84	86
86	76	78	79	81	82	84
84	75	76	78	79	80	82
82	73	75	76	77	79	80
80	72	73	75	76	77	78
78	71	72	73	74	75	76
76	70	71	72	73	74	75

Temperature humidity index = temperature in degrees Fahrenheit − {0.55 − [0.55 × (relative humidity/100)]} × (Tdbf − 58).
Normal = <74, alert = 75–78, danger = 79–83, emergency = >84
Source: Mader et al. 2000.

be determined not only by the environment but also by body condition, hair coat length, coat color, and plane of nutrition (Mader et al. 2000).

Estimating Forage Intake under Grazing Situations

As noted earlier, forage intake is affected by many factors (Cordova et al. 1978; Paterson et al. 1994). Often when the actual animal performance is not accurately predicted by diet formulations, the animal's intake has been poorly estimated. For example, what is the estimated intake of a 1150-lb cow grazing on low-quality grain sorghum crop residue? Assuming adequate protein is supplied through a supplement, the cow will consume about 1.8% of her body weight in forage dry matter. Multiplying 0.018 by 1150 lb, a dry-matter intake of 20.7 lb is calculated. If the crop residue is assumed to be 90% dry matter, the "as-fed" intake is 20.7 divided by 0.90, which equals 23 lb of crop residue per day plus the protein supplement. Again, Table 6.6 can be used as a general guide to estimate the intake of any forage available free choice.

Protein Supplementation

Energy is generally considered to be the most common nutrient that is deficient in beef cow diets. In reality, rumen-degradable protein, referred to as DIP, is the real culprit. Proper protein supplementation of poor, low-quality forages will increase forage intake and, in turn, increase energy consumption. Thus, to maximize profitability, it is essential to optimize protein supplementation. The consequences of inadequate protein supplementation are low forage intake, low forage digestibility, and poor performance. This reduction in performance may be expressed as weight loss, a decline in body condition, lower milk production, lower antibody transfer to newborn calves through colostrum, or numerous health problems resulting from lowered resistance to disease.

Meeting the protein requirements of beef cattle is complicated by the fact that the microorganisms in the ruminant digestive system can use many sources of nitrogen to make bacterial proteins. In addition, the digestive characteristics of feeds made from plant and animal sources are variable. The following section is an attempt to define and explain some terms used in describing types of protein and to give guidelines for practical application of these concepts.

DIP and UIP

The protein fraction of the diet can be divided into two components: DIP and UIP. Feedstuffs used for protein supplementation can be classified into three categories according to their DIP-to-UIP ratios (Chalupa 1975; Satter et al. 1982; Agricultural Research Council 1980). Feeds with less than 40% UIP include soybean meal and peanut meal. Medium bypass feeds, or those feeds with between 40% and 60% UIP, are cottonseed meal, dehydrated alfalfa meal, corn grain, and brewers dried grains. The last category comprises those feeds high in UIP content (greater than 60%). Examples are corn gluten meal, feather meal, and fish meal. These estimates do not consider feed processing or condition and other factors that may affect the rumen environment.

The DIP can be consumed and broken down by rumen microflora and, thus, used for their growth and for the digestion of dietary fiber. Supplementing low-quality forages with DIP has been shown to increase forage digestion and intake, thus increasing energy intake (McCollum and Gaylean 1985). Because the UIP fraction is not available to the rumen microorganisms, it has essentially no effect on forage utilization. The UIP fraction can be a direct supply of amino acids to the host animal, or it can go undigested and be expelled from the body.

From these general statements, one can appreciate that, for most forage-based diets, supplementation programs should focus on the inclusion rates of DIP in the diet. Most research would indicate that the amount of DIP required to maximize forage use appears to be about 10%–11% of the digestible organic matter (TDN is nearly the same as digestible organic matter). Therefore, to correctly balance diets to meet foraging cattle's requirements, one needs to know the source of supplemental protein as well as an estimate of forage intake, digestibility, and nutrient composition.

Nonprotein Nitrogen

Cattle have the ability to use either "natural" protein, such as that contained in plant and animal-sourced feedstuffs, or various other nitrogen sources. Sources other than natural protein are generally referred to as nonprotein nitrogen (NPN) sources. Common NPN sources used in cattle diets are urea, biuret, and ammonia hydroxide.

The rumen microflora can convert NPN into protein if they are supplied with adequate energy and carbon skeletons. Forage-based diets usually do not supply excessive amounts of either, thus limiting the quantity of NPN that can be fully converted by the rumen. Therefore, the use of NPN in cow rations has been discouraged in the past in the United States, although it is widely used in some other countries. Most NPN sources are considered to be 100% DIP in nature.

It appears that NPN can replace limited amounts of DIP in supplements. It seems that an inclusion level of 15% of the total crude protein, or about 20% of the DIP as NPN, can be used without jeopardizing livestock performance (Koster et al. 1997). Because of the severe energy limitations of low-quality forage diets, NPN is less potent as an energy source than it would be with, for example, high-quality forage diets or high-concentrate diets. Some research has indicated that in supplements containing significant amounts of NPN, only those containing 50% or less of the NPN will be used by the cows. As NPN inclusion rates increase, NPN utilization can steadily decline from 50% to as low as 20%–25%. Therefore, cattle producers need to carefully analyze their protein supplementation programs and only include those amounts of NPN that will be optimum for their operations.

Feed tags must display the NPN content of commercial protein supplements. The tag commonly shows the total protein concentration, followed by the amount of protein coming from NPN. For example, a tag on a range supplement might show the figures "20-10" for protein. This means that the total protein content of the supplement is 20%, with the equivalent of 10 percentage units—or 50%—of the protein coming from NPN.

For use with beef cows grazing dormant range or crop residues in late winter, the 20-10 supplement should be considered, at best, a 15% supplement (10% from natural sources plus 50% of the remaining 10% from NPN). Again, this is allowing for 50% utilization of the NPN, as in a "best case" scenario. If the actual rate of utilization is only 20%, as could be the case, for example, with crop residues in late winter, this supplement may only give performance equal to a 12% all-natural supplements.

Biuret is another commonly used NPN in beef cattle supplements. It is essentially two urea molecules chemically bonded together. Biuret is somewhat safer than urea, although a review of research comparing urea to biuret indicates that the performance of cattle treated with either item has been similar.

During the last several decades, ammoniation of roughages has gained in popularity. Ammoniation is a method for improving the forage intake and digestibility of low-quality forages that contain a low amount of soluble sugars. Examples of forages ideal for ammoniation are wheat straw, mature grass hay, and other crop residues. Feed analyses routinely show an increase in the crude protein content of wheat straw by four to six percentage points. Thus, a straw at 4% crude protein before ammoniation often tests at 9% crude protein after ammoniation. However, this rise in protein is in the form of ammonia hydroxide, an NPN source. Supplementing ammoniated forages with natural proteins has resulted in increased animal performance (Beck et al. 1991). This indicates that the primary reason to ammoniate low-quality forages is to increase fiber, energy intake, and digestibility, rather than to meet the entire protein requirement of the beef cow.

To ammoniate forage properly, the hay supply should be arranged so that a plastic cover can be used to trap the anhydrous ammonia within the haystack. After sufficiently sealing the hay pile, the anhydrous ammonia should be slowly introduced until 3% by weight has been added to the stack. The stack should remain covered for several days to several weeks, depending on the environmental temperature. Daytime temperatures of greater than 90°F will allow the chemical reactions to proceed at a rapid pace, and curing will take place in about 1 week. Colder temperatures, less than 50°F, may prolong reaction times to greater than 30 days. Extreme caution should be used when working around anhydrous ammonia. Even low concentrations can easily cause injury or death.

Forages containing significant amounts of soluble sugars should not be ammoniated. Brome, small grain, forage sorghum, sudans, and other high-quality grass hays are not recommended to be treated with anhydrous ammonia. The high soluble carbohydrate content of such forages when treated with ammonia appears to produce imidazole compounds in some cases. These compounds can produce extreme hyperactivity, convulsions, and even death in cattle, especially when treated forages make up most of the ration or the entire ration.

Mineral Supplementation

Mineral requirements of beef cattle can be grouped into the major and trace mineral categories. Major minerals are required in large enough quantities that the requirements are listed as a percentage of the diet. However, trace minerals are needed in much lower concentrations and are usually designated by parts per million units. Salt, calcium, phosphorus, and magnesium are major minerals, and copper, zinc, cobalt, selenium, and manganese are examples of trace minerals. Table 6.16 lists several of the important minerals and their suggested daily requirements.

Calcium, phosphorus, and salt are minerals required at significant levels by beef cows. Thus, major consideration should be given to them during diet formulation. Most forages contain fairly high levels of calcium, so it is rarely lacking in diets typically fed to beef cows. However, if cows are consuming diets high in concentrates, calcium levels should be monitored closely. Conversely, forages are relatively poor sources of phosphorus. This means this mineral should be the first considered in a supplementation program. Unfortunately, phosphorus is relatively expensive, and producers are advised to meet animal requirements but to avoid oversupplementation.

Cattle have a definite requirement and appetite for sodium. Because most feedstuffs are deficient in this mineral, cattle should be offered supplements containing salt at all times.

Trace minerals are extremely important nutrients, even though they are required in small amounts. Inadequate trace mineral consumption has been shown to affect reproductive performance, disease resistance, growth, and thriftiness. Much attention has been given to the relationship between trace minerals and immune functions. Research (National Research Council 2000) has illustrated the potential improvement of animal health with dietary fortification of trace minerals. Because clinical signs of trace mineral deficiencies are usually not apparent when the immune system is compromised, one of the first signs of deficiency may be an impaired immune response. After the deficiencies worsen, clinical signs typically associated with trace mineral deficiencies (growth impairment, infertility, and outward appearance) may become more apparent at a herd level.

Of the numerous minerals present in the body, six to eight trace minerals are believed to be the most likely to influence the production of grazing cattle: copper, cobalt, iodine, selenium, zinc, and manganese. Iron and molybdenum also should be monitored because of their antagonistic behavior in the rumen; and iron, molybdenum, and sulfur can negatively affect copper utilization. Deficiencies of these three elements, however, very seldom affect animal performance. Rather, excessive amounts of molybdenum or iron (and, in some cases, sulfur) are usually present.

Because mineral-deprived animals have been observed chewing on bones, eating dirt, and engaging in pica, it was assumed that they could control their own mineral intakes. That assumption, however, is inaccurate. Today practitioners recognize that cattle do not have the ability to balance their own diets. Thus, food animal nutritionists and producers need to understand the mineral needs of beef herds and to correctly formulate dietary supplementation accordingly.

Corah and Dargatz (1996) compiled the mineral profiles of 352 forage samples from 18 states and

Table 6.16. Mineral requirements and maximum tolerable levels in beef cattle.

		Requirements			
		Calves	Cows		
Mineral	Unit	Growing and Finishing	Gestating	Lactating	Maximum Tolerable Concentration
Calcium	%	See Table 6.19	See Table 6.19	See Table 6.19	See Table 6.19
Chlorine	%	—	—	—	—
Chromium	ppm	—	—	—	1000
Cobalt	ppm	.10	.10	.10	10
Copper	ppm	10	10	10	100
Iodine	ppm	.50	.50	.50	50
Iron	ppm	50	50	50	1000
Magnesium	%	.10	.12	.20	.40
Manganese	ppm	20	40	40	1000
Molybdenum	ppm		—	—	5.0
Nickel	ppm	—	—	—	50
Phosphorus	%	See Table 6.19			
Potassium	%	.6	.6	.7	3.0
Selenium	ppm	.10	.10	.10	2.0
Sodium	%	.06–.08	.06–.08	.10	—
Sulfur	%	.15	.15	.15	.40
Zinc	ppm	30	30	30	500

Source: NRC 2000.
Dash indicates that no requirement was reported in the text reference.

Table 6.17. Trace mineral classification of forage samples.

Trace Element	Adequate (%)	Deficient (%)	Marginal (%)	High (%)	Marginal Antagonist Levels (%)	Very High Antagonist Levels (%)
Copper	36.0	14.2	49.7	—	—	—
Manganese	76.0	4.7	19.3	—	—	—
Zinc	2.5	63.4	34.1	—	—	—
Cobalt	34.1	48.6	17.3	—	—	—
Selenium	19.7	44.3	19.3	16.7	—	—
Iron	62.8	8.4	—	—	17.0	11.7
Molybdenum	42.2	—	—	—	48.6	9.2

Dash indicates that no requirement was reported.

327 cooperators. Table 6.17 summarizes the results of this study. A large portion of the samples demonstrated that fortifying trace minerals in beef cattle diets should be a major concern for producers. In particular, zinc, cobalt, and copper need to be monitored for low levels, and iron and molybdenum for antagonistic levels. The practice of supplementing trace minerals is often considered an insurance policy against poor herd performance. Table 6.18 is provided as a guide to an acceptable cattle mineral supplement.

Certain production situations deserve special mineral considerations. Many producers must supplement with magnesium in the early spring to prevent grass tetany. This is especially true with cool-season forages such as brome, fescue, rye, and wheat pastures. Supplementation should start about 3 weeks before the initiation of grazing for best results. The recommended intake of magnesium oxide is at least 0.03 lb per day, which means that it must make up 15%–20% of a free-choice mineral supplement. A

Table 6.18. Characteristics of an acceptable complete free-choice cattle mineral supplement.

Final mixture containing a minimum of 6%–8% total phosphorus. In areas where forages are consistently lower than 0.20% phosphorus, mineral supplements in the 8%–10% phosphorus range are preferred.
A calcium-to-phosphorus ratio not substantially over 2:1.
A significant proportion (e.g., about 50%) of the trace mineral requirements for cobalt, copper, iodine, manganese, and zinc.
For most regions, inclusion of selenium, unless toxicity problems have been observed. Iron should be included in temperate region mixtures, but often iron and manganese can be eliminated for acid soil regions. In certain areas where parasitism is a problem, iron supplementation may be beneficial.
In known regions of trace mineral deficiency, 100% of specific trace minerals provided.
High-quality mineral salts that provide the best biologically available forms of each mineral element and avoidance or minimal inclusion of mineral salts containing toxic elements. As an example, phosphates that contain high concentrations of fluoride should be avoided or formulated so that breeding cattle would receive no more than 30–50 ppm fluoride in the total diet. Fertilizers or untreated phosphates could be used to a limited extent for feedlot cattle.
A formula sufficiently palatable to allow close-to-adequate consumption in relation to requirements.
Reputable manufacturer backing with quality control guarantees as to the accuracy of the mineral supplement label.
An acceptable particle size that allows adequate mixing without the settling of smaller-sized particles.
A formula for the area involved, level of animal productivity, and environment (temperature, humidity) in which it will be fed, providing the mineral element as economically as possible.

McDowell 1997.

good "home mix" for this period would be 15% magnesium oxide, 50% dicalcium phosphate, 25% salt, and 10% molasses or soybean meal as a flavoring agent to increase intake, as magnesium oxide is unpalatable. Commercial mineral mixtures used during early spring should be at least 10% magnesium.

Chelated minerals are trace minerals that are attached to an organic molecule (i.e., an amino acid or protein). Most research has shown that chelated minerals have greater bioavailability (higher absorption potential) than some inorganic forms, and especially oxides, used in common mineral supplements.

However, inorganic forms are much less expensive than organic or chelated trace minerals. At present, the question of whether chelates are beneficial or not is one of the most controversial areas in beef cattle nutrition. This controversy results from the fact that even though absorption is greater, there is little, if any, research demonstrating improved animal performance. Until further research demonstrates an economic advantage, producers should be advised to use highly available inorganic forms of trace minerals either alone or in a mixture with chelated minerals to minimize costs while assuring nutritional adequacy.

Human nature dictates that when trying to determine the cause of poor reproductive performance, exotic causes are considered, rather than the most logical ones. In troubleshooting cowherd nutrition programs, keep in mind that a very high percentage of problems, including poor reproduction or low weaning weights, can be directly attributed to inadequate energy or protein intake, and not to a trace mineral deficiency or lack of an unknown growth factor. Only after the adequacy of energy and protein has been established should the focus turn to mineral and vitamin nutrition.

Vitamin Requirements

Beef cattle require vitamins, in common with other mammals. The rumen microbes produce an abundance of the water-soluble vitamins, so they rarely need to be added to cattle diets. However, fat-soluble vitamins do need to be included in cow diets, and vitamin A specifically must be supplied in some form. (Requirements for vitamin A are included in Table 6.19.) Cattle are capable of storing large quantities of vitamin A in the liver during periods of high intake. Therefore, during the spring, summer, and fall months, cattle will store excessive vitamin A that will be needed in winter. Drought, poor-quality forages, unfavorable forage harvesting, and disease will affect the efficiency and amount of vitamin A stored.

It is generally advised to supplement winter cow diets with vitamin A. Supplementing vitamin A is very inexpensive and can be accomplished by any of the following: feeding forages high in vitamin A such

Table 6.19. Daily nutrient requirements of breeding beef cattle.

Wt (lbs)[a]	Gain (lbs)[b]	DMI (lbs)[c]	TDN (lbs)	TDN (% Dry Matter)	NEm (Mcal)	NEm (Mcal/lb)	NEg (Mcal/lb)	Protein (lbs)	Protein (% DM)	Calcium (% DM)	Phosphorus (% DM)	Vit A (1000 U)[d]
Pregnant yearling heifers—last third of pregnancy												
700	0.9	15.3	8.5	55.4	7.95	0.52	NA	1.3	8.4	0.27	0.20	19
700	1.4	15.8	9.6	60.3	7.95	0.60	0.34	1.4	9.0	0.33	0.21	20
700	1.9	15.8	10.6	67.0	7.95	0.70	0.43	1.5	9.8	0.33	0.21	20
800	0.9	16.8	9.2	54.8	8.56	0.51	NA	1.4	8.2	0.28	0.20	21
800	1.4	17.4	10.4	59.6	8.56	0.59	0.33	1.5	8.8	0.33	0.21	22
800	1.9	17.5	11.6	66.1	8.56	0.69	0.42	1.6	9.3	0.35	0.21	22
900	0.9	18.3	9.9	54.3	9.15	0.51	NA	1.5	8.1	0.26	0.20	23
900	1.4	19.0	11.3	59.1	9.15	0.58	0.32	1.6	8.5	0.30	0.21	24
900	1.9	19.2	12.5	65.4	9.15	0.68	0.41	1.7	9.0	0.32	0.21	24
Dry pregnant mature cows—middle third of pregnancy												
900	0.0	16.7	8.2	48.8	7.00	0.42	NA	1.2	7.0	0.18	0.18	21
1100	0.0	19.5	9.5	48.8	8.13	0.42	NA	1.4	7.0	0.19	0.19	25
1300	0.0	22.0	10.8	48.8	9.22	0.42	NA	1.5	6.9	0.20	0.20	28
Dry pregnant mature cows—last third of pregnancy												
900	0.9	18.2	9.8	54.0	9.1	0.50	NA	1.5	8.0	0.27	0.21	23
1000	0.9	19.6	10.5	53.6	9.7	0.50	NA	1.6	7.9	0.26	0.20	25
1200	0.9	22.3	11.8	52.9	10.8	0.49	NA	1.7	7.8	0.26	0.21	28
1400	0.9	24.9	13.1	52.5	11.9	0.48	NA	1.9	7.6	0.26	0.21	32
2-year-old heifers nursing calves—first 3–4 months postpartum—10 lbs milk/day												
700	0.5	15.9	10.3	65.1	9.20[e]	0.67	0.40	1.8[f]	11.3	0.36	0.24	28
800	0.5	17.6	11.2	63.8	9.81[e]	0.66	0.39	1.9[f]	10.8	0.34	0.24	31
900	0.5	19.2	12.0	62.7	10.40[e]	0.64	0.37	2.0[f]	10.4	0.32	0.23	34
1000	0.5	20.8	12.9	61.9	10.98[e]	0.62	0.36	2.1[f]	10.0	0.31	0.23	37

(Continued)

Table 6.19. Daily nutrient requirements of breeding beef cattle. (*Continued*)

Wt (lbs)[a]	Gain (lbs)[b]	DMI (lbs)[c]	TDN (lbs)	TDN (% Dry Matter)	NEm (Mcal)	NEm (Mcal/lb)	NEg (Mcal/lb)	Protein (lbs)	Protein (% DM)	Calcium (% DM)	Phosphorus (% DM)	Vit A (1000 U)[d]
Cows nursing calves—average milking ability—first 3–4 months postpartum—10 lbs milk/day												
900	0.0	18.8	10.8	57.3	10.40[e]	0.55	NA	1.9[f]	9.9	0.28	0.22	33
1100	0.0	21.6	12.1	56.0	11.54[e]	0.54	NA	2.0[f]	9.4	0.27	0.22	38
1300	0.0	24.3	13.4	55.1	12.63[e]	0.52	NA	2.2[f]	9.1	0.27	0.22	43
Cows nursing calves—superior milking ability—first 3–4 months postpartum—20 lb milk/day												
900	0.0	18.7	13.1	69.8	13.81[e]	0.74	NA	2.4[f]	12.9	0.41	0.28	33
1100	0.0	22.3	14.5	65.2	14.94[e]	0.67	NA	2.6[f]	11.9	0.38	0.27	40
1300	0.0	25.3	15.9	62.6	16.03[e]	0.64	NA	2.8[f]	11.2	0.36	0.26	45

NEm = net energy for maintenance, NEg = net energy for gain, NA = not applicable.

Source: NRC 1984.

[a]Average weight for a feeding period.

[b]Approximately 0.9 + 0.2 lb weight gain/day over the last third of pregnancy is accounted for by the products of conception. Daily 2.15 Mcal of NEm and 0.1 pound of protein are provided for this requirement for a calf with a birth weight of 80 lb.

[c]Dry-matter intake (DMI; consumption) should vary depending on the energy concentration of the diet and environmental conditions. These intakes are based on the energy concentration shown in the table and assuming a thermoneutral environment without snow or mud conditions. If the energy concentrations of the diet to be fed exceed the tabular value, limit feeding may be required.

[d]Vitamin A requirements per pound of diet are 1273 IU for pregnant heifers and cows and 1773 IU for lactating cows and breeding bulls.

[e]Includes 0.34 Mcal NEm/pound of milk produced.

[f]Includes 0.03 pound protein/pound of milk produced.

Table 6.20. Estimated daily water intake of cattle.

Month	Temperature (°F)	Non-lactating Cows (gallons)[a]	Lactating Cows (gallons)	Nonlactating Bulls (gallons)
January	36	11.0	6.0	7.0
March	50	12.5	6.5	8.6
May	73	17.0	9.0	12.0
July	90	16.5	14.5	19.0
September	78	17.5	10.0	13.0
November	52	13.0	6.5	9.0

Source: Guyer (1988).
[a]Cows nursing calves during the first 3–4 months after parturition (peak milk production period).

as high-quality alfalfa hay that has been in storage for less than 6–8 months, including vitamin A in mineral mixtures or other supplements, and injecting vitamin A. When choosing the latter, injecting 2 million IU should provide sufficient vitamin A for 80–100 days.

Water Requirements

Normally, water is easily supplied to cattle and is therefore taken for granted. Table 6.20 is provided to estimate water consumption at various production stages and times of the year. Water should always be provided in ample amounts: Clean, uncontaminated water is the basis of all good nutritional programs. Filthy water is often high in solid contents that can have excessive amounts of minerals that are antagonistic to many of the essential trace minerals. If water is limited, feed intake will be depressed, resulting in subpar performance.

GUIDELINES FOR MANAGING BEEF COW DIETS

First, feeding cattle is dynamic. Animal, environmental, and diet factors must all be considered when formulating correct beef cow diets.

Second, a cow's nutrient requirements (energy, protein, and mineral) will increase about 30%–40% with calving. Forage intake will generally increase about 30% with calving.

Third, a positive response to providing a supplement with high protein concentrations is most likely when forage crude protein is less than 7%. The first-limiting nutrient in low-quality forages is DIP. Therefore, the best approach for increasing total protein and energy supply is to supplement with DIP.

Fourth, the NPN inclusion rate in supplements for forage-based diets should be monitored closely, as only up to 15% of the total dietary crude protein should be NPN in nature. Including NPN at levels that are too high may result in animal refusal to consume the supplement.

Fifth, starches can negatively affect forage intake and fiber digestion. Supplementing low-quality forage diets with corn, grain sorghum, and other cereal grains can actually decrease energy intake. The key is to meet the dynamic protein requirements of the rumen. As starch is added to the rumen, the TDN content of the diet is increased. This increases the crude protein requirement of the animal. The proper TDN to protein ratio of a diet is between 4 and 6 to 1.

Sixth, when alfalfa hay or a low-protein supplement is fed at levels exceeding 0.5% of body weight, intake of low-quality forage is going to be reduced by 0.5 lb per 1 lb of supplement.

Last, only minor differences in performance are evident for cattle supplemented every other day or three times weekly, as compared with those supplemented daily.

REFERENCES

Ames, D.R. 1987. Effects of cold environment on cattle. *Agri-Practice* 8(1):26–29.

Agricultural Research Council. 1980. The Nutrient Requirements of Ruminant Livestock. Farnham: Agricultural Research Council, Commonwealh Agricultural Buraux.

Beck, T.J., D.D. Simms, R.T. Brandt, Jr., R.C. Cochran, G.L. Kuhl. 1991. Supplementation of ammoniated wheat straw in wintering diets of gestating beef cows. Ag. Exp. Stat., Kansas State University, Manhattan. Rep. Of Progress 623:101.

Blasi, D.A., R.J. Rasby, I.G. Rush, C.R. Quinn. 1999. Cow body condition scoring management tool for monitoring nutritional status of beef cows. *Beef Cattle Handbook*. University of Wisconsin-Extension. BCA-5405.

Chalupa, W. 1975. Rumen bypass and protection of proteins and amino acids. *Journal of Dairy Science* 58:1198.

Corah, L.R., D. Dargatz. 1996. Forage Analyses from Cow/Calf Herds in 18 States. US Department of Agriculture, Animal and Plant Health Inspection Service, Veterinary Services, National Animal Health Monitoring System.

Corah, L.R., R.P. Lemenager, P.L. Houghton, D.A. Blasi. 1991. *Feeding Your Cows by Body Condition*. C-842. Manhattan, KS: Kansas State University.

Cordova, F.J., J.D. Wallace, and R.D. Pieper. 1978. Forage intake by grazing livestock: a review. *Journal of Range Management* 31:430–438.

Duble, R.L., J.A. Lancaster, E.C. Holt. 1972. Forage characteristics limiting animal performance on warm-season perennial grasses. *Agronomy Journal* 63:795.

Ewing, S.A., L. Smithson, D. Stephens, D. McNutt. 1966. Weight loss patterns of beef cows at calving. *Oklahoma State University Feeding and Breeding Tests with Sheep, Swine and Beef Cattle Progress Report, 1965–1966* MP-78:64.

Guerrero, J.N., B.E. Conrad, E.C. Holt, H. Wu. 1984. Prediction of animal performance on bermudagrass pasture from available forage. *Agronomy Journal* 76:577.

Guyer, P.Q. 1977. Water Requirements for Beef Cattle. Coop. Ext. Inst. Agric. & Natural Resources. University of Nebraska–Lincoln. G-77-372-A. Electronic version, 1966. www.ianrpubs.unl.edu/beef/g372.

Hibberd, C.A., T.A. Thrift. 1992. Supplementation of forage-based diets: Are results predictable? *Journal of Animal Science* 70(Suppl. 1):181.

Institut National de la Recherche Agronomique. 1989. *Ruminant Nutrition*. Montrouge: Libbey Eorotext.

Iowa Beef Center. 2002. *Grazing: Making Extended Grazing Work in Nebraska*. IBC-22. Ames, Iowa: Iowa Beef Center.

Koster, H.H., R.C. Cochran, E.C. Titgemeyer, E.S. Vanzant, T.G. Nagaraja, K.K. Kreikemeier, G. St Jean. 1997. Effect of increasing proportion of supplemental nitrogen from urea on intake and utilization of low-quality, tallgrass-prairie forage by beef steers. *Journal of Animal Science* 75:1393–1399.

Ludwig, C., S.A. Ewing, L.S. Pope, D.F. Stephens. 1967. The cumulative influence of level of wintering on the lifetime performance of beef females through seven calf crops. *Oklahoma State University Feeding and Breeding Tests with Sheep, Swine and Beef Cattle Progress Report 1966–1967*. p. 58. Stillwater: Oklahoma State University Cooperative Extension Service.

Lusby, K.S., V. Stevens, K. Apple, M. Scott, R. Bartling, F. Bates. 1985. *Supplement the Cow Herd. Oklahoma State University OSU Ext. Facts No. 3010*. Stillwater: Oklahoma State University Cooperative Extension Service.

Mader, T., D. Griffin, L. Hahn. 2000. *Managing Feedlot Heat Stress*. University of Nebraska, Nebraska Cooperative Extension, G00-1409-A. Lincoln: Nebraska Cooperative Extension.

McCollum, F.T., M.L. Gaylean. 1985. Influence of cottonseed meal supplementation on voluntary intake, rumen fermentation and rate of passage of prairie hay in beef steers. *Journal of Animal Science* 60:570.

McDowell, L.R. 1997. *Minerals for Grazing Ruminants in Tropical Regions*, 3rd edition. Gainesville: University of Florida, Cooperative Extension Service.

Momont, P.A., R.J. Pruitt. 1998. Condition scoring of beef cattle. *Cow-Calf Management Guide, Cattle Producer's Library, Cow-Calf Section CL720*. Moscow: University of Idaho.

National Research Council. 1984. *Nutrient Requirements of Beef Cattle, Sixth Revised Edition*. Washington, DC: National Academy Press.

National Research Council. 2000. *Nutrient Requirements of Beef Cattle, Update 2000*. Washington, DC: National Academy Press.

Paterson, J.A., R.L. Belyea, J.P. Bowman, M.S. Kerley, J.E. Williams. 1994. The impact of forage quality and supplementation regimen on ruminant animal intake and performance. In *Forage Quality, Evaluation, and Utilization*, edited by G.C. Fahey, Jr. Madison, WI: American Society of Agronomy.

Satter, L.D. 1982. A metabolizable protein system keyed to ruminal ammonia concentration—the Wisconsin system. In *Protein Requirements for Cattle: Proceedings of an International Symposiom*, edited by F.N. Owens. MP-109. Stillwater: Division of Agriculture, Oklahoma State University.

Stevenson, J.M. 1954. Diet and survival. In *Transactions of the Third Conference on Cold Injury*, edited by M.I. Ferrer. New York: Josiah Macy Jr. Foundation.

Ugherughe, P.O. 1986. Relationship between digestibility of *Bromus inermis* plant parts. *Journal of Agronomy and Crop Science* 157:136.

Wagner, J.J. 1985. Carcass composition in mature Hereford cows: Estimation and influence on metabolizable energy requirements for maintenance during winter. Ph.D. thesis. Stillwater: Oklahoma State University.

Whitman, R.W., E.E. Remmenga, J.N. Wiltbanks. 1975. Weight change, condition and beef-cow reproduction. *Journal of Animal Science* 41: 387.

7
Behavior and Handling

Temple Grandin

INTRODUCTION

Understanding how behavior affects ease of cattle handling will make working cattle easier and safer. Many serious injuries to both people and cattle occur when animals become agitated. There are also increasing concerns in modern society about animal welfare. Practices that may have been considered acceptable in the past may not be acceptable today. Restaurant companies and supermarkets now have standards for the humane treatment of animals. Good cattle handling practices will improve welfare, safety, and productivity.

Careful, quiet handling of cattle will also help improve profitability. Research in our laboratory has shown that cattle that become agitated during handling and restraint will have lower weight gains and tougher meat (Voisinet et al. 1997a, 1997b). Cattle that run wildly out of the squeeze chute will also have poorer performance (Fell et al. 1999). Progressive livestock producers have found that learning behavioral principles of animal handling helps to reduce sickness and improve productivity. Cattle will settle down and go back onto feed more quickly after quiet handling. Research studies done over 20 years ago clearly demonstrated the bad effect of handling stresses on cattle productivity (Hixon et al. 1981). Further studies also show that handling restraint and transport stresses are detrimental to immune, reproductive, and rumen function (Galyean et al. 1981, Kelly et al. 1981, Blecha et al. 1984).

When animals become agitated during handling, it is usually a result of fear. Fear is a very strong stressor (Boissey 1995). Reducing the animal's fear will make handling easier for both the producer and the animal. Australian researcher Paul Hemsworth has conducted many studies that show that farm animals that fear people are less productive (Hemsworth and Coleman 1998). Reducing negative interactions (such as hitting) between people and dairy cows improved production (Hemsworth et al. 2002).

IMPORTANCE OF GOOD MANAGEMENT FOR PROPER HANDLING

Why do some people continue to handle animals roughly when so much research shows the detrimental effects of stressful handling practices? The author has observed that people often are more willing to buy technology such as a new chute system rather than adopt better management practices. Management requires continuous effort, whereas buying technology is a one-time investment.

Managers need to constantly monitor handling to prevent rough practices, such as excessive electric prod use, from recurring. Standards need to be upheld to prevent "bad" from becoming "normal." One of the author's biggest frustrations has been teaching people how to handle cattle quietly, and then seeing them a year later revert back to screaming, yelling, and whips. To prevent this, different parts of handling procedures must be measured against an objectively defined standard. People manage the things they measure. Measurement of the percentage of animals that vocalize, fall, or have been electrically prodded has been successfully used to greatly improve handling in slaughter plants (Grandin 1998a, 2001). Each variable is measured on a yes/no basis for each animal. In a plant with well-designed facilities and trained staff, 95% of the cattle can be moved without an electric prod, and three or fewer animals per hundred (3%) will vocalize (moo or bellow).

On feedlots and ranches, the best variables to measure are percentage of animals goaded with an electric prod, percentage that fall down, and exit speed out of the squeeze chute such as walking, trotting, or running. In a cattle operation that worked hard to improve handling, it was possible to move 99% of the cattle without an electric prod, with 0% falling down and 90% exiting from the squeeze chute at a walk or trot. In most cases, slow is actually faster because of fewer holdups, and injuries to both people and cattle will be reduced. Regular audits of handling with objective scoring will help prevent people from reverting back to old, rough methods.

Preventing Losses

Injections in the muscle cause extensive meat damage, and this damage is long lasting. Injections given to calves at branding or weaning can result in meat damage at slaughter (George et al. 1995a, 1995b). George et al. (1995a) found that injections in the muscle can cause a persistent sphere of toughness the size of a baseball. Quiet handling is important because it is easier to properly give subcutaneous injections when cattle are standing still in a squeeze chute. When agitated, cattle struggle, making it more difficult to administer injections in the correct position on the neck. Agitated cattle that hit the headgate too hard can sustain extensive injuries that may destroy a large portion of the shoulder meat. Careful, quiet handling will also help prevent the toe abscesses that occur when excited cattle scuff their toes during handling.

Importance of Reducing Fear during Handling

A calm animal is much easier to handle than a fearful, agitated animal. Cattle are easier to handle in corrals if they are allowed to settle down for 15–30 minutes after they have been brought into the corral. It takes 20 minutes for the heart rate to return to normal. Groups of fearful, excited animals are more difficult to separate and sort because scared animals will stick together in a bunch.

Fear is a universal emotion in the animal kingdom (LeDoux 1996, Rogan and LeDoux 1996). It motivates animals to avoid predators and survive in the wild. Scientists have learned that the amygdala is the brain's fear center (Davis 1992); it is the central fear system that is involved both in fear behavior and in learning to fear certain things or people. Stimulating the amygdala elicits responses in the nervous system that are similar to fear in humans (Redgate and Faringer 1973). A more complete review of this topic may be found in Grandin 1997a. "Fear learning" takes place in a subcortical pathway, and extinguishing a learned fear response is difficult because it requires the animal to suppress the fear memory via an active learning process. A single, very frightening or painful event can produce a strong learned fear response, but eliminating this fear response is much more difficult (LeDoux 1996). The animal may develop fear memories that are difficult to eliminate. The stubborn cow that refuses to enter the squeeze chute probably had a bad previous experience such as getting hit hard on the head with the headgate. Ranchers have reported that yearling bulls that had been electroejaculated by a careless veterinarian, who applied excessive amounts of stimulation, refused to reenter the squeeze chute.

Good First Experiences Are Important

Observations by the author on cattle ranches have shown that, to prevent cattle from becoming averse and fearful of a new squeeze chute or corral system, painful or frightening procedures that cause visible signs of agitation should be avoided the first time the animals enter the facility (Grandin 1997a). It is important that an animal's first experience with a new corral, trailer, or restraining chute be a good one. Progressive ranchers walk their calves through the corrals and the squeeze chute. After the calves exit the squeeze chute, they are given some tasty feed. First experiences with new things make a big impression on animals. When an animal is first brought onto a new ranch, make its first experiences pleasant by feeding it and giving it time to settle down. Nonslip flooring is essential, because slipping and falling in the new facility may create a fear memory.

Cattle struggle in chutes when they get scared. Hitting or yelling at the cow just makes her more scared. The cow needs to be taught that, when she cooperates, the handler will reward her by releasing pressure. For example, if pressure is applied to her tail to get her in the squeeze chute, the handler must release pressure the instant she takes the first step toward the squeeze chute. If continuous pressure is applied, then she is still being punished when she is starting to do what the handler wants. The handler should instead reward her by releasing pressure. Most breeding cows will quickly learn that if they get in the squeeze, they can avoid pressure on the tail.

Genetics and Handling

Cattle that have an excitable temperament are more fearful than are more placid cattle. There are differences in the level of fearfulness both within and between breeds of cattle. On average, Herefords are less fearful than are Salers, although much variation can occur within each breed. An animal that is more fearful is more likely to panic during handling. More excitable, fearful cattle need to be introduced more gradually and gently to new experiences; these cattle are also more aware and vigilant of their surroundings. This fear trait helps animals to survive in the wild and avoid predators. Physical traits in cattle are also linked to temperament: Cattle with fine leg bones with a small diameter tend to be more "flighty" than are more heavily boned cattle, and cattle with spiral hair whorls above the top of the eyes are more likely to struggle in a squeeze chute than are cattle with a spiral hair whorl below the eyes (Grandin and Deesing 1998).

The National Non-Fed Beef Quality Audit indicated that losses caused by lameness significantly increased between 1994 and 1999 (Roeber et al. 2001) when approximately 25% of beef cows arrived at the packing plant lame. The author has recently observed that at this time, increasing numbers of cull cows and feedlot steers arriving at packing plants have poor structural conformation of the feet and legs, and fracturing of the forelegs of feedlot steers and heifers has been increasingly observed at packing plants. Veterinarians should urge their clients to breed cattle that are structurally sound in addition to having good expected progeny differences. It is likely that genetics is a major cause of the leg problems increasingly being encountered.

Cattle should also be selected to have a calm temperament. Excitable animals are more prone to quality problems such as dark cutters (Voisinet et al. 1997b) and damaged meat from bruising. The best approach to temperament selection is to cull the crazy cows instead of selecting for the most placid and calm ones, as overselection for any single trait often results in problems. Breeders need to look at the whole animal: The ideal animal for the feedlot may not be the ideal animal for rugged range. There is some evidence that extremely placid animals are less likely to move long distances and graze in rugged country. (Herefords tend to prefer lowland grazing, and Tarentaise and Salers are more likely to graze in the hills; Bailey et al. 2001).

EFFECTS OF VISION ON HANDLING

Contrary to popular belief, cattle can see color (Arave 1996). Research indicates that cattle are dichromats, with eyes that are most sensitive to yellow–green (553–555 nm) and blue–purple light (444–455 nm; Jacobs et al. 1998). This means that grazing animals may have partial color blindness similar to that of a human dichromat. They do not have black and white, total color blindness. Dichromatic vision may provide better vision at night and make the animal more sensitive to seeing motion. Dichromatic vision may also partially explain why cattle and other grazing animals are easily spooked and frightened by sudden movements and by high contrasts such as shadows. This theory may explain why animals will often refuse to walk over objects that have high contrast, such as sparkling reflections in a puddle, drain grates, or a shadow or bright spot of sunlight on the floor.

All grazing animals have wide-angle vision because their eyes are located on the sides of their head. Wide-angle vision enables a grazing animal to see all around itself and to detect predators while grazing. Their visual field is over 300° (Prince 1977), with a small blind spot immediately behind the animal's rear.

Animals notice small details and distractions that people do not notice, such as moving objects or shadows with harsh contrasts of light and dark. Even a small, swinging piece of chain in the entrance of a chute may cause animals to balk and refuse to enter (Grandin 1996). Animals should move through a facility easily, but quiet handling is impossible if animals constantly balk, back up, or turn back. Instead of resorting to greater force to move the animal in this case, the distractions that cause balking and backing up should be located and removed. Handlers need to be very observant to identify what is causing an animal to balk (Grandin 1998b). A calm animal will look right at the thing it is afraid of, showing you the distraction that needs to be removed. If removal is not possible, the animal should be allowed to slowly investigate the distraction. Most animals will be willing to move forward after they have looked at or sniffed a distraction such as a shadow or drain grate on the floor.

To find distractions that impede animal movement, a handler needs to walk through the chutes and pens to observe what the animals are seeing. It is important that the handler bend down and get his or her eyes at the same level as the animal's eyes because, for example, reflections on wet floors often cause animals to refuse to move, and a reflection that is visible at the animal's eye height may not be visible to a person who is standing up.

Anything that makes rapid movement is also likely to cause animals to stop. Look for swinging chains, people moving outside the chute, vehicles, shiny metal that jiggles, a flapping piece of plastic, or fan blades turning in the wind. Air blowing into the faces of approaching animals will also cause them to balk or back up, so ventilation systems should be designed so that air does not blow into the faces of approaching animals.

Managing Sensitivity to Contrasts of Light and Dark

All grazing animals are sensitive to harsh contrasts of light and dark. Facilities should be painted a single solid color to prevent balking at contrasts of light and dark. Shadows will often cause animals to stop, and balking caused by shadows is often worse on a bright sunny day. Changing shadows can cause a facility to work well at one time of day and poorly at another time.

Drain grates in the middle of the floor will often cause animals to balk because the cattle are sensitive to changes in flooring type or texture. They may refuse to move from a dirt floor onto concrete, and balking can often be reduced by putting some dirt on the concrete floor. Cattle will also balk if they see a high-contrast object. A coat flung over a chute fence or the shiny reflection off a car bumper will cause balking.

Effects of Light on Cattle Movement

Cattle have a tendency to move toward the light, but they will not approach blinding light. If livestock must be handled at night, it is strongly recommended that indirect lighting that does not glare in the animals' faces be positioned inside the truck or building. Loading ramps and squeeze chutes should face either north or south, because animals may balk if they have to look directly into the sun. Sometimes it is difficult to persuade animals to enter a darker-roofed working area. Persuading animals to enter a dark building from bright sunlight in an outdoor pen is also often difficult. In some cases, incoming animals will enter if doors on the opposite side of the building are opened up so they can see daylight through the other side. Movement into a building can also be improved by installing white translucent panels in the walls or roof that let in lots of shadow-free light so that the animals will be attracted to the light (Grandin 1998b). Problems with getting animals to move into a darker building are most likely to occur on a bright sunny day. A facility that works poorly on a bright sunny day may work well at night or on a cloudy day. The ideal illumination inside a building for moving animals should resemble a bright but cloudy day.

Solid Sides and Vision Blockage

Solid sides are recommended on cattle handling facilities because cattle behavior is strongly influenced by what the animal sees (Grandin 1987, 2000). Solid sides prevent animals from seeing distractions outside the chute such as moving people or vehicles. Grazing animal behavior is controlled by the animal's vision. Adding solid sides to a chute will often help keep animals quieter. Solid sides are especially important when animals with a large flight zone are handled. They should be placed on a single-file chute that leads to the squeeze chute, crowd pen, crowd gate, and loading ramp. A solid crowd gate prevents the animals from turning back to the pens where they came from. Solid sides have little effect if animals are completely tame and have no flight zone. A tame, halter-broken cow does not need solid sides. A basic principle is that the bigger the flight zone and the wilder the animal, the more you need a visually solid barrier. Wild cattle can be calmed by placing a mask of completely opaque material over their eyes (Andrade et al. 2001).

Dark Box Chutes and Vision Blockage

Cows can be easily restrained for artificial insemination or pregnancy testing in a dark box chute that has no headgate or squeeze (Figure 7.1; Canada Plan Service 1984). This is an example of using behavior instead of force to restrain an animal. Even the wildest cow can be restrained in a dark box chute with a minimum of excitement.

A dark box chute can be easily constructed from plywood or steel. It has solid sides, top, and front. When the cow is inside the box, she is inside a quiet, snug, dark enclosure. A chain or bar is latched behind her rump to keep her in. After insemination, the cow is released through a gate in either the front or the side of the dark box. If wild cows are being handled, an extra long, dark box can be constructed. A tame cow that is not in heat is used as a "pacifier" and is placed in the chute in front of the cow to be bred. Even a wild cow will stand quietly and place her head on the pacifier cow's rump. After breeding, the cow is allowed to exit through a side gate, while the pacifier cow remains in the chute.

HEARING

Cattle are more sensitive to high-pitched sound than are people. The auditory sensitivity of cattle is greatest at 8000 Hz (Heffner and Heffner 1983).

People working around large animals should speak softly with a low tone of voice. Observations by the author indicate that cattle remain calmer and handling is easier when handlers refrain from yelling and loud whistling.

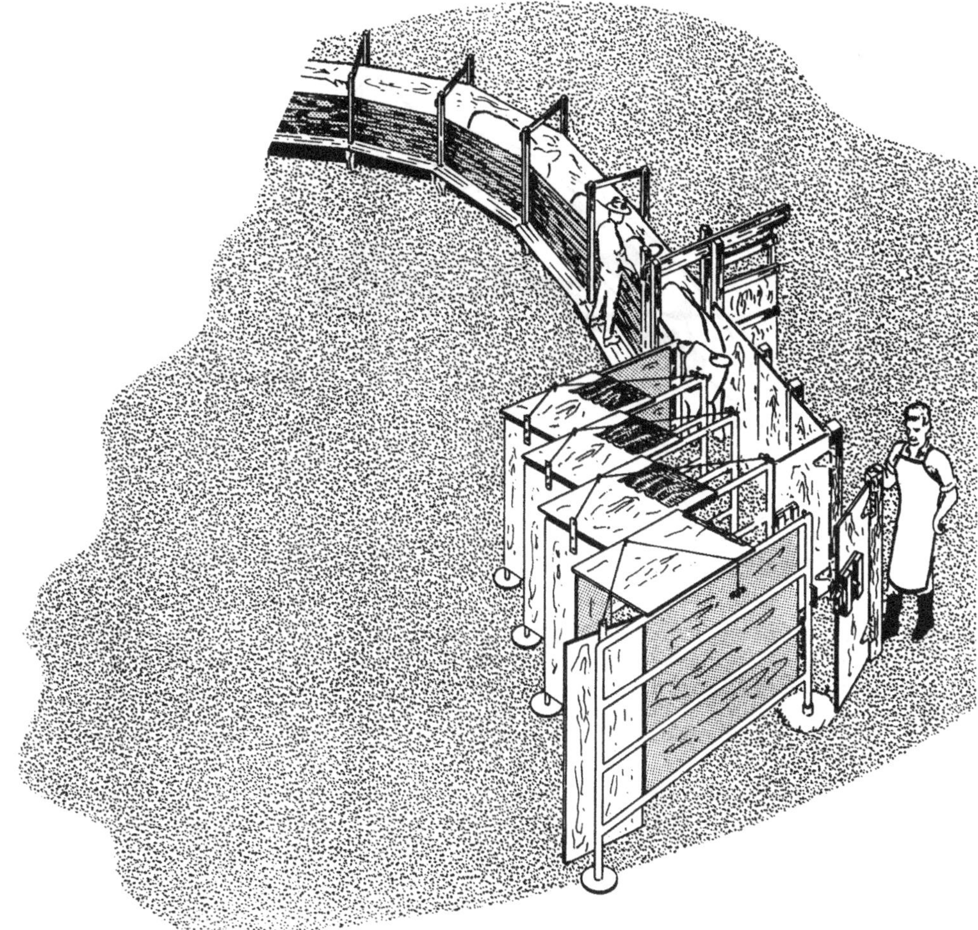

Figure 7.1. Dark box chutes keep cows calm for pregnancy testing or artificial insemination because the cow's vision is blocked by solid sides, top, and front. To entice the cow to enter the box, a small window 8 in (20 cm) by 3 in (8 cm) may need to be cut in the door at cow eye height.

Research indicates that yelling and loud whistling at cows is very aversive and increases an animal's heart rate more than does the sound of a gate slamming (Waynert et al. 1999). Cattle with a nervous temperament that became agitated in an auction ring were more sensitive to the sound of a person yelling and to sudden movement than were calmer cattle (Lanier et al. 2000).

BEHAVIORAL HANDLING PRINCIPLES

All people handling cattle need to understand the flight zone. The flight zone is the animal's safety zone, and its size varies depending on how wild or tame the animal is (Figure 7.2; Grandin 1987, Hedigar 1968). The flight zone may vary from a few feet to 100 yards or more: A show steer or a riding horse has no flight zone, but cattle that seldom see people will have a large zone. If a person is outside the animal's flight zone, it will turn and look at him or her, but when a person enters the flight zone, the animal will turn away. The size of the flight zone is determined by three interacting factors: genetic traits (excitable vs. calm), amount of contact with people (sees them every day or only twice a year), and quality of their contact with people (aversive vs. positive). Cattle with large flight zones may become fearful and agitated when a person deeply penetrates their zone while they are in a confined space and unable to move away. Cattle rearing up in squeeze chutes or single-file chutes have caused many accidents. Wild

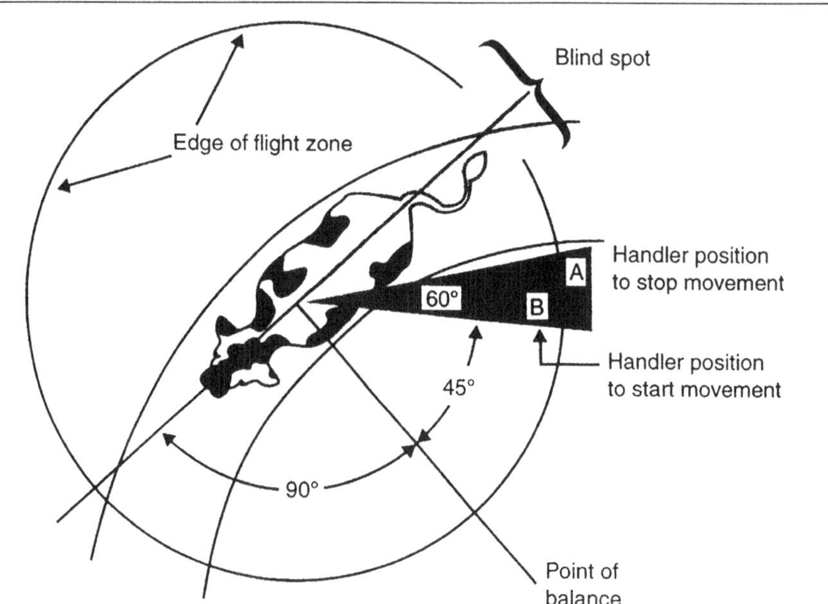

Figure 7.2. Handler should work on the edge of the flight zone and avoid the blind spot behind the animal's rear. The cow will go forward if the handler is behind the point of balance at the shoulder and backward when the handler is in front of the point of balance.

cattle may do this because they are attempting to escape from a person who is deep in their flight zone. If an animal rears, a person should back up and remove himself or herself from the animal's flight zone. When the person backs away, the animal will often settle back down. Handlers should be instructed to never attempt to push a rearing animal back down. This is likely to increase its agitation and may cause injuries to either the animal or its handlers.

Cattle that have a large flight zone will move more quietly with less agitation if the handler works on the edge of the zone (Figure 7.2). The handler penetrates the edge of the flight zone to make the animal move and retreats outside the flight zone to induce the animal to stop moving. Excited, agitated animals will have a larger flight zone than will calm animals. The flight zone will be larger when a person faces an animal and smaller if he or she turns sideways. The flight zone enlarges when the handler makes himself or herself look bigger and more intimidating. Flight zone principles may not work on completely tame animals, which should be led or trained to move.

Handlers also need to understand the "point of balance." The point of balance is an imaginary line at the animal's shoulders. To induce the animal to move forward, the handler must be behind the point of balance. To make it move backwards, he or she must be in front of the point of balance. Cattle will move forward when a handler walks inside the flight zone, past the point of balance, in the direction opposite that of desired movement (Figure 7.3; Kilgour and Dalton 1984, Grandin 1987, 1998b). The handler must walk quickly past the point of balance. If the handler moves too slowly, the animal will back up. Progressive cattle handlers have been able to almost eliminate the use of electric prods by implementing these movement patterns.

On most ranches and feedlots, 99% of the cattle can be moved quietly and efficiently without electric prods. In large slaughter plants, 15 minutes of instruction on flight zones and movement patterns resulted in a reduction of electric prodding of beef cattle from 83% of the animals to 17% (further reduction of electric prod use required modifications of the facilities). In addition, the workers were able to keep pace with the slaughter line even with reduced prodding. The very best plants with good equipment had to use an electric prod on only 5% of the cattle. There are a few animals that refuse to move unless the electric prod is used. In this situation, the electric prod is preferable to hitting the animal or twisting the tail to make it move.

For both welfare and safety reasons, the use of electric prods should be avoided as much as possible.

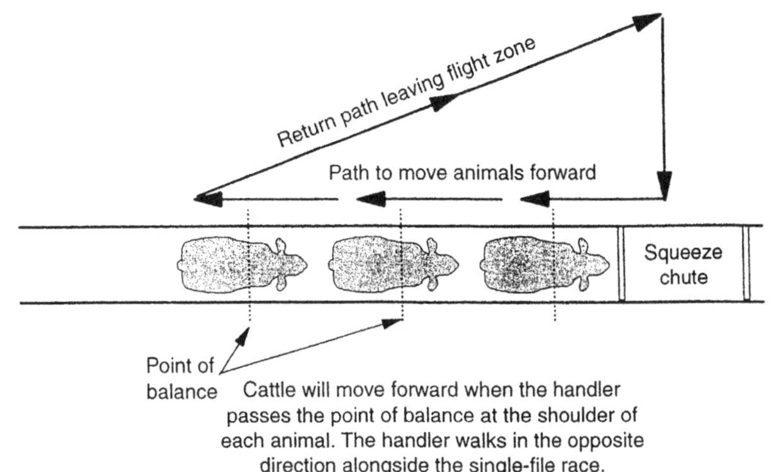
Return path leaving flight zone
Path to move animals forward
Squeeze chute
Point of balance
Cattle will move forward when the handler passes the point of balance at the shoulder of each animal. The handler walks in the opposite direction alongside the single-file race.

Figure 7.3. An animal can be moved into a squeeze chute easily if the handler quickly walks past the point of balance at the shoulder.

Nonelectric driving aids such as plastic paddles or a flag on the end of a stick should be used as the primary driving tools. Plastic streamers or a flag on the end of a stick can also be used to quietly turn animals. Be careful not to be too aggressive with a flag. Use it to guide an animal and do not wildly wave it. Handlers should be careful to avoid scaring the animals. It is important to get electric prods out of people's hands as much as possible. Observations by the author indicate that the attitude of the people toward the cattle improved when they stopped carrying electric prods.

Handling in Crowd Pens

The number one mistake made by handlers is putting too many animals in the crowding pen that leads to a single-file chute or to a loading ramp. Animals need room to turn. When cattle are handled, the crowd pen should be filled only one-half to three-quarters full (Figure 7.4). Do not push the crowd gate up tight against the animals. Cattle should be moved in small groups, which will help keep them calm.

Cattle will move through the crowd pen into a loading ramp or single-file chute more easily if the handler takes advantage of their natural following behavior. When a truck is being loaded, do not allow animals to stand in the crowd pen. They should walk through the pen and immediately go up the ramp before they have a chance to turn around. When animals are being moved into a single-file chute, the crowd pen should not be filled until there is space in the chute. In fact, the crowd pen should be renamed the "passing-through pen." The animals should walk through without stopping on their way to either the loading ramp or a single-file chute. If the animals balk or turn back, distractions must be removed.

Cattle will often become very agitated and stressed when a lone animal is separated from the herd. A single animal that is frantically attempting to rejoin its herdmates is highly stressed. The author has observed that a lone bovine left behind in a crowd pen or alley has on several occasions caused serious accidents when it jumped a fence or ran over a person. A person should never get into a confined space, such as a crowd pen, with a single, agitated, large animal. Either the cow should be released or more animals should be put in with it. However, it is safe for experienced handlers to be in a larger pen or alley with a group of cattle. In this situation, there is sufficient room for the animals to move away. The handler is not constantly standing inside the flight zone. Calm animals can be sorted out at a gate by facing and staring at the animals you want to hold in the pen and turning sideways and looking away from the animals you want to move through the gate.

Habituation to Handling

Cattle will usually habituate to being quietly moved through a squeeze chute (Grandin and Deesing 1998). If a bovine is moved through a squeeze chute every day for several days, it will usually become calmer on each successive day because it learns that going in the squeeze chute will not hurt it. Animals with a calm temperament will habituate to a series of forced, nonpainful procedures. For example, cortisol levels

Figure 7.4. These people are doing a good job, and the crowd pen is only half full even when the crowd gate is advanced. Cattle need room to turn so that they can be guided with paddle sticks or flags.

in cattle decreased after they were moved through the squeeze chute a number of times over a period of days (Crookshank et al. 1979, Alam and Dodson 1986, Andrade et al. 2001).

A review of several studies showed that wild beef cattle become more stressed during handling than do tamer animals that remain quiet during handling (Grandin 1997a). However, extremely flighty, excitable animals may not habituate. Instead, they often react explosively to a forced-handling procedure and severely injure themselves. Rather than becoming less and less fearful with each successive pass through the chute, they may become **increasingly** fearful. They are likely to be injured when they rear, jump out of a facility, or violently struggle. Animals with flighty, excitable genetics are more likely to panic when they are suddenly confronted with a novel experience. The cattle may be calm and docile at the home ranch, when they are with familiar people, but they may become very agitated when brought to an auction. Cattle that are genetically calmer are less likely to panic when they are brought to new places such as auctions or feedlots.

Effects of Previous Experiences on Handling

Previous experiences affect how animals will behave during handling. Cattle have excellent memories. They remember painful or aversive experiences, and they will be more reluctant to reenter a facility in which an aversive event occurred. The author has observed that cattle that have had previous experiences with rough handling will have bigger flight zones and become more agitated during restraint when they are handled in the future.

Calves that have been reared in close association with people will usually be easier to handle and have a smaller flight zone when they mature. The author has also observed that cattle reared in the colder parts of the United States, where they are fed every day during the winter, have a smaller flight zone than do cattle raised in southern states. Some southern cattle are handled only a few times each year, and they are not fed during the winter because grass grows year-round in that region. There is a tendency for southern cattle to become more agitated in squeeze chutes compared to northern cattle, which are exposed to people feeding them all winter.

Australian researchers conducted some of the first training experiments of extensively raised beef calves that had large flight zones. These experiments were conducted to determine whether training calves would make the animals easier to handle when they matured. They found that walking quietly among the calves and moving them quietly through the chutes produced calmer adult animals (Fordyce 1987, Fordyce et al. 1988). Extensively reared Zebu calves handled (petting the calves in a single-file chute) ten times at 1–2 months of age were calmer and less likely to jump fences in the future than were those calves who had not been handled (Becker and Lobato 1997).

It is important to calmly and quietly train calves to handling both on foot and on horseback. Cattle that have never seen a person on foot may become fearful when they see a person walking in a pen. In cattle with excitable genetics, this can cause serious handling problems when the cattle go to a feedlot. Observations by the author have indicated that cattle that originated from ranches where they had become accustomed to people both on foot and on horseback were calmer and easier to handle at the feedlot. Frequent gentle handling and contact with people also reduces cortisol levels and stress associated with restraint in cattle (Hopster et al. 1999). Training young cattle to the quiet presence of people walking amid them will produce calmer adult animals.

Animals need to be habituated to a variety of vehicles and people. They need to learn that some variation in their routine will not hurt them. This is especially important for cattle with flighty, excitable genetics. Over 40 years ago, Reid and Mills (1962) found that exposing sheep to variations in routine helped to reduce their stress when they were exposed to change.

Moving Large Groups on Pasture

Cows can be easily taught to come when called. They will often learn to associate a truck with feed. It is best to teach the animals that tooting the horn means feed rather than teaching them that the sight of the truck means feed. This will prevent the cattle from running after vehicles and leaving their calves behind. When cows with young calves are being moved, it is important to control the movement of the cows so that the young calves stay with their mothers. This will help reduce stress on the calves. Handlers must not chase stragglers when cows are moved. Allow the motion of the herd to attract them back to the herd. When animals are moved out of a pen, the handler should stand near the gate and control their movement out of it. In feedlots, a lead horse should be used to prevent cattle from running down the alley.

Large groups of extensively raised cattle can be gathered by triggering the animals' natural tendency to bunch. This method will not work on very tame cattle with little or no flight zone: Tame cattle should be led, not driven. To induce bunching, the handler walks on the edge of the collective flight zone in a windshield wiper pattern (Figure 7.5; Grandin 2000, Smith 1998). The handler walks quietly back and forth until the bunching instinct is triggered. The bunching instinct must be triggered before any attempt is made to move the cattle. Attempting to move the cattle before bunching will cause them to scatter.

Derek Bailey at Montana State University and Floyd Reed with the U.S. Forest Service have worked with ranchers to use herding methods to keep cows away from fragile, riparian areas near streams without the use of fences. Following are some tips:

- Move cows away from the riparian area later in the day to allow adequate time for them to bed down their calves at the new pasture. Cows tend to stay where their calves bed down.
- Move the herd quietly and do not allow cows and calves to get separated.
- Cull problem cows and "bunch quitters."
- Cows prefer the forage grasses on which they were raised as calves. Cows raised on forages away from streams will be easier to keep off riparian areas.
- In hilly country, breeds such as Salers that were originally developed in mountainous regions may be easier to keep off the lowlands than are breeds such as Herefords that were developed in flatter country. However, within a breed, some cows are lowland dwellers and others are highland dwellers (Bailey et al. 2001).
- Tasty supplements will help keep the cows at the new grazing site.
- Practice the principles of pressure and release. When cows go where you want them to, back up and reduce pressure on the collective flight zone. Continuous pressure on the flight zone will cause cows to run.

BULL BEHAVIOR

A bull often attacks a person because he perceives the person as a conspecific (herdmate) that he attempts to dominate. This is true aggression and it is not motivated by fear. Why is the dairy bull more dangerous

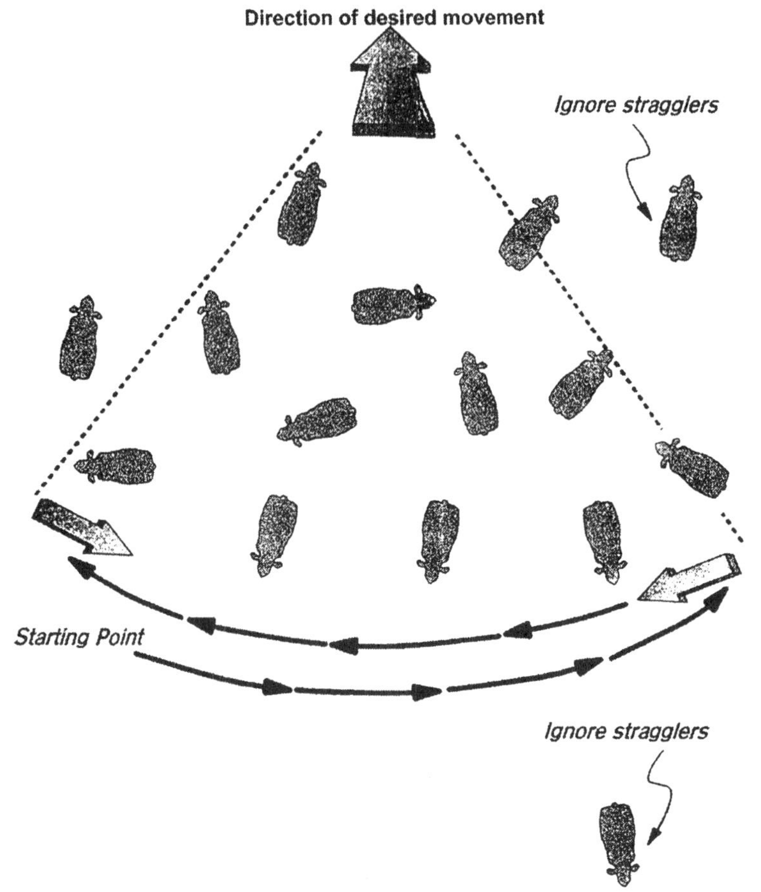

Figure 7.5. Windshield-wiper pattern to induce the bunching instinct in pastured cows. The handler quietly walks back and forth on the edge of the collective flight zone until the cows start to line up. Do not circle around the cows. Do not attempt to move the cows until they line up.

than are most beef bulls? The difference in beef and dairy bull behavior may be partly explained by differences in rearing methods. Beef bull calves are reared on their mothers, and most dairy bull calves are bucket-fed by people. Research by Edward Price and his associates at the University of California found that Hereford bulls reared in groups were less likely to attack people than were bulls bucket-fed in individual pens (Price and Wallach 1990). Seventy-five percent of the individually reared bulls threatened handlers. In 1000 dam-reared bulls, only one bull attacked. Bulls that grow up with other cattle learn that they are bulls. Individually reared bulls may think they are people, and when they become sexually mature, they may challenge a person to exert dominance. Bull calves reared with other cattle usually direct their challenges toward other bulls rather than toward people.

A bull will challenge another bull or a person by making a broadside threat (Smith 1998), facing sideways and flexing his neck to show how big he is. The bull will often not look at a person he is threatening. The threat will be made before the bull charges. Bulls that make broadside threats toward people should be culled because they may charge people. More information on bull behavior can be found in Smith 1998.

To reduce risk on beef cattle operations, an orphan bull calf should be either castrated or placed on a nurse cow. Castration at a young age will reduce aggression toward people. Practical experience in dairies indicates that 6-week-old bull calves reared together in large groups are less likely to attack people. After a short period of individual rearing, putting 6-week-old bull calves together for the rest of the rearing process can help reduce risks. Orphan male calves, llamas, and buck deer have also been known to attack people. The problem of attacking bulls is not caused by tameness, or a lack thereof, but, rather, appears to be caused by the bull mistaking a human for his own species. A basic principle is that in most grazing animals, intact males will have less of a tendency to attack people if they are reared by their own species. This effect is especially strong very early in life.

Cattle form social dominance hierarchies. Bulls will fight to determine the boss bull. Dominant bulls will generally breed more cows than will subordinate bulls. However, if the boss bull has poor semen quality, conception rates for the entire herd can be lower. A recent genetic analysis of sires of calves indicated that some bulls breed many cows and others breed very few. Veterinarians need to examine dominant bulls to make sure that they are fertile and free of defects that could affect breeding.

REDUCING WEANING STRESS

Both research and practical experience has shown that fenceline weaning will reduce stress on both the cows and the calves. Vocalization and constant pacing are greatly reduced. Research by Edward Price at the University of California indicated that calves weaned with fenceline contact with their mother also had better weight gains than did calves that were totally separated. A five-wire barbed wire fence covered with woven wire mesh will keep the cows and calves separated. An electric fence wire can be placed 12 in (30 cm) to 15 in (38 cm) out from the fence on the calf side. On many ranches the electric wire will not be needed. The calf is more interested in seeing its mother than in nursing. To further reduce stress, weaning and vaccination should be done 35–45 days before shipment off the ranch.

If a stocker operator receives a truckload of bawling calves, the use of experienced trainer cows may help to induce them to eat. The calves may be more likely to eat when they see the trainer cows eating. Constant pacing of the calves can be reduced by a person walking in the same direction as the calves are moving to help slow the pacing. When the calves slow down, they will start eating.

CHUTE DESIGN

For cattle, a single-file curved chute leading up to a squeeze chute or truck works better than does a straight chute for three reasons. First of all, it prevents the animal from seeing the truck, the squeeze chute, or people until the last moment. A curved chute also takes advantage of the animal's natural tendency to circle around the handler (Grandin 1987, 1997b). When a handler enters a pen of cattle or sheep, the animals will turn and face the handler but will maintain a safe distance. As the handler moves through the pen, the animals will keep looking at the handler and will circle around as he or she moves. A curved chute also takes advantage of the natural tendency of cattle to go back to where they came from.

A well-designed, curved, single-file chute has a catwalk beside the inner radius for the handler to use (Figure 7.6). Working along the inner radius puts the handler in the best position to move the animals. Another design that is becoming popular is making the outer radius of the single-file chute completely solid to block vision; the inner radius has a four-foot-high (1.2 m) solid side that allows the cattle to see out. The catwalk is eliminated, and the handler works on the ground by penetrating and then retreating from the flight zone. Walking back by the point of balance also works well in this type of facility. Another alternative is to block the opening along the top of the chute with a flexible rubber curtain made from conveyor belting. If an animal stops moving, a handler on the ground can lift up the curtain to help promote movement.

The crowd pen, or "tub," should have a solid fence that prevents the cattle from seeing out. Circular crowd pens should be laid out to take advantage of the animal's natural tendency to go back to where they came from (Figure 7.7). This will require at least a half circle. The ideal radius (gate length) for round crowd pens is 12 ft (3.5 m). Animals will be more difficult to handle if the pen is either too small or too big. Catwalks should never be placed overhead: Having people walking above them frightens animals. The best catwalk design is to locate the catwalk 42 in (76 cm) from the top of the fence. The top of the chute or crowd pen fence should be at waist height on the average person. Using single-file chutes and making animals stand in a queue works well for cattle, as lining up in a single file is natural for these species. Cattle will stand quietly in a queue as long as they can see the next animal in front of them fewer than 3 ft (1 m) away.

USE OF OPEN BACKSTOP GATES TO REDUCE BALKING

Many cattle facilities have too many backstop gates in the single-file chute to prevent animals from backing up. These gates can increase balking because the

Figure 7.6. A curved chute with solid sides is more efficient because cows will back up less. Curves work because cattle have an instinct to go back to where they came from.

animals may refuse to walk through the devices. If cattle constantly back up, this is a symptom of a problem that needs to be corrected. The distractions discussed previously should be removed. If cattle balk at a backstop gate at the single-file chute entrance, the gate should be either tied open or equipped with a remote control rope so that it can be held open for the cattle. In a well-designed beef cattle facility with a curved, single-file chute, the only backstop gate that is really needed should be located two body lengths behind the squeeze chute. This will prevent the leaders from backing out.

Problems with balking tend to come in bunches: When one animal balks, the tendency to balk seems to spread to the next animals in line (Grandin 1980). When an animal is being moved through a single-file chute, the animal must never be urged forward unless it has a place to go. Once it has balked, others will start balking.

Dead-End Curved Chutes and Balking

Cattle will balk if the entrance to a single chute appears to be a dead end (Grandin 1987, 1998b). Sliding and one-way gates in the single-file chute must be constructed so that the animals can see through them; otherwise, the animals will balk. This is especially important at the junction between the single-file chute and the crowd pen. However, palpation gates for pregnancy checking or artificial insemination should be solid so that approaching animals do not see a person standing in the chute.

When a curved chute is built, it must be laid out properly so that it does not appear to be a dead end. A cow standing in the crowd pen must be able to see a minimum of two body lengths up the chute. Animals will balk if the chute is bent too sharply at the junction between the crowd pen and the single-file chute. This is one of the worst design mistakes in

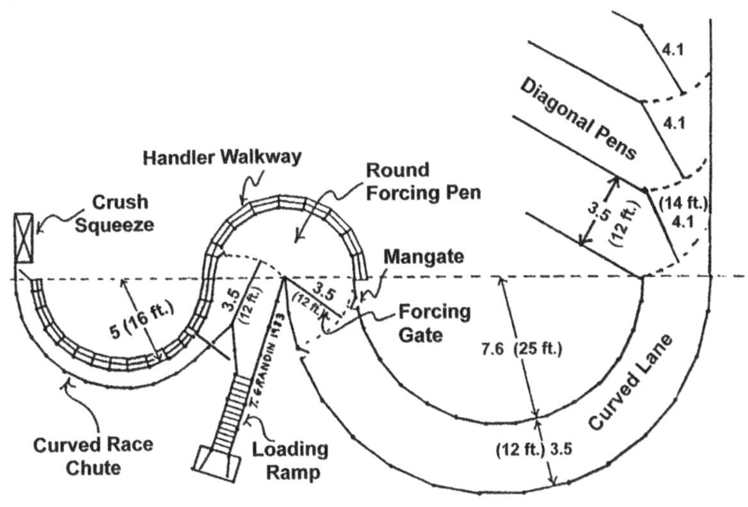

Figure 7.7. Curved handling system for a ranch. The correct layout is essential. The most critical part of this design is the junction between the curved chute and the round forcing pen. Round forcing pens work best when they are laid out so the cow thinks she is going back to where she came from.

an animal handling facility. Figure 7.7 illustrates an efficient, curved facility that is easy to lay out. The round crowd pen (tub) in this facility works efficiently because cattle moving through the tub think they are going back to where they came from. The entire facility consists of three half circles laid out along a line (Figure 7.7; Grandin 1998b).

CATTLE RESTRAINT PRINCIPLES

Cattle sometimes become severely stressed and agitated in a conventional squeeze chute. This is caused by deep invasion of the animal's flight zone by the operator and other people that the cattle can see through the open-barred sides. Agitation and struggling can be reduced by installing solid sides or rubber louvers on the open-barred sides. Louvers 6–8 in (15–20 cm) wide and made from rubber conveyor belting can be installed on the drop-down bars. The strips of belting are installed on a 45° angle. The bars can still be opened, but incoming animals cannot see out as they enter. Producers can experiment by covering up the sides of the squeeze chute with cardboard. The most important part to cover is the back half nearest the tailgate.

The use of a complete squeeze chute is strongly recommended for most ranches. Restraint of the body will prevent the animal from fighting the headgate. The best squeeze chutes squeeze evenly on both sides. Chutes that squeeze on only one side tend to throw the cow off balance, sometimes causing them to struggle more. On hydraulic chutes, the pressure relief valve must be adjusted to prevent excessive squeeze pressure. The animal must be able to breathe normally and must show no signs of straining. Excessive pressure can cause severe injuries such as a ruptured diaphragm or broken bones. If an animal strains or vocalizes (moos or bellows) when the valve lever is held down until the valve releases pressure, the pressure setting is too high. Devices for restraining the head also must be designed to avoid distress. If the animal struggles or vocalizes at the moment the device is applied, it will need to be redesigned or the pressure may have to be reduced.

To prevent choking in a headgate with curved stanchion bars, the squeeze sides must be adjusted so that the V shape of the sides prevents the animals from lying down. Pressure exerted by the headgate on the carotid arteries can kill the animal. If an animal collapses while held in a headgate, the headgate must be released immediately to avoid death. Some veterinarians prefer a chute that does not pinch the feet together at the bottom. If a squeeze chute with straight sides is used, it must be equipped with a straight bar stanchion headgate to prevent choking. An animal can safely lie down in a straight bar stanchion because no pressure is exerted on the underside of the neck. Care must be taken with self-catching headgates. Cattle can be injured if they run into the self-catcher at a high speed. Severe injuries can also occur if a self-catcher is adjusted too wide and the animal's shoulder passes part way through the closed gate. Self-catchers are usually not recommended for wild, horned cattle. It is also essential to adjust the self-catcher for the size of the cattle. **Never** leave an animal unattended in any restraint device.

Latches and ratchet locks must be kept well maintained to prevent accidents to people. If a ratchet device becomes worn, replace it immediately. Friction-type latches must never be oiled: Oiling will destroy the ability of a friction latch to hold. On self-catching headgates, the mechanism must be kept maintained to prevent an animal from getting stuck part way through the closed gate.

Painful restraint methods such as nose tongs should be avoided whenever possible. Cattle that have been restrained with nose tongs become more and more difficult to restrain repeatedly. They also remember aversive experiences such as being accidentally banged on the head with the headgate of a squeeze chute. The use of a halter is recommended. Practical experience has shown that cattle become ever easier to restrain for blood testing from the jugular vein if a halter is used. Calves and polled cattle held in a headgate can be easily restrained for blood testing by a person who physically pushes the animals' heads to one side by using his or her rear end. The person's rear covers the eyes and an animal will remain calmer. The animal will usually cooperate if steady pressure is applied and sudden jerky movement is avoided.

Electrical immobilization that paralyzes an animal must never be used as a substitute for well-designed restraint equipment. Scientific studies clearly show that electrical immobilization is highly aversive and detrimental to animal welfare (Lambooy 1985, Grandin et al. 1986, Pascoe 1986). Electrical immobilization with a small current should never be confused with the electrical stunning that is used in slaughter plants. Electrical stunning uses a high-amperage current that induces instantaneous unconsciousness.

NONSLIP FLOORING

Low-stress animal handling is impossible if animals constantly slip on the floor. Even slight slipping can make an animal become agitated and nervous. The two most common facility problems that make quiet handling difficult are slick floors and lighting problems that cause animals to balk. In many cases, older facilities that are not state of the art are often adequate provided that they are well maintained, have nonslip flooring, and do not have the lighting problems discussed previously.

Existing slick concrete floors can be roughened with a concrete-grooving machine. Grooves made by a concrete-grooving machine are suitable for many different species. Rubber mats are also available to provide nonslip footing. In high-traffic areas in beef cattle facilities, a grating constructed by placing steel bars that have a 1-in diameter in 12 × 12 in (30 × 30 cm) squares can be used. The bars must be welded flush so that all bars lie flat against the floor. Woven rubber mats made from old car tires are another alternative.

In new facilities, concrete should be deeply grooved for beef cattle. Make an 8 × 8 in (20 × 20 cm) pattern of 1-in deep V grooves. A less rough pattern should be used in dairy cattle facilities, where cattle are walking on the floor every day, as a floor that is too rough can damage their feet.

TROUBLESHOOTING HANDLING PROBLEMS

To solve a handling problem, one must first determine the cause of the problem. Do you have a facility problem or a problem caused by the way people are handling the animals? Difficulties can arise from any one or more of the following factors:

- Facility design problem such as a dead-end chute (or race)
- Small, easily corrected distractions that cause balking
- Too many animals being placed in the crowd pen, which should only be half full
- Handlers who get the animals agitated, excited, and scared
- Animal temperament problems caused by flighty, excitable genetics
- Problems with lighting and a chute entrance that is too dark.

You must determine whether you have a basic design problem, a small distraction that can be easily fixed, or an animal or handling technique problem.

LOW-STRESS TRANSPORT

The single most important factor that affects the condition of cattle during and after transport is the condition of the animal when it is placed on the truck.

Roeber et al. (2001) reported that timely marketing of cull cows is still a big issue. A cow should be shipped when she is still fit for transport. An emaciated, weak cow is likely to become a downer if she is shipped. These cattle should be euthanized on the ranch. Only physically fit cattle should be transported.

It is the author's opinion that freshly weaned calves that are still vocalizing (bawling) for their mother are not fit for transport. Both practical experience and research shows that freshly weaned calves are more likely to get sick than are calves that are preweaned and vaccinated before transport.

Truck drivers should be trained in cattle handling and be aware of how their driving will affect the cattle. Sudden stops and starts must be avoided. Drivers and shippers must follow all of the recommended guidelines for stocking densities in trucks. Guidelines are available from the National Institute for Animal Agriculture in Bowling Green, Kentucky (http://www.animalagriculture.org); the American Meat Institute (http://www.meatami.org); and Grandin (2000). Overloading trucks will increase the number of downed or injured cattle.

Practical experience has shown that air-ride suspensions will reduce stress on cattle and provide a smoother ride. Many new cattle trailers are equipped with air ride. Another important factor during transport is nonslip flooring. When aluminum trailers get older, the nonslip flooring wears off, resulting in slipping and stress. On older trailers, small rods may need to be welded to the floor to prevent slipping.

Effects of Weather Conditions

Drivers must also be aware of how weather and temperature can affect the cattle. Freezing rain can be deadly to cattle because it destroys the ability of the coat to insulate the animal. Observations by the author indicate that more deaths occur during freezing rain than occur during dry conditions that are much colder. During hot weather, trucks must be kept moving because heat builds up rapidly in a stationary vehicle. Charts that show heat stress and wind chill factors are available from the National Institute for Animal Agriculture and the American Meat Institute.

The ability of an animal to tolerate either very cold or very hot conditions depends on many factors. There is some evidence that where a calf is born will affect its tolerance to either heat or cold. Cattle that are acclimated to the local conditions will handle environmental stressors more easily; for example, cattle from a northern area that have heavy winter coats are more likely to become heat stressed than are cattle raised in a hotter climate. Another big factor is genetics. Braham-cross cattle can tolerate more heat than can English breeds.

Loading Ramp Design

Loading ramps should have a level dock at the top, which provides a flat surface for the cattle to walk on when they get off the truck. The best cattle ramps have concrete stair steps with a 3.5-in (10 cm) rise and a 12-in (30 cm) tread. The steps should be deeply grooved. If cleats are used, there should be 8 in (20 cm) of space between the cleats. Hardwood 2 × 2s make good cleats. On permanent ramps, the slope should not exceed 20°, and adjustable ramps should not exceed 25°. Many people mistakenly make the ramp too wide. The best width is 30 in (76 cm). The sides of the ramp and the crowd pen should be solid. Man gates should be installed for the easy access and safety of people.

CONCLUSIONS

Animals that are calm and have low fear of humans will be more productive animals. When cattle become agitated in a squeeze chute, it is usually as a result of fear. Calm cattle are easier and safer to handle than are agitated, fearful cattle. Handlers who use behavioral principles will be able to handle cattle more efficiently. Research has shown that yelling at cattle is very stressful for the animals. Cattle remember painful or aversive experiences, and they become ever easier to handle when they become habituated to quiet handling. The use of low-stress fenceline weaning, an understanding of bull behavior, and good transport principles can improve welfare and improve both productivity and safety.

REFERENCES

Alam, M.G.S., H. Dodson. 1986. Effects of various veterinary procedures on plasma concentrations of cortisol, luteinizing hormone and prostaglandin E2 metabolite in the cow. *Veterinary Record* 118:7–10.

Andrade, O., A. Orihuela, J. Solano, C.S. Galina. 2001. Some effects of repeated handling and use of a mask on stress responses in zebu cattle during restraint. *Applied Animal Behaviour Science* 71:175–181.

Arave, D.R. 1996. Assessing sensory capacity of animals using operant technology. *Journal of Animal Science* 74:1996–2009.

Bailey, D.W., D.D. Kress, D.C. Anderson, D.L. Boss, E.T. Miller. 2001. Relationship between terrain use and performance of beef cows grazing foothill rangeland. *Journal of Animal Science* 79:1883–1891.

Becker, B.G., J.F.P. Lobato. 1997. Effect of gentle handling on the reactivity of zebu cross calves to humans. *Applied Animal Behaviour Science* 53:219–224.

Blecha, F., S.L. Boyles, J.G. Riley. 1984. Shipping suppresses lymphocyte blastogenic responses in Angus and Braham x Angus feeder calves. *Journal of Animal Science* 59:576.

Boissey, A. 1995. Fear and fearfulness in animals. *Quarterly Review of Biology* 70:165.

Canada Plan Service. 1984. *Herringbone AI Breeding Chute Plan 1819.* Ottawa: Agriculture Canada.

Crookshank, H.R., M.H. Elissalde, R.G. White, D.C. Clanton, H.E. Smalley. 1979. Effect of transportation and handling of calves upon blood serum composition. *Journal of Animal Science* 48:430–435.

Davis, M. 1992. The role of the amygdala in fear and anxiety. *Annual Review of Neuroscience* 15:353.

Fell, L.R., I.G. Colditz, K.H. Walker, D.L. Watson. 1999. Associations between temperament, performance and immune function in cattle entering a commercial feedlot. *Australian Journal of Experimental Agriculture* 39:795–802.

Fordyce, G. 1987. Weaner training. *Queensland Agricultural Journal* 6:323–324.

Fordyce, G., R.M. Dodt, J.R. Wythes. 1988. Cattle temperaments in extensive herds in northern Queensland. *Australian Journal of Experimental Agriculture* 28:683–688.

Galyean, M.L., R.W. Lee, M.W. Hubbert. 1981. Influence of fasting and transit on rumen blood metabolites in beef steers. *Journal of Animal Science* 53:7.

George, M.H., P.E. Heinrich, D.R. Dexter, et al. 1995a. Injection site lesions in carcasses produced by cattle receiving injections at branding and weaning. *Journal of Animal Science* 73:32–35.

George, M.H., J.B. Morgan, R.D. Glock, et al. 1995b. Injection site lesions, tissue histology, collagen concentration and muscle tenderness in beef rounds. *Journal of Animal Science* 73:3510–3518.

Grandin, T. 1980. Observations of cattle behavior applied to the design of cattle-handling facilities. *Applied Animal Ethology* 6:19–31.

Grandin, T. 1987. Animal handling. In *Veterinary Clinics of North America*, Vol. 3, edited by E.O. Price, p. 323. Philadelphia: W.B. Saunders.

Grandin, T. 1996. Factors that impede animal movement at slaughter plants. *Journal of the American Veterinary Medical Association* 209:757–759.

Grandin, T. 1997a. Assessment of stress during handling and transport. *Journal of Animal Science* 74:249.

Grandin, T. 1997b. The design and construction of facilities for handling cattle. *Livestock Production Science* 49:103–119.

Grandin, T. 1998a. Objective scoring of animal handling and stunning practices in slaughter plants. *Journal of the American Veterinary Medical Association* 212:36–93.

Grandin, T. 1998b. Handling methods and facilities to reduce stress on cattle. *Veterinary Clinics of North America: Food Animal Practice* 14:325–341.

Grandin, T. 2000. *Livestock Handling and Transport.* Wallingford: CAB International.

Grandin, T. 2001. Welfare of cattle during slaughter and the prevention of nonambulatory (downer) cattle. *Journal of the American Veterinary Medical Association* 219:1377–1382.

Grandin, T., M.J. Deesing. 1998. Genetics and behavior during handling, restraint and herding. In *Genetics and the Behavior of Domestic Animals*, edited by T. Grandin. San Diego, CA: Academic Press.

Grandin, T., S.E. Curtis, T.M. Widkowski, J.C. Thurman. 1986. Electro-immobilization versus mechanical restraint in an avid choice test. *Journal of Animal Science* 62:1469–1480.

Hedigar, H. 1968. *The Psychology and Behavior of Animals in Zoos and Circuses.* New York: Dover Publications.

Heffner, R.S., H.E. Heffner. 1983. Hearing in large mammals: Horses (*Equus caballus*) and cattle (*Bos Taurus*). *Behavioral Neuroscience* 97(2):299–309.

Hemsworth, P.H., G.J. Coleman. 1998. *Human-Livestock Interactions: The Stockperson and the Productivity of Intensively Farmed Animals.* Wallingford: CAB International.

Hemsworth, P.H., G.J. Coleman, J.C. Barnett, S. Borg, S. Dowling. 2002. The effect of cognitive behavioral interventions on the attitude and behavior of stockpersons and the behavior and productivity of commercial dairy cows. *Journal of Animal Science* 80:68–78.

Hixon, D.L., D.K. Kesler, T.R. Troxel. 1981. Reproductive hormone secretions and first service conception rate subsequent to ovulation control with Synchromate-B. *Theriogenology* 16:219.

Hopster, H., J.T. vanderWert, J.H. Erkens, H.J. Blokhuis. 1999. Effects of repeated jugular puncture of plasma cortisol concentrations in loose housed dairy cows. *Journal of Animal Science* 77:708–714.

Jacobs, G.H., J.F. Deegan, J. Neitz. 1998. Photopigment basis for dichromatic colour vision in cows, goats and sheep. *Visual Neuroscience* 15:581–584.

Kelly, K.W., C. Osborn, J. Eermann, S. Parish, D. Hinrichs. 1981. Whole blood leukocytes vs. separated mononuclear cell blastogenesis in calves, time dependent changes after shipping. *Canadian Journal of Comparative Medicine* 45:249.

Kilgour, R., D.C. Dalton. 1984. *Livestock Behaviour.* Sydney: University of New South Wales Press.

Lambooy, E. 1985. Electro-anesthesia or electro-immobilization of calves, sheep and pigs with Feenix Stockstill. *Veterinary Quarterly* 7:120–126.

Lanier, J.L., T. Grandin, R.D. Green, K. McGee. 2000. The relationship between reaction to sudden intermittent movements and sounds to temperament. *Journal of Animal Science* 78:1467–1474.

LeDoux, J.E. 1996. *The Emotional Brain.* New York: Simon and Schuster.

Pascoe, P.J. 1986. Humaneness of electrical immobilization unit for cattle. *American Journal of Veterinary Research* 10:2252–2256.

Price, E.O., S.J.R. Wallach. 1990. Physical isolation of hand reared Hereford bulls increases their aggressiveness towards humans. *Applied Animal Behavior Science* 27:263–267.

Prince, J.H. 1977. The eye and vision. In *Dukes' Physiology of Domestic Animals*, edited by M.J. Swenson, pp. 696–712. New York: Cornell University Press.

Redgate, E.S., E.E. Faringer. 1973. A comparison of pituitary adrenal activity elicited by electrical stimulation of preoptic amygdaloid and hypothalamic sites in the rat brain. *Neuroendrocrinology* 12:334.

Reid, R.L., S.C. Mills. 1962. Studies of carbohydrate metabolism of sheep. XVI. The adrenal response to physiological stress. *Australian Journal of Agricultural Research* 13:282–294.

Roeber, D.L., P.D. Mies, C.D. Smith, et al. 2001. National Market Cow and Bull Beef Quality Audit, 1999. A survey of producer related defects in market cows and bulls. *Journal of Animal Science* 79:658–665.

Rogan, M.T., J.E. LeDoux. 1996. Emotion: Systems, cells and synaptic plasticity. *Cell* 85:369–475.

Smith, B. 1998. *Moving 'Em: A Guide to Low Stress Animal Handling.* Kamuela, HI: Graziers Hui.

Voisinet, B.D., T. Grandin, J.D. Tatum, S.F. O'Connor, J.J. Struthers. 1997a. Feedlot cattle with calm temperaments have higher average daily weight gains than cattle with excitable temperaments. *Journal of Animal Science* 75:892–896.

Voisinet, B.D., T. Grandin, S.F. O'Connor, J.D. Tatum and M.J. Deesing. 1997b. *Bos indicus* cross feedlot cattle with excitable temperaments have tougher meat and a higher incidence of borderline dark cutters. *Meat Science* 46:367.

Waynert, D.E., J.M. Stookey, J.M. Schwartzkopf-Gerwein, C.S. Watts, C.S. Waltz. 1999. Response of beef cattle to noise during handling. *Applied Animal Behaviour Science* 62:27–42.

8
Replacement Heifers

R. L. Larson

INTRODUCTION: DEVELOPING HEIFERS THAT CALVE AT 24 MONTHS

Replacement heifer development is a critically important area in which veterinarians should offer production medicine advice to their beef-producing clients, because of the importance of having a high percentage of heifers pregnant to calve by 24 months of age and then to retain a high percentage of replacement heifers as pregnant mature cows in subsequent pregnancies. In a survey conducted by the National Animal Health Monitoring System, Cow/Calf Health and Productivity Audit, of 35 farms that completed a standardized performance analysis, only 25% of the farms that were profitable calved their heifers at greater than 24 months of age, whereas 50% of unprofitable farms calved their heifers at greater than 24 months of age (U.S. Department of Agriculture 1996). Development costs for preparing a heifer to calve at 24 months of age average about 31% of her lifetime expenses, whereas delaying first calving to 30 or 36 months increases the cost of development to 42% or 46% of her lifetime costs, even if her lifetime productivity is extended by 6–12 months (R.L. Larson and V.L. Pierce, unpublished data). Because of the high fixed and maintenance costs required for cattle production, decreasing the percentage of the herd that is producing income and increasing the number of nonproductive animals by delaying first calving is not likely to improve economic return and should not be recommended without a comprehensive analysis of available options.

Not only does the group of replacement heifers need to calve at a mean of 24 months, but the distribution of calving should result in most if not all of the heifers calving early in the first calving season (Lesmeister et al. 1973). To reach this goal, a producer's heifer development program should result in most heifers in the replacement pool reaching puberty at least 42 days before the start of breeding because the percentage conceiving at first service is lower on the pubertal estrus as compared to the third estrus (Byerley et al. 1987, Perry et al. 1991).

A fact that puts additional pressure on producers to have heifers reach puberty at a young age is that many producers breed heifers 3–4 weeks earlier than they breed the mature cow herd. The stress of calving is greater on heifers than on older cows and is more likely to be accompanied by calving difficulty. Thus, breeding replacement heifers essentially one heat cycle earlier than breeding the mature cows allows the producer to concentrate on the heifers at calving. In addition, the length of time from calving to the resumption of cycling is longer in heifers than in cows (Short et al. 1990). Therefore, calving heifers earlier gives the heifers the extra time they need to return to estrus and be cycling at the start of the subsequent breeding season. A heifer development program that is designed to start breeding replacements 28 days earlier than the mature herd, and that strives to have a high percentage of heifers reaching puberty 42 days before the start of breeding, needs to have the group reaching puberty by 12.5 months of age.

Because one goal of proper heifer development is to improve second parity pregnancy percentage, a beef producer may ask what the effect of higher pregnancy percentages during the second breeding season on costs and income is. Table 8.1 displays the effect of changing pregnancy percentage for first-calf heifers in five–percentage point increments on the percentage of the herd that must be replaced each year and on the average age of the herd. In general, given the assumptions in the table, for every five–percentage point improvement in first-calf heifer pregnancy percentage, the number of replacements needed for the

Table 8.1. Effect of second parity pregnancy percentage on herd replacement percentage and average cow age at the start of the breeding season.

	Scenario 1	Scenario 2
Pregnancy percentage for third parity	90%	88%
Pregnancy percentage for fourth through eighth parity	95%	92%
Pregnancy percentage for ninth through eleventh parity	90%	88%
Pregnant replacements needed, [percentage of mature herd; second parity and older (average age of cow herd, in years)]		
Second parity pregnancy percentage		
95	14.1 (6.08)	15.7 (5.87)
90	14.9 (6.09)	16.6 (5.88)
85	15.8 (6.10)	17.6 (5.78)
80	16.7 (6.11)	18.7 (5.90)
75	17.9 (6.12)	19.9 (5.91)
70	19.1 (6.13)	21.4 (5.92)

Assumptions: All cows greater than eleventh parity (13 years) are culled, all replacements are pregnant

herd decreases by about one percentage point, and average cow age increases by 0.01 years.

HEIFER SELECTION

CONFORMATION

Selection begins at birth. Heifer calves from early-maturing cows requiring minimal nutritional supplementation to conceive early in the calving season should be identified as possible replacements. These heifers should be from dams that have excellent udder, foot, and leg conformation. Structural correctness is critical in female selection because of the importance of longevity on cowherd efficiency and profitability (Rogers 1972, Tanida et al. 1988, Arthur et al. 1993).

BREED TYPE

Although great differences in fertility and growth occur within breeds, there are differences among breeds of beef cattle that should be considered when selecting replacement heifers (Cundiff 1986). Table 8.2 summarizes expected differences between breeds with regard to mature size, milking ability, and age at puberty. Whether examining within-breed or across-breed differences, producers should concentrate on matching the metabolic demands of the mature cows with available forage and environment. Mature cow size and milking ability are important considerations in matching breed and type to production environment. Producers should choose breeds and biological types that will optimize milk production without sacrificing reproductive efficiency or increasing nutritional requirements above those provided by available grazed forages.

In general, faster-gaining breeds that mature at a larger size (e.g., Charolais, Chianina) reach puberty at an older age than do slower-gaining breeds with a smaller mature size (e.g., Hereford, Angus; Martin et al. 1992). Researchers have also shown that breeds selected for milk production (e.g., Gelbvieh, Brown Swiss, Simmental, Braunvieh, Red Poll, Pinzgauer) reach puberty at younger ages than do breeds of similar size not selected for milk production (e.g., Charolais, Chianina, Limousin, Hereford; Gregory et al. 1991, Martin et al. 1992).

Researchers have also found that *Bos indicus* (Brahman-derivative) breeds and breed crosses are older at puberty than are British-breed heifers (Gregory et al. 1979, Stewart et al. 1980, Martin et al. 1992). British-breed heifers reach puberty at lighter weights than do Brahman × British heifers (Reynolds et al. 1963, Gregory et al. 1979). However, once *Bos indicus* heifers reach puberty, their percentage conceiving does not differ from that of *Bos taurus* heifers. In addition, *Bos indicus* cows have been shown to have longevity that is greater compared to

Table 8.2. Breed characteristics for mature size, milk production, and age at puberty.

Breed	Mature Size	Milk Production	Age at Puberty
Hereford-Angus	Small	Low	Moderate
Red Poll	Small	Moderate	Early
Devon	Small	Low	Moderate
Tarentaise	Moderate	Moderate	Early
Pinzgauer	Moderate	Moderate	Early
Brangus	Moderate	Low	Late
Santa Gertrudis	Moderate	Low	Late
Gelbvieh	Large	High	Early
Simmental	Very large	High	Moderate
Maine-Anjou	Very large	Moderate	Moderate
Limousin	Moderate	Very low	Late
Charolais	Very large	Very low	Late
Chianina	Very large	Very low	Late

Adapted from Cundiff 1986.

that of purebred *Bos taurus* cows (Rohrer et al. 1988, Bailey 1991). Therefore, the slow onset of puberty seen in *Bos indicus* heifers does not extend to decreased fertility as cows.

For commercial operators, crossbred heifers should be preferred because of their inherent hybrid vigor and greater fertility (Wiltbank et al. 1966, Cundiff et al. 1974, Steffan et al. 1985), longevity (Rohrer et al. 1988, Bailey 1991, Núñez-Dominguez et al. 1991, Arthur et al. 1993), and lifetime production (Cundiff et al. 1992). In the U.S. Gulf Coast and in other less temperate environments, some influence from Brahman, Brahman-derivative (e.g., Beefmaster, Brangus), or other heat-tolerant (e.g., Senepol, Tuli) cattle may be needed for heat tolerance and parasite resistance. Care should be taken to limit the percentage of Brahman in the feeder cattle progeny to 25% to maintain acceptable meat tenderness (Wheeler et al. 1994, Sherbeck et al. 1995).

Expected Progeny Differences

Expected progeny differences (EPDs) are a prediction of the transmitting ability of a parent animal, or how a bull's or cow's progeny will compare to another animal's progeny for various traits (Pollack 1992). EPDs allow producers to make valid comparisons among purebred animals of the same breed raised in different herds, and even under differing environmental and management conditions. The traits measured vary slightly among breeds but generally include birth weight, weaning weight, yearling weight, and a prediction of daughter's milking ability. Some breeds include a measurement for calving ease. EPDs are expressed in units of the trait of interest. For example, birth weight EPD is expressed in pounds of birth weight, and milk EPD is expressed in pounds of weaning weight resulting from milk production of the dam. Some breeds have expanded their evaluation programs to include traits such as scrotal circumference, mature size, and carcass characteristics.

Calculation of EPDs considers the performance data of the animal, its relatives, and its offspring as compared to other members of those animals' contemporary groups. As the amount of information on an animal and its relatives increases, the accuracy of these predictions also increases. Older bulls used for artificial insemination (AI) in many herds for several years have the most accurate EPDs. Young bulls from sires with low-accuracy EPDs can less reliably be compared to other bulls. However, these low-accuracy EPDs are still much more useful for across-herd comparisons than are adjusted weights or ratios.

Producers should use EPDs to select sires that will add the optimum level of growth, milk production, and other economically important traits. Scrotal circumference EPD, in particular, may be of special interest when selecting sires of replacement heifers. A valuable objective measurement of a bull's ability to produce daughters that reach puberty at a young age is the bull's yearling scrotal circumference. Scrotal circumference is a moderately to highly heritable

Table 8.3. Mature weights and target-weights for heifers to reach puberty for different frame sizes.

Frame Score	Expected Mature Weight (lb)[a]	55% of Mature Weight (lb)	60% of Mature Weight (lb)	65% of Mature Weight (lb)
2	953	524	572	619
3	1027	565	616	668
4	1100	605	660	715
5	1173	645	704	762
6	1247	686	748	811
7	1320	726	792	858

[a]Mature weights are from Fox et al. 1992.

trait (Bourdon and Brinks 1986, Smith et al. 1989, Kriese et al. 1991, Keeton et al. 1996). Heifers sired by bulls with high scrotal circumference EPD have been shown to reach puberty at significantly earlier ages than do daughters of bulls with low scrotal circumference EPD (Moser et al. 1996).

Geneticists are finding that the heritability of reproductive traits is higher than previously assumed ($h^2 \approx 0.20$) and that using EPDs for selection for heifer fertility will allow herds to make genetic progress toward greater pregnancy proportions for heifers (Evans et al. 1999, Doyle et al. 2000). In response to this information, some breed associations are either currently reporting or plan to report an EPD for heifer pregnancy percentage.

Selection Criteria at Weaning

A rigorous selection standard should be set at weaning time for prospective replacements based on available records and visual appraisal. Complete records of calf, dam, and sire performance are ideal; however, selection pressure can be applied to the herd simply by knowing a potential replacement's weaning weight, week of birth, and dam's identity. Heifers identified at birth as unsuitable replacements because of either sire or dam shortcomings should not be allowed in the selection pool.

Producers should select heifers born early in the calving season, as older females are more likely to have reached puberty by the start of the breeding season and consequently have a higher pregnancy percentage than do heifers born late in the calving season (Bergman and Hohenboken 1992). The rate of gain needed to reach the target weight that coincides with puberty by the start of the breeding season is less for older heifers as compared with that for younger calves in the same herd. These older calves thus allow greater feeding and management flexibility than do lighter, younger heifers.

Selection Criteria at Yearling Age

Yearling Weight

The target-weight concept (Dziuk and Bellows 1983, Wiltbank et al. 1985) is based on the fact that *Bos taurus* breed heifers such as Angus, Charolais, or Limousin are expected to reach puberty at about 60% of mature weight. Dual-purpose breed heifers such as Braunvieh, Gelbvieh, or Red Poll tend to reach puberty at about 55% of mature weight. *Bos indicus* heifers, most commonly Brahma or Brahma-cross, are older and heavier at puberty than are the other beef breeds, at about 65% of mature weight (Laster et al. 1972, 1976, 1979, Stewart et al. 1980, Sacco et al. 1987). One can determine the target weight for heifers by knowing the average mature weight of the cow herd, or by knowing the frame score and predicting mature weight (Table 8.3). Yearling weight should approach the target weight to have a high percentage of a group of heifers pubertal by the start of the breeding season.

Reproductive Tract Scores

Onset of puberty can be determined fairly closely in a laboratory setting by measuring blood progesterone levels from samples taken every 10 days. Of course, this method is not practical for production herds. Another method for determining onset of puberty is the reproductive tract scoring system, developed to subjectively classify pubertal status using size of the uterus and ovaries as estimated by palpation per rectum (Anderson et al. 1991). The system assigns a score to each heifer using a 5-point scale, where a score of 1 is considered an immature tract and 5 is considered a cycling tract (Table 8.4).

A reproductive tract score (RTS) of 1 is used to describe heifers with infantile reproductive tracts that are estimated to not be near the onset of puberty when palpated. These heifers have small, flaccid tracts and small ovaries with no significant structures. Heifers

Table 8.4. Reproductive tract score.

Score	Uterine Horns	Ovarian Length (mm)	Ovarian Height (mm)	Ovarian TCHWidth (mm)	Ovarian Structures
1	Immature, <20 mm diameter, no tone	15	10	8	No palpable follicles
2	20–25 mm diameter, no tone	18	12	10	8-mm follicles
3	25–30 mm diameter, slight tone	22	15	10	8–10-mm follicles
4	30 mm diameter, good tone	30	16	12	>10-mm follicles, corpus luteum possible
5	>30 mm diameter, good tone, erect	>32	20	15	>10-mm follicles, corpus luteum present

Source: Anderson et al. 1991

assigned a RTS of 1 are assumed to be either too young to fit into the breeding season being planned or too light to reach their target weight and not able to express their genetic potential for reaching puberty. Heifers assigned an RTS of 2 have slightly larger uterine diameter, but tone is still lacking and the ovaries have very small follicles. Heifers described as having an RTS of 3 have some uterine tone and larger uterine diameter than do heifers with more immature scores. These heifers are close to cycling and many will begin cycling within 6 weeks. Heifers assigned a score of either 4 or 5 are considered to be cycling, as indicated by good uterine tone and size and by easily palpable ovarian structures. A RTS of 4 is assigned to heifers that, although they have large follicles present, do not have a palpable corpus luteum (CL), because either they are in their pubertal cycle or they are in a stage of the estrous cycle in which a CL is absent. Heifers with an RTS of 5 are similar in uterine and ovarian size, tone, and structure when palpated per rectum as RTS 4 heifers, except that in heifers with a RTS of 5, a CL is palpable.

Studies have reported sensitivity of ovarian palpation per rectum for the presence of corpora lutea to be between 70% and 90%. Specificity has been reported to be between 50% and 84% (Boyd and Munro 1979, Watson and Munro 1980, Mortimer et al. 1983, Ott et al. 1986, Rosenkrans and Hardin, 2003). False positives may have several explanations. A developing CL may be palpable between days 1 and 4 of the estrous cycle and may be mistaken for a mature CL, though not yet producing large quantities of progesterone (Hansel et al. 1973). Regressing corpora lutea may be palpable well into the next cycle, even though they cease to produce progesterone beyond day 17 of the estrous cycle (Hansel et al. 1973). Kelton et al. (1991) suggested that the high false-negative percentage may be caused by the difficulty in palpating luteal tissue deeply embedded in ovarian stroma, or by small, progesterone-producing corpora lutea being mistaken as atretic (recognizing that there is no correlation between size of CL and level of progesterone secretions; Watson and Munro 1980).

Rosenkrans and Hardin (2003) report that results achieved by the RTS system are repeatable both within (same palpator evaluating the same heifer twice a few hours apart) and between palpators. Substantial agreement ($\kappa = 0.6$–0.8) was found within palpator, and moderate agreement ($\kappa = 0.4$–0.6) was found between palpators in determining individual tract scores. By convention, κ values of 0.0–0.2 equal slight, 0.2–0.4 fair, 0.4–0.6 moderate, 0.6–0.8 substantial, and 0.8–1.0 almost perfect agreement between tests (Sackett 1992).

The Rosenkrans study demonstrated that the RTS system can be used as a screening test for herds. However, because the negative predictive value of transrectal reproductive tract palpation for classifying pubertal status appears to be 70%–80%, then 20%–30% of the heifers classified as prepubertal may actually be pubertal. The positive predictive value of reproductive tract classification to characterize puberty appears to be 75%–80%. Thus 20%–25% of heifers classified as pubertal would actually be prepubertal. The "cost" of false-positive and false-negative classification for an individual operation may dictate the value of using the RTS system as a basis for the culling of individual heifers.

The scores assigned with the RTS system are retrospectively correlated with reproductive performance of yearling heifers, especially for pregnancy percentage both to synchronized breeding and to pregnancy percentage at the end of the breeding season. Heifers with more mature reproductive tracts as yearlings had a higher pregnancy percentage and calved earlier (Patterson and Bullock 1995).

Heifers should be evaluated for tract score about 6–8 weeks before the breeding season. If deficiencies are found, management changes instituted this far ahead of the breeding season can result in an increased number of heifers reaching puberty by the start of the breeding season. If the heifers are evaluated too far ahead of the breeding season (more than 8 weeks), the heifers are likely to be young and to therefore have lower tract scores than what truly reflect their potential to reach puberty before the breeding season.

If a low percentage of heifers are cycling at the time of RTS evaluation and many of the heifers are scored as having a RTS of 2, management changes must be instituted immediately. These changes may include increasing the plane of nutrition so that increased weight gain will allow the heifers to reach their target weight by the start of the breeding season, increasing the plane of nutrition and delaying the start of the breeding season by several weeks, holding the heifers over to breed 6 months later to calve in the fall (for spring-calving herds), and marketing the heifers for feeder cattle and finding another source of replacements.

Selection Criteria at 1 Year of Age: Pelvic Area Measurement

The use of pelvic measurement at 1 year of age as a tool to decrease the incidence of dystocia has been described extensively since the early 1980s (Neville et al. 1978, Holtzer and Schlote 1984, Deutscher 1985). Veterinarians have used pelvic area measurements of yearlings because the major cause of dystocia is a disproportionately large calf compared with the heifer's pelvic area. The correlation between yearling and 2-year-old pelvic areas is 0.70 (Neville et al. 1978); therefore, measuring a heifer's pelvic area as a yearling is beneficial for predicting pelvic size at the time of parturition. Pelvic area is moderately to highly heritable (0.4–0.61), so after a few years of measuring replacement heifers and bulls used to produce replacements, producers can increase average pelvic size of the herd (Benyshek and Little 1982, Holtzer and Schlote 1984, Morrison et al. 1984).

Critics of using pelvic area measurements to decrease dystocia point out that pelvic area is also positively correlated with mature cow size and calf birth weight (Laster 1974, Price and Wiltbank 1978). If producers place selection pressure on heifers for pelvic area by selecting for increasingly larger pelvic area, calf birth weight will also increase, and the occurrence of dystocia is not likely to decrease (Basarab et al. 1993). A number of researchers have shown that selection based on pelvic area alone did not significantly reduce the incidence of dystocia in groups of heifers (Naazie et al. 1989, Van Donkersgoed et al. 1990, Whittier et al. 1994).

Rather than using pelvic area measurement to select for maximum pelvic size, it should be used to set a minimum pelvic size (such as 150 cm^2 at a year of age) as a culling criterion without assigning preference for heifers that exceed the minimum. In addition, by including mature weight as a selection criterion, heifers with a genetic predisposition for small pelvic area are culled without increasing mature size.

Selection Criteria After the Breeding Season

The final culling of prospective replacement heifers is done once pregnancy status is determined, soon after the end of the breeding season. By selecting only those heifers that conceive by a proven AI sire or by natural service during a short breeding season, producers can be assured of selecting for females that reach puberty at a young age and that conceive early in the breeding season. Lesmeister et al. (1973) showed that heifers that conceive early in their first breeding season have greater lifetime productivity than do their counterparts that conceive later in their first breeding season.

HEIFER HEALTH PROGRAM

Biosecurity is the attempt to keep infectious agents (bacteria, virus, fungi, parasites, and so forth) away from a herd. One aspect of biosecurity is a vaccination program that improves the immunity of cattle against the infectious agents they may contact. Not all diseases of cattle have commercial vaccines available, and no vaccine is completely effective at preventing disease in all situations. Therefore, other aspects of disease prevention and biosecurity are at least as important as a vaccination program. A vaccination program should be tailored for specific risk factors and then rigorously applied to the herd.

Commercial vaccines are not available for all pregnancy-wasting infectious agents. Other diseases have vaccines manufactured for their control, but these vaccines are not adequately efficacious or a primary concern for a particular area or herd, and thus are not always worth giving. For most beef herds, the potential list of diseases in a vaccination program would include brucellosis, vibriosis (campylobacteriosis), leptospirosis, infectious bovine rhinotracheitis (IBR), and bovine viral diarrhea (BVD). Other diseases for which vaccines are available include *Haemophilus somnus* and trichomoniasis.

As brucellosis is a zoonotic disease, its control in animals is especially important to the human population. *Brucella abortus* strain RB51 vaccine is a live bacterial product and confers long-term, cell-mediated protection in healthy animals vaccinated properly. For many areas of the country, brucellosis has been eradicated, and in those areas, many herds are no longer using official calfhood vaccination. Whether or not to continue with a brucellosis vaccination program should be determined only after considering interstate movement, risk of exposure, and legal responsibility.

Commercially available vaccines exist for two viral pregnancy-wasting diseases (IBR and BVD). To decrease the risk of pregnancy wastage from these viral diseases, nonpregnant heifers should be given modified live vaccines two or more times after weaning to 6 weeks before breeding. Although modified live IBR/BVD vaccines do not require a booster to induce a protective response, it is recommended that vaccinations be repeated two or more times because it is not possible to know when maternal antibody interference with active immunization wanes or whether nutritional or host factors interfering with immunization are present. Multiple vaccinations allow the maximum number of heifers to develop active immunity to the vaccination. An open herd with a high level of risk may benefit from having higher levels of circulating IgG subsequent to annual IBR/BVD booster immunizations.

Leptospirosis is a zoonotic bacterial disease that causes pregnancy wastage primarily in the last trimester of gestation. Leptospiral organisms cause latent infection in the kidneys of host animals, and the organisms are excreted in urine for a variable time depending on serovar and age of the host. Leptospires survive in wet environments for up to 30 days. Infection of susceptible cattle occurs through mucous membranes and abraded or water-softened skin, or by sexual contact. *Leptospira* sp has many serovars, grouped into 19 serogroups. Each serovar is adapted to a particular maintenance host, although they can each cause disease in any mammalian species.

In the United States, *L. borgpetersenii* serovar hardjo type hardjo-bovis (formerly *L. interrogans* hardjo hardjo-bovis) has a "maintenance host" relationship with cattle. A maintenance host relationship is characterized by high susceptibility to infection, endemic transmission within the host species, relatively low pathogenicity of the serovar for its host, tendency to cause chronic rather than acute disease, and low efficacy of vaccination for prevention of infection (Prescott 1993). Infertility may follow localization of leptospires in the uterus and oviduct of maintenance host hardjo carriers. Historically, vaccination against hardjo infection with vaccines available in the United States (*L. interrogans* serovar hardjo type hardjoprajitno) did not prevent kidney establishment, urinary shedding, or fetal infection after conjunctival infection with type hardjo-bovis (Bolin et al. 1989). In contrast, a monovalent *L. borgpetersenii* serovar hardjo hardjo-bovis vaccine (recently introduced to the United States) protected previously seronegative heifers against renal colonization and urinary shedding when challenged 4 months after vaccination with *L. borgpetersenii* serovar hardjo hardjo-bovis (Bolin and Alt 2001).

In contrast, an "incidental host" relationship is characterized by relatively low susceptibility to infection but high pathogenicity for the host, with a tendency to cause acute, severe disease, sporadic transmission within the incidental host species and acquisition of infection from other species, and good efficacy of vaccination in preventing infection (Prescott 1993). Serovar *pomona* (*kennewicki*) is a common incidental pathogen of cattle, and the maintenance host is swine. Leptospira strains maintained by nondomestic animals such as skunks, raccoons, opossums, foxes, beavers, mice, deer, and others can infect cattle herds that are exposed to environmental contamination such as at urine-contaminated water holes.

A protocol to enhance immunity against the Leptospira organisms should include primary immunization of heifers two or three times at monthly intervals prebreeding, followed by another booster in midgestation of the first pregnancy. *Leptospira interrogans* bacterins apparently produce immunity of fairly short duration (at most, a few months) for controlling clinical disease. The length of protection in the genital tract against abortion may be even shorter than for clinical disease. Because of these limitations, annual (preferably in midgestation) or twice-annual boosters should be given. Methods other than vaccinations for reducing risk of exposure to leptospirosis should also be implemented. These would include having a closed herd and fencing cattle away from water sources that can be contaminated by other herds, swine, or nondomestic animals (Prescott 1993).

Campylobacter fetus ss *venerealis* is an infertility-inducing venereal disease causing early embryonic mortality. After transmission to susceptible females, the organism can be found initially in the vagina, cervix, uterus, and oviducts. Infection of the uterus and oviducts persists for up to 2 months, but thereafter it is progressively eliminated, and by the end of the third month, it is usually confined to the cervix and vagina (Duffy and Vaughan 1993). Management factors that minimize risk include using artificial insemination with semen from noninfected bulls, using bulls younger than 3 years of age (as they tend to be difficult to infect when exposed to the organism), treating or culling infected females, and initiating an immunization program.

Campylobacter fetus ss *venerealis* is an unusual bacterial pathogen in that infection is limited to the genital tract, and only local immunity results from infection. Systemic immune response, as indicated by antibodies against *C. fetus* in serum, is not helpful in diagnosing vibriosis because titers do not change before or after infection (Corbeil et al. 1980). Both local humoral and local cellular immune responses are responsible for clearing the organism from the uterus and oviducts following a natural infection. Once the organism is cleared from the uterus and oviducts, the female regains fertility. Immunoglobulin G is the primary Ig class active in the uterus following infection, whereas IgA predominates in the vagina. Immunoglobulin G acts to immobilize and opsonize *C. fetus*, allowing intracellular killing by neutrophils and macrophages present on the endometrium. In contrast, IgA produced in the vagina immobilizes the bacteria, preventing uterine infection, but it does not opsonize and likely blocks the opsonizing effect of IgG, thereby preventing the complete clearance of organisms from the vagina. After eliminating the uterine infection, convalescent females are resistant to further *C. fetus* colonization of the uterus, but colonization of the vagina often occurs and may persist for 6–24 months (Corbeil et al. 1980). This apparent immunity to disease, combined with a vaginal carrier state, may be an adaptation that keeps protective antibody levels high by providing constant antigenic stimulation (Corbeil et al. 1980).

Protection of a herd from *C. fetus*–induced disease by parenteral vaccination apparently violates the assumption that to create a protective immune response, the induced response should be of the same character as the natural infection. In natural infection, local humoral- and cell-mediated responses clear the organism and then confer protective immunity for 2–4 years. In contrast, parenteral vaccination induces a systemic humoral response consisting primarily of IgG_1 and IgG_2. However, this route has proven to be effective in preventing the clinical infertility syndrome, and in fact, systemic vaccination can be used to cure, as well as prevent, infection in both males (Clark et al. 1968) and females (Schurig et al. 1975, Winter et al. 1980). Because the IgG produced in response to parenteral vaccination is found in both uterine and vaginal secretions, the systemic immunity induced by the bacterin is successful in clearing both the uterine and the vaginal (carrier state) infection (Winter et al. 1980). This vaccine-induced clearance of the carrier state is contrary to the usual dogma on the use of vaccines.

To induce an immune response to vibriosis, for the primary immunization, heifers should be vaccinated two or three times at monthly intervals after they reach 6 months of age. Annual boosters should be given 30 days before each breeding season if risk of exposure is present. Because of the curative ability of the vaccination, all bulls brought into a herd should be vaccinated a minimum of twice, at monthly intervals, with the last vaccination taking place 30 days before the breeding season. If risk of exposure to carrier males or females is present, annual vaccinations should be used to boost immunity.

Haemophilus somnus can cause vulvitis, vaginitis, male and female infertility, and rarely, sporadic abortion. Transmission for the abortion syndrome is uncertain but is most likely ingestion. Data showing the ability of vaccination to protect against abortion are lacking (Little 1993), although the development of a systemic IgG_2 antibody response may provide local immunological protection in the uterus (Butt et al. 1993). Vaccination-induced protection from clinical disease is probably short-lived at best; therefore, a minimum of two primary vaccinations given at monthly intervals and at least annual boosters would be needed to provide even theoretical protection in those herds in which the disease has been demonstrated.

Trichomonas fetus infection is a protozoal venereal disease, and transmission occurs during coitus. In cows, the parasite is confined to the reproductive tract. Trichomonads produce cytotoxic factors that damage host tissue, cause inflammation of the uterus, and invade placental and fetal tissue, resulting in early embryonic death. After a variable period of infertility following the initial exposure, cows are usually able to clear the infection, although persistently infected females have been reported. On subsequent exposure to infected bulls, cows appear to be less susceptible to infection.

In bulls, trichomoniasis is asymptomatic. The organisms are located on the surfaces of the penis and penal sheath, where they cause little damage. Bulls younger than 4 years of age appear to recover spontaneously or to be refractory to infection (Kimsey 1986). Control of trichomoniasis outbreaks involves management practices including use of artificial insemination, use of only bulls younger than 4 years of age, culling females that do not conceive in a short breeding season, and continued surveillance of the herd by culturing bulls and culling carriers.

Vaccination programs for females exposed to *Trichomonas*-infected bulls would appear to be beneficial in controlling outbreaks. The program should include two vaccinations 4 weeks apart for primary vaccination, and annual boosters thereafter. Researchers have shown that although an immunization program did not prevent *Trichomonas fetus* infection, it did decrease the duration and the incidence of infection (Kvasnicka et al. 1989).

Preventing the introduction of trichomoniasis into a herd in endemic areas includes eliminating common pastures and examination of smegma samples from replacement bulls three times at weekly intervals before the start of the breeding season. Smegma samples may be placed in sterile saline and examined directly under the microscope for *Trichomonas* organisms, or they may be placed in Diamond's medium for culturing.

Because most infectious agents cannot live very long outside or off of an animal, and because most do not travel great distances through the air, a method to keep other animals and people away from a herd nearly accomplishes the goal of keeping infectious agents away. Keeping a closed herd is one method of biosecurity. A closed herd is one where no cattle enter the farm and no cattle on the farm have contact with cattle from other farms. A herd is not closed if cattle share a fence with cattle from a different farm, cattle are purchased (bulls, replacement heifers, replacement cows, stocker cattle), cattle return to the herd after being at a performance evaluation (i.e., bull test station) or show, bulls are borrowed or loaned, or cattle are transported in a vehicle that transports other cattle. Using this definition, one can agree that it is difficult (and maybe not desirable from a production standpoint) to have a completely closed herd. However, it is a good practice to keep the herd as closed as possible to minimize exposure to infectious agents.

In open herds, additions (replacement females and bulls) should only be purchased from herds about which you know the health status and that have a known, effective vaccination and disease testing and diagnosis program. Producers should avoid purchasing animals from unknown sources or animals that have been mixed with other cattle before sale. Also, additions to the herd should be isolated for at least 1 month before introduction to the resident herd. Isolated cattle should not share feeders, waterers, or airspace (distance depends on wind velocity and direction). During the isolation period, the additions should be tested for and vaccinated against transmissible diseases.

Equipment and animals other than cattle can carry infectious diseases. Rodents, birds, cats, and dogs should all be limited in their exposure to cattle. Rodents and birds are primarily a problem when cattle are confined, and professional exterminators may be needed to devise an effective control plan. Salmonellosis, cryptosporidiosis, and other diseases can be passed by dogs and cats; therefore, keeping pet animals away from cattle is an important aspect of biosecurity.

During the quarantine period, animals should be screened to identify those persistently infected with BVD by using an immunohistochemistry (immunoperoxidase) test on a skin biopsy sample or by polymerase chain reaction, virus isolation, or antigen-capture enzyme-linked immunosorbent assay of tissue. Some herds may also screen for Johne's disease and BLV.

REPLACEMENT HEIFER MANAGEMENT

Puberty

Puberty in the beef heifer is reached when she is able to express estrous behavior and ovulate a fertile oocyte. The maturing of the neuroendocrine system that induces the maturation and ovulation of the first oocyte, as well as the hormonal changes that induce the first expression of behavioral estrus, are the result of a gradual increase in gonadotropic (luteinizing hormone, or LH, and follicle stimulating hormone, or FSH) activity. This increased gonadotropic activity is caused by a decreased negative feedback of estradiol on the hypothalamic secretion of gonadotropin-releasing hormone (Niswender et al. 1984, Foster et al. 1985). The gradually increased secretion of LH initiates ovarian production of steroid hormones and gametes, resulting in follicle maturation and ovulation. The first ovulation is usually not accompanied by external indications of estrus. It is generally believed that a certain amount of progesterone is needed during a period preceding estrus to induce estrus behavior and for the following cycle to be of normal length (Dodson et al. 1988). Once the heifer has gone through a cycle with CL development or has been exposed to sufficient progesterone levels from other endogenous sources, the following cycles are normal (Gonzalez-Padilla et al. 1975b).

The onset of puberty is primarily influenced by age and weight within breed (Wiltbank et al. 1969, Oeydipe et al. 1982, Nelsen et al. 1985). Other factors can also have some influence on the onset of puberty and include exposure to bulls (Pennel et al. 1986, Roberson et al. 1991), time of year (Schillo et al. 1983), and exposure to progestagens (Gonzalez-Padilla et al. 1975a, Short et al. 1976, Spitzer 1982). Although important, weight is not the only controlling factor for the onset of puberty, as a minimum age requirement must also be reached (Nelsen et al. 1982). Age at puberty can be decreased by selecting for breeds with a younger age at puberty, selecting within a breed for younger age at puberty, or crossbreeding with another breed that has a similar or younger age at puberty.

Some studies have indicated that exposing prepubertal heifers to bulls (Pennel et al. 1986, Roberson et al. 1991) or bull urine (Izard and Vandenbergh 1982) decreased the age at puberty. However, other studies have not shown the same results (Berardinelli et al. 1978, Roberson et al. 1987). The proposed mechanism of action of bull exposure involves stimulation of the hypothalamic-pituitary axis so that LH or FSH secretion is increased (Patterson et al. 1992), as has been shown in mice (Bronson and Desjardins 1974) and ewes (Poindron et al. 1980, Martin et al. 1983). Exposure of prepubertal heifers to mature cows (Nelsen et al. 1985) or cycling heifers (Roberson et al. 1983) did not decrease age or weight at puberty.

Although cows are considered a nonseasonal animal, season does have some effects on bovine reproductive performance. Heifers born in the fall reach puberty at a younger age than do heifers born in the spring (Schillo et al. 1983). This difference may be the result of photoperiod differences during the maturation of the neuroendocrine system. Work in ovariectomized cows (Stumpf et al. 1988) and in intact and castrated bulls (Stumpf et al. 1993) indicates that either mean concentration of circulating LH or amplitude of LH pulses is increased at the spring equinox as compared with at the fall equinox. Concentrations of LH from blood samples taken from prepubertal heifers between 6 and 7 months of age were higher in the spring (fall-born heifers) than in the fall (spring-born heifers; Schillo et al. 1983). If heifers are kept in environmental chambers and exposed to spring, followed by summer, temperatures and photoperiods from 6 to 12 months of age, no difference in age at puberty exists between spring-born and fall-born heifers, supplying evidence that photoperiod differences during the second 6 months of life influence age at puberty (Schillo et al. 1983).

Progestagens

Progesterone and synthetic progestagens induce puberty in heifers, and management systems that capitalize on this result have been developed. Short et al. (1976) showed that more prepubertal heifers (8.5 months old and 249 kg) given a progesterone implant for 6 days plus an injection of estradiol-17β 24 hours after implant removal showed estrus and ovulated within 4 days than did heifers treated with estradiol-17β alone. Gonzalez-Padilla et al. (1975b) also used progesterone or norgestomet (a synthetic progestagen), in conjunction with estradiol valerate, to induce estrus in prepubertal beef heifers in a series of experiments. Gonzalez-Padilla et al. (1975b) were able to induce estrus in approximately 93% of heifers treated either with a 9-day, 6-mg norgestomet implant coupled with an injection of 3 mg norgestomet plus 5 mg estradiol valerate at the time of implant insertion, or with daily intramuscular injections of 20 mg progesterone for 4 days plus 2 mg estradiol-17β 2 days after the last progesterone injection. Pregnancy percentage ranged from 43% to 73%.

In a study by Spitzer (1982), treatment with a norgestomet implant and injectable norgestomet plus estradiol valerate induced estrus in a percentage of lightweight peripubertal heifers, but pregnancy percentages were low and a large proportion of the heifers that failed to conceive to the induced estrus failed to continue cycling regularly. Short et al. (1976) reported that although treatment with a norgestomet implant and injectable norgestomet plus estradiol valerate resulted in estrous expression in both prepubertal heifers and cycling heifers, the pregnancy proportion to the treatment-induced estrus was lower for the prepubertal heifers. The authors speculated that some of the estrous behavior may have been induced by the 5-mg estradiol valerate injection because estrus can be induced in ovariectomized cows with as little as 500 μg estradiol-17β. This was also confirmed when 55%–60% of ovariectomized cows or heifers treated with the same protocol responded by exhibiting estrus behavior, obviously without concurrent development or ovulation of a fertile oocyte (McGuire et al. 1990, Larson and Kiracofe 1995).

A progesterone-impregnated intravaginal device known as a CIDR is now available in the United States. The CIDR is a T-shaped device with a nylon spine covered by a progesterone-impregnated silicone skin. On insertion, blood progesterone concentration rises rapidly. Maximal concentration is reached within an hour. Progesterone concentration is maintained at a relatively constant level during the 7 days the insert is in the vagina. On removal of the insert, progesterone concentration in the bloodstream drops quickly. This product is labeled for use to cause suckled beef cows to show estrus sooner after calving and will cause replacement heifers to express heat at a younger age and weight than do nontreated animals. Research using CIDR in beef heifers and cows was conducted over several years at a number of universities; those trials indicate that use of CIDR did not decrease fertility compared with untreated females and that it was successful in inducing almost 50% of noncycling females in the herds tested to show signs of a fertile heat following removal of the CIDR (Lucy et al. 2001).

Another commercially available synthetic progestagen is melengestrol acetate (MGA). Research has also demonstrated the ability of MGA to induce puberty in heifers, especially heifers near the age and weight requirements for spontaneous induction of puberty. Percentage pregnant at first service for heifers that attained puberty while being treated with MGA (administered orally for 14 days followed by prostaglandin $F_{2\alpha}$ intramuscular injection 17 days after the final day of MGA feeding) was not different from that of control heifers that attained puberty during the same period (Jaeger et al. 1992).

Ionophores

Ionophores were originally cleared for use to improve the feed efficiency of feedlot cattle on high-concentrate diets (Raun et al. 1976) and to improve pasture cattle gains (Oliver 1975, Potter et al. 1976). Now ionophores are cleared for use in replacement heifers. Inclusion of ionophores in heifer diets has been shown to increase the number of heifers that had reached puberty by the start of the breeding season (Moseley et al. 1977), decrease the age at puberty (Moseley et al. 1982, Sprott et al. 1988, Purvis and Whittier 1996), decrease the weight at puberty (Purvis and Whittier 1996), increase the corpora lutea weight, and increase the amount of progesterone produced (Bushmich et al. 1980). The decrease in age at puberty was independent of improved average daily gain and increased body weight. Moseley et al. (1982) speculate that changes in ruminal fermentation patterns to favor propionic acid production produce an endocrine response that influences the mechanisms regulating puberty (Purvis and Whittier 1996).

Growth Implants

Implanting suckling calves with anabolic growth promoters is a highly profitable practice used by cow/calf operators to increase weaning weights of calves intended for slaughter. Research on the effect of implanting potential replacement heifers on percentage cycling and conceiving has been somewhat inconsistent, with results ranging from negative (Rusk et al. 1992, King et al. 1995) to positive (Whittier et al. 1991). When nutritional levels are adequate to sustain the anabolic effects on weight gain, implants have been reported to have no negative effects (Deutscher 1991, Larson and Corah 1995). Negative results were most likely to occur when implants were placed at birth, or when heifers were implanted with anabolic agents three times between birth and puberty (Rusk et al. 1992, King et al. 1995).

However, a paper revealed possible negative effects of a progesterone and estradiol implant that is approved for use in heifers that are intended to be retained as heifers. Bartol et al. (1995) implanted some heifers according to label directions at 45 days of age. Other heifers in the experiment were implanted at birth or at 21 days of age or remained unimplanted controls. All implanted heifers had reduced uterine weight, decreased myometrial area, decreased endometrial area, and reduced endometrial gland density compared with the control heifers. The effects were greatest in heifers that were implanted at birth.

Numerous studies have shown that heifers implanted with anabolic growth promoters at 2–3 months of age have a larger pelvic area as yearlings than do controls without implants (Lawrence et al. 1985, Carpenter and Sprott 1991, Whittier et al. 1991, Rusk et al. 1992, Hancock et al. 1994, Larson and Corah 1995). This increase ranged from 10 to 29 cm^2. A few studies have followed the heifers to calving at 2 years of age to determine whether the larger pelvic areas were maintained (Whittier et al. 1991, Rusk et al. 1992, Larson and Corah 1995, Hancock et al. 1994). These studies showed that much of the advantage for implanted heifers seen as yearlings was lost by the time they were ready to calve; the advantage was only 3–9 cm^2 when compared with controls with no implants.

Some implants are approved for use in suckling heifers that are to be retained as replacements, but the author does not recommend implanting calves that can be identified at a young age as likely replacements. There are no benefits to implanting replacement heifers because producers do not benefit economically from maximum growth. Instead, economic benefits from replacement heifers occur because of early onset of puberty, high fertility, and a long productive life in the cowherd.

Anthelmintic Treatment

Internal parasites can have a negative effect on virtually all production characteristics of beef cattle, including gains from weaning through the first pregnancy (Williams et al. 1989, Wohlgemuth et al. 1989, Larson et al. 1995). The presence of internal parasites affects nutrient utilization and possibly alters metabolism in infected animals. Minimizing the negative effect of internal parasites with the use of broad-spectrum anthelmintics that are able to kill inhibited stages of *Ostertagia ostertagi* improves the efficiency of gain for replacement heifers. Improved gain increases body weight, and hence the number of heifers cycling at the beginning of the breeding season (Larson et al. 1995, Purvis and Whittier 1996, Lacau-Mengido et al. 2000). It is interesting to note, however, that improvements in reproductive response in replacement heifers treated with anthelmintics may not be solely caused by these heifers reaching target weights faster than nontreated heifers. It is noteworthy that Larson et al. (1995) found that the correlations between weight gain or prebreeding heifer weight and puberty in ivermectin-treated heifers approached zero, indicating that the gain response does not fully explain the earlier onset of puberty. Purvis and Whittier (1996) also showed that decreased age and weight at puberty in ivermectin-treated heifers compared to controls was not caused by improved average daily gains. Therefore, other pathways affecting onset of puberty, in addition to weight gain, are being stimulated because of treatment with ivermectin and possibly other anthelmintics.

MATERNAL NUTRITION AND SUBSEQUENT PERFORMANCE

Breeding through Midgestation

Overfeeding protein during the breeding season and early gestation, particularly if inadequate energy is supplied to the rumen, may be associated with a decline in fertility (Canfield et al. 1990, Elrod and Butler 1993). The mechanism for this decline may work by decreasing uterine pH during the luteal phase in cattle fed high levels of degradable protein (Elrod and Butler 1993). The combination of highly digestible protein and low energy concentrations on an as-fed basis in early growth, cool-season grasses may explain the lower-than-expected fertility seen in females placed on such pastures near the time of breeding.

The target-weight concept can be applied when planning the nutritional requirements through pregnancy (Figure 8.1). A heifer should weigh 80%–85% of her mature weight at the time of calving when she is 2 years old. By the following calculation, one finds that heifers need to gain between 1.3 and 1.5 lb per day depending on mature weight. Heifers that will mature at heavier weights will need to have a higher daily gain.

Energy and protein requirements (National Research Council estimates) for growing heifers during midgestation should be used to formulate rations that allow heifers to maintain body condition and progress toward target calving weight. As long as the environmental temperature remains above the critical point, and the levels of pathogen exposure, mud, or other stressors remain low, the nutrient requirements for this period of time during heifer development can often be met with fairly low cost

Target weight at parturition = 85% of mature weight
Weight of pregnancy = 2× calf birth weight
Wt. gain during pregnancy = (Target wt. at parturition + wt. of pregnancy) - wt. at breeding
Daily gain during pregnancy = Weight gain during pregnancy ÷ 283

- For a heifer with a mature weight of 1100 lb having a 75 lb. calf, the equation would be as follows:
  [(935 + 150)- 660] ÷ 283 = 1.5 lb. per day daily gain during pregnancy.

Figure 8.1. Calculation of target weight of heifers.

Table 8.5. Estimated critical temperatures for cattle.

Coat Description	Critical Temperature (°F)
Summer coat	59
Fall coat	45
Winter coat	32
Heavy winter coat	18
Wet coat	59

forages. For example, for an 885-lb heifer (body condition score [BCS] 6, with a mature weight of around 1100 lb) on day 130 of gestation, the requirements are about 9.9 Mcal/day and 476 g metabolizable protein. We can expect her to consume 2.0%–2.5% of her body weight on a dry-matter basis of an average-quality brome hay (net energy for maintenance = 1.18 Mcal/kg; net energy for gain = 0.61 Mcal/kg; metabolizable protein = 7.34%). Consuming 17 lb (dry-matter basis) of this hay will leave a deficit of 0.97 Mcal net energy for gain and is adequate for protein. One and one-half pounds of cracked corn (as fed) will supply the needed energy.

If, however, the wind chill is below the critical temperature (Table 8.5), energy requirements increase about 1% for every degree Fahrenheit drop in wind chill. Protein requirements are not appreciably increased because of temperatures below the critical point. For example, if the same 885-lb, midgestation heifer is subjected to several October days of rain and 40°F temperature, her NE requirement increases by 19% to 11.7 Mcal/day during the period of inclement weather. Because of the importance of environmental temperature, the expected environmental conditions for a locality should be considered when planning rations and purchasing supplements.

Economic considerations may favor limited weight gain or even weight loss during midgestation in mature beef cows. As long as deficits are made up before calving, midgestation weight loss is acceptable for mature cows, but because of higher nutrient demands of heifers, little or no decrease in body condition should occur during the first pregnancy.

Last 60 Days of Gestation

The nutritional demands of pregnancy increase as gestation progresses. These demands occur not only because of fetal growth but also because of uterine/placental growth and metabolism involved with the fetal/maternal interaction and exchange of nutrients and waste.

Heifers calving at a BCS of 4, 5, or 6 had calves with progressively heavier birth weights, but the dystocia score was not influenced by BCS at calving (Spitzer et al. 1995). Heifers with greater weight gains prepartum had calves with heavier actual and 205-day adjusted weaning weights than did heifers with moderate weight gains (Spitzer et al. 1995). Greater BCS at calving resulted in more heifers in estrus and more heifers pregnant by 40 and 60 days of the subsequent breeding season (Spitzer et al. 1995). Thin females should be fed at levels during the last third of their pregnancy to achieve a targeted BCS greater than 6 at calving, whereas those in moderate-high to high-body condition at 90 days prepartum should be fed at levels intended to maintain body reserves.

When body weight or condition loss occurred during the middle third of pregnancy, increased nutrient intake 1–3 months before calving substantially improved pregnancy percentage when compared with cows that continued to lose weight (Selk et al. 1988). However, cows that maintained weight throughout the last half of pregnancy had higher pregnancy proportions than did those that lost weight and had to gain it back later, even though precalving BCSs were similar between the two groups (Selk et al. 1988).

Although there is disagreement over the effect of level of nutrition and BCS changes after calving in cows and heifers that calve in good condition, most research clearly demonstrates that body condition at calving is a dominating factor in postpartum fertility. Higher BCSs or greater levels of supplemental energy during late gestation improved the percentage of cows showing estrus by 60 days after calving and subsequent pregnancy proportions (Ames 1985, Selk et al. 1988). Heifers that calve in poor body condition have lighter–birth weight calves (Spitzer et al. 1995), a longer postpartum interval to return to estrus, and a lower pregnancy percentage during the following breeding season (Selk et al. 1988, Spitzer et al. 1995).

First 80 Days of Lactation

During the first 80–100 days following parturition, the heifer must continue to grow at about 0.5 lb per day, support lactation for a suckling calf, resume estrous cyclicity, and conceive for her second pregnancy. The maintenance requirement for lactating heifers averages about 20% higher than that for nonlactating heifers, but maintenance requirements are greatly affected by milk production potential. In beef cattle, peak lactation occurs at approximately 60 days postpartum, and maximum yield has been reported to range from 9 to 30 lb/day (National Research Council 1996).

It is clear that energy and protein requirements postcalving greatly exceed those of midgestation heifers, and even late-gestation heifers. These higher demands make it difficult to add body condition to heifers once they begin lactation. Because postcalving condition score and energy balance control ovulation (Wright et al. 1992) and condition scores of 6 or greater are required for high conception percentages in heifers (DeRouen et al. 1994) both body condition at calving and level of nutrition postpartum are critical control points (CCPs) affecting pregnancy.

Marston et al. (1995) illustrate the importance of adequate body condition at calving, in that postpartum supplementation of energy or protein after calving had little effect on subsequent pregnancy percentage. The period of time between calving and rebreeding is fairly short (only 82 days to maintain a 365-day calving interval), and during this time the cow has her highest nutritional demand as a result of lactation. Because of these factors, weight gain or body condition increase is difficult to achieve in the early postpartum cow. Lalman et al. (2000) found that feeding thin heifers high-energy diets postpartum reduced the negative effects of prepartum nutrient restriction but did not completely reverse those effects. However, increasing dietary energy intake was associated with a curvilinear increase in milk yield and percentage milk fat and with a linear increase in energy available for milk production; as a consequence, a high-energy diet, rather than a moderate- or low-energy diet, was necessary to improve the energy status of thin heifers. Marston and Lusby (1995) also show that in grazing cattle, it is difficult to increase the intake of energy once protein requirements are met. For cows grazing forage or consuming poor- to average-quality hay that is deficient in protein, supplementing the diet with protein will increase dry-matter digestibility and intake and, subsequently, energy intake. However, after protein deficiencies are corrected, additional protein or energy merely replaces forage, rather than supplementing it. These findings do not decrease the importance of postpartum nutrition, they only illustrate the constraints placed on the postpartum period nutritionally if body condition is not adequate at calving.

By recognizing the importance of body condition at calving, one should not assume that if condition is adequate or good at calving that postpartum nutrition is less critical. Wiltbank et al. (1962) found that regardless of prepartum energy regimens, more cows that were fed greater amounts of energy diet after calving became pregnant than did cows fed a reduced-energy diet postpartum. Dunn et al. (1969) showed that although precalving energy level exerted a strong influence on the early postpartum anestrous period, pregnancy percentage at 120 days postcalving were directly related to postcalving energy levels. Prepartum nutrition does, to some extent, influence early postcalving ovarian function, but data from Rakestraw et al. (1986) as well as the results of Wiltbank et al. (1962) and Dunn et al. (1969) support the concept that good body condition at calving does not guarantee optimal rebreeding unless nutrition during early lactation is adequate.

HEIFER BREEDING MODEL: CCPS IN HEIFER DEVELOPMENT

A CCP can be defined as a point in the production process at which a value or values can be measured that are the direct result of previous management and that affect the success of the remaining production process. A number of CCPs are possible in heifer development (Table 8.6). Not all of these points are important as control points for every farm. How well a farm can manage earlier or later CCPs in the production process affects the necessity of measuring each of the following potential control points. These CCPs are based on a production system in which replacement heifers are moved into the replacement pool at weaning, estrous synchronization and AI are used for the first breeding opportunity when the heifer is approximately 14 months of age (followed by exposure to bulls for the remainder of a 60-day breeding season), and natural service only is used for the second breeding season, at approximately 27 months of age.

The following potential heifer development CCPs start with those closest to the end of heifer development as the author has defined it (bred for second calf) and work backward to selection of heifers for the replacement pool.

SECOND BREEDING

Pregnancy Percentage by 20-Day Periods

The overall success of the heifer development system for a farm is determined at 40 days after breeding season (late summer/early fall for spring-calving herds). The goal is to have a high percentage of heifers that were selected to enter the herd become pregnant at 13–15 months of age and to be pregnant for their

Table 8.6. Calendar for heifer development critical control points.

	Veterinary Visit or Data Collection				
Age Group	September–November	December–February	April	May	July
Weaning to first breeding	Weight, WDA, structure, EPDs of sire. Begin vaccination program.	Mid-development weight (ADG). Continue vaccination program.	BSE of heifers (RTS, PA, YW). BSE of bulls used to breed heifers. Continue vaccination program.	Estrous response to synchronization	Early palpation to determine artificial insemination pregnancy percentage
First breeding to first calving	Pregnancy determination for entire 60-day breeding season. BCS of heifers going into winter. Continue vaccination program.	Midwinter weight (ADG) and BCS	Record calving distribution and calving ease scores. BSE of bulls used to first-calf heifers. BCS of first-calf heifers going into breeding season.		
First-calf heifers and adult cows	Pregnancy determination for entire breeding season. BCS of cows going into winter.	Midwinter BCS	Record calving distribution and calving ease scores. BSE of bulls to breed cows. BCS of cows going into breeding season.		

WDA = weight per day of age, BSE = breeding soundness exam, YW = yearling weight, EPD = expected progeny differences, RTS = reproductive tract score, ADG = average daily gain, PA = pelvic area.

second calves, and also to have become pregnant early in the breeding season. A pregnancy percentage for first-calf heifers over 90%–95% is an achievable goal for farms that have met previous CCPs.

If the farm's heifer development system was meeting the goals for all CCPs before this point, but did not meet the rebreeding goals of the farm, bull fertility and the nutritional program for first-calf heifers would be areas for further investigation.

Reasons that rebreeding percentage could be below farm goals in a system in which earlier CCPs were not met could include heifers becoming pregnant late in the breeding season, and consequently not reaching the producer's goal for days between calving and rebreeding, or first-calf heifers being underconditioned going into the breeding season, thereby reducing the number that are cycling at the start of the breeding season.

BCS/Body Weight at Palpation for Pregnancy Determination

The adequacy of the nutrition/forage program on the farm for the late portion of the breeding season can probably be determined by BCS at the time of pregnancy determination if palpation is not done too late in gestation. BCS at this time is also a good indicator of whether the herd's genetic potential for growth and milk production matches forage quality and quantity. BCS at this time is not a good indicator of the nutritional program going into the previous breeding season. First-calf heifers in their first lactation should weigh at least 85% of their mature weight at this time and should be near a BCS of 5, with no individuals falling below a BCS of 4.

If earlier CCPs for body condition were met, but the heifers do not meet farm goals at this measurement, the producer needs to determine whether unusual weather caused forage production to be greatly decreased compared with its normal variation, if milk production or growth rate caused energy requirements to exceed that supplied by forage within normal yearly variation, or if the farm should plan to supplement energy given to first-calf heifers during lactation either routinely or when forage production falls below a given quantity and quality. If BCS is lower than desired, calves should be weaned and the diet should be adjusted so that condition can be added before winter and late gestation, when maintenance requirement increases.

BCS/Body Weight before the Start of the Second Breeding Season

Because body condition at the start of the breeding season (late spring for spring-calving herds) is a good predictor of breeding season success, it is important that the farm's goal be met for BCS (5–6) and body weight (82%–84% of mature weight). Adding body condition to a growing heifer that is lactating is very difficult; therefore, meeting this CCP is largely dependent on having the heifer in adequate condition at the time of calving and then providing either adequate quality and quantity of forage to meet her needs for maintenance, lactation, and growth or implementing a supplementation strategy that adds adequate calories to the forage base to meet her energy needs during early lactation.

If earlier CCPs for body weight and condition were met, but the herd failed to meet this CCP, a supplementation strategy or a change in forage quality and quantity for future groups of first-calf heifers early in their first lactation is needed. If body condition goals were not met at earlier CCPs, and energy intake did not increase to the degree necessary to replace deficient condition and maintain lactation, BCSs at this CCP are likely to be below the farm's standards.

Breeding Soundness Examination of Bulls

Only bulls that pass a Breeding Soundness Examination (BSE; given in late spring for spring-calving herds) should be considered for use in the breeding pasture. The need for a BSE of bulls is based on the fact that many prospective breeding bulls are infertile, subfertile, or unable to copulate. However, as producers evaluate the use of BSE, they should also recognize the limitations of the process. It should be remembered that the BSE reflects an animal's breeding soundness only on the date tested. A BSE does not reflect the bull's soundness in the past; neither does it definitely define the bull's ability to cause conception in the future. However, the overall effect of BSE is to eliminate many infertile bulls and to improve the genetic base for fertility within the herd.

BSEs consist of a complete physical, a scrotal measurement as an indication of testicular size, and a semen evaluation. Once the evaluator has collected all the available information, he or she can determine whether the bull is a satisfactory breeder on the day tested. The veterinarian may also give an indication of the severity of any abnormality and a prognosis for the bull's recovery and use as a breeder in the future.

First Calving

Calving Distribution by 1-, 2-, or 3-week Intervals; Calving Difficulty Scores; Pregnancy Loss between Pregnancy Determination via Palpation and Calving

Data collected at calving (in the spring for spring-calving herds) is very valuable for determining whether earlier CCPs adequately monitored and guided the farm's heifer development program. Sixty-five percent or more of the calves should be born in the first 3 weeks of the calving season, with greater than 80% being born in the first 6 weeks. The prediction of AI pregnancy percentage should be compared to the percentage of calves born in the first 2 weeks of the calving season. In addition, the percentage of heifers that are confirmed to be pregnant but that fail to calve should not exceed 2%. Calving ease scores should reflect less than 15% of heifers having dystocia, with levels exceeding that goal causing one to examine both the growth of the heifers and the birth weight EPDs of the bulls used.

Failure to have a high percentage of calves born when predicted by palpation will allow the palpator to recalibrate his or her criteria for fetal aging and to determine the stage of pregnancy at which he or she is most accurate to improve future predictions of calving date. A high gestational loss of viable pregnancies should cause the herd's veterinarian to focus on biosecurity and vaccination protocols for diseases that cause pregnancy wastage. Excessive occurrence and severity of dystocia indicate either that heifers were underdeveloped or, more likely, that calf birth weight was excessive because of genetic predisposition by either the dam or sire. Because each sire will affect many calves, accurate predictions of the sires' influence on birth weight by using EPDs is critical to avoiding excessive dystocia.

BCS/Body Weight at Start of Calving Season

Heifers should have a BCS of 6 at the start of the calving season (spring for spring-calving herds) and should weigh 80%–85% of mature weight after parturition. Failure to have adequate energy reserves will negatively affect lactation, weaning weight, and rebreeding percentage. Appropriate use of supplements after calving will be necessary to meet breeding season goals if this CCP is not met.

First Breeding

Pregnancy Percentage to First Breeding Season by 20-Day Periods

Determining pregnancy percentage for the entire breeding season (determined 40 days after breeding season in the late summer/early fall for spring-calving herds) allows producers to identify heifers that are not pregnant and to determine the best marketing plan for those animals. In addition, if more pregnant heifers are available than are needed as replacements, selection criteria can be used to market them to farms needing pregnant animals. A goal of 90%–95% of heifers in the replacement pool being pregnant at the end of the breeding season is achievable if CCPs earlier in the heifer development process are used to remove heifers that do not meet the farm's standards. If CCPs were not measured or met earlier in the heifer development process, lower pregnancy percentages should be expected.

If pregnancy percentages do not meet farm goals even though earlier CCP criteria were met, bull fertility, nutrition during the breeding season, and the presence of pregnancy-wasting disease should all be investigated. If the number of pregnant heifers available will not supply the necessary replacements for the herd, additional replacements must be located and purchased.

BCS/Body Weight at Palpation for Pregnancy Determination

Heifers should be gaining about 1.3–1.5 lb per day weigh 80%–85% of mature weight at the time of calving as 2-year-olds. Failure to meet farm goals at this CCP indicates that forage quality and quantity do not meet the nutritional demands of growing, pregnant heifers in this herd. If BCS is lower than desired, the diet should be adjusted so that condition can be added before winter and late gestation, when maintenance requirement increases. If low BCS and weight are not caused by an unusually decreased forage quality or quantity, as compared to normal annual variations, either the nutritional strategy or the genetic potential for growth for the herd should be modified.

Pregnancy Percentage to AI Breeding

This section concerns the first 40–60 days after AI breeding, which is midsummer for spring-calving herds. If heifers are synchronized and bred AI, bulls should be held out of the breeding pasture for 2 weeks following the last day of AI breeding, so that AI pregnancy percentage can be accurately determined early in gestation via fetal aging by palpation. Sixty-five percent to 70% or more of the heifers identified in estrus and bred artificially should become pregnant with AI. Failure to meet farm goals could indicate inaccurate determination of estrus, poor semen delivery by the AI technician, poor semen quality, or poor condition of the females (stress, high environmental temperature, losing weight).

Identification of AI-sired pregnancies has value if more pregnant heifers are available than are needed as replacements and if those heifers are either given preference to enter the herd or are marketed at a higher value than are natural service–sired pregnancies. By doing an early pregnancy determination, not only does one know the AI pregnancy percentage but because the bulls are still in the breeding pasture, they can be left longer than planned if the AI pregnancy percentage is less than expected, to ensure that every heifer has several opportunities to become pregnant.

Estrous Response to Synchronization System

If heifers have reached puberty and the synchronization system was applied appropriately (at the start of the breeding season; late spring for spring-calving herds), 70%–80% or more of heifers should display estrus within the time window predicted by the synchronization system. If results do not meet this goal, the percentage of heifers that are pubertal, the accuracy of estrous detection, and the success of administering the synchronization system should all be investigated. If estrous response to synchronization is poor, alternate or additional synchronization systems can be implemented, the period of estrous detection and AI can be extended, or the date for the start of the natural breeding season can be altered.

Breeding Soundness of Bulls

All bulls used to breed heifers should pass a BSE (in the late spring for spring-calving herds), and their EPDs for birth weight and or calving ease should be consistent with the farm's goals.

Breeding Soundness of Heifers: RTS, Pelvic Area, Yearling Weight

This section refers to heifers tested at 30–60 days before the start of the breeding season, which is early- to midspring for spring-calving herds. Heifers should be 55%–65% of mature weight, depending on breed, before the start of the breeding season to ensure that most animals in the group have already reached puberty. Palpation of the reproductive tract aids in determining the percentage of the group that has reached puberty and identifies individuals that have abnormal or infantile reproductive tracts. Heifers with pelvic areas that are abnormally small (when considering age, weight, and maturity) should be identified.

Sixty percent of heifers should already have reached puberty 30–60 days before the start of the breeding season. If the herd does not meet this criteria, the energy content of the diet can be increased so that weight gain is increased to the necessary level to meet target weight, the start of the breeding season can be delayed, or the breeding season can be abandoned, with the heifers either being sold as feeder cattle or being held over for a later breeding season.

Body Weight during Weaning-to-Breeding Period

This section refers to weaning, taking place in the winter for spring-calving herds. Because weight is a primary factor determining the onset of puberty, ensuring that the nutritional program is meeting average daily gain requirements for the period from weaning to breeding is critical for a successful heifer development program. If weight gain is not as projected, the energy content of the diet can be increased so that target weight will be met. In addition, the use of ionophores, progestagens, and anthelmintics will help ensure that heifers reach target weights and puberty before the start of the breeding season.

Vaccination Protocol to Enhance Herd Immunity to Pregnancy-Wasting Diseases

For most beef herds, the potential list of diseases in a vaccination program would include brucellosis, IBR, BVD, vibriosis (campylobacteriosis), and leptospirosis. Other diseases for which vaccines are available include *H. somnus* and trichomoniasis.

Weight per Day of Age at Weaning, Sire EPDs for Milk Production, Growth, and Calving Ease/Birthweight, and Structural Soundness

Successful heifer development starts with the selection of candidates for the replacement pool that are likely to reach puberty before the start of the breeding season, become pregnant early in the breeding season, have little calving difficulty, and rebreed early in the second breeding season (selected for in the fall for spring-calving herds). Because puberty is age and weight dependent, only heifers whose age and weight at weaning are compatible with being old enough and heavy enough before the start of breeding to reach puberty should be selected. EPDs for the sires and dams (when available) of individual heifers should be examined to find those heifers that are predicted to meet herd goals for mature size, growth rate, milking ability, and calving ease. Heifers with undesirable structural conformation of the feet and legs should not be included in the replacement pool, nor should heifers with a familial history of genetic defects such as vaginal prolapse be included.

If subsequent CCPs have high failure rates, reexamining the selection criteria may be necessary to identify those traits most likely to be correlated with failure later in the heifer development system.

SUMMARY

A system of CCPs to evaluate and plan intervention for heifer development is necessary to ensure that a high percentage of heifers in the replacement pool become pregnant early in their first breeding season, continue to grow adequately during gestation, have a healthy calf unassisted at 24 months of age, and rebreed for a second pregnancy early in the second breeding season. Proper selection of replacement candidates, adequate nutritional development to reach target weights, and utilization of commercially available ionophores, anthelmintics, and progestagen-containing estrous synchronization systems (MGA) will ensure that a high percentage of heifers are pubertal and available for breeding at the start of the breeding season. In addition, a herd biosecurity program that includes stringent vaccination and quarantine protocols for replacements will minimize the risk of pregnancy-wasting diseases.

REFERENCES

Ames, D.R. 1985. Managing cows during the winter. *Proceedings, the Range Beef Cow Symposium IX* December 2–4, 1985, Chadron, NE, pp. 11–16.

Anderson, K.J., D.G. Lefever, J.S. Brinks, K.G. Odde. 1991. The use of reproductive tract scoring in beef heifers. *Agri-Practice* 12(4):19.

Arthur, P.F., M. Makarechian, R.T. Berg, R. Weingardt. 1993. Longevity and lifetime productivity of cows in a purebred Hereford and two multibreed synthetic groups under range conditions. *Journal of Animal Science* 71:1142–1147.

Bailey, C.M. 1991. Lifespan of beef-type *Bos taurus* and *Bos indicus* × *Bos taurus* females in a dry, temperate climate. *Journal of Animal Science* 69:2379–2386.

Bartol, F.F., L.L. Johnson, J.G. Floyd, et al. 1995. Neonatal exposure to progesterone and estradiol alters uterine morphology and luminal protein content in adult beef heifers. *Theriogenology* 43:835–844.

Basarab, J.A., L.M. Rutter, P.A. Day. 1993. The efficacy of predicting dystocia in yearling beef heifers: I. Using ratios of pelvic area to birth weight or pelvic area to heifer weight. *Journal of Animal Science* 71:1359–1371.

Benyshek, L.L., D.E. Little. 1982. Estimate of genetic and phenotypic parameters associated with pelvic area in Simmental cattle. *Journal of Animal Science* 54:258–263.

Berardinelli, J.G., R.L. Fogwell, E.K. Inskeep. 1978. Effect of electrical stimulation or presence of a bull on puberty in beef heifers. *Theriogenology* 9:133–141.

Bergman, J.A.G., W.D. Hohenboken. 1992. Prediction of fertility from calfhood traits of Angus and Simmental heifers. *Journal of Animal Science* 70:2611–2621.

Bolin, C.A., D.P. Alt. 2001. Use of a monovalent leptospiral vaccine to prevent renal colonization and urinary shedding in cattle exposed to *Leptospira borgpetersenii* serovar hardjo. *American Journal of Veterinary Research* 62:995–1000.

Bolin, C.A., A.B. Thiermann, A.L. Handsaker, J.W. Foley. 1989. Effect of vaccination with a pentavalent leptospiral vaccine on *Leptospira interrogans* serovar hardjo type hardjo-bovis infection of pregnant cattle. *American Journal of Veterinary Research* 50:161–165.

Bourdon, R.M., J.S. Brinks. 1986. Scrotal circumference in yearling Hereford bulls: Adjustment factors, heritabilities and genetic, environmental and phenotypic relationships with growth traits. *Journal of Animal Science* 62:958–967.

Boyd, H., C.D. Munro. 1979. Progesterone assays and rectal palpation in pre-service management of a dairy herd. *Veterinary Record* 104:341–343.

Bronson, F.H., C. Desjardins. 1974. Circulating concentrations of FSH, LH, estradiol, and progesterone associated with acute, male-induced puberty in female mice. *Endocrinology* 94:4658–4668.

Bushmich, S.L., R.D. Randel, M.M. McCartor, L.H. Carroll. 1980. Effect of dietary monensin on ovarian response following gonadotropin treatment in prepubertal heifers. *Journal of Animal Science* 51:692–697.

Butt, B.M., T.E. Besser, P.L. Senger, P.R. Widders. 1993. Specific antibody to *Haemophilus somnus* in the bovine uterus following intramuscular immunization. *Infection and Immunity* 61:2558–2562.

Byerley, K.G., R.B. Staigmiller, J.G. Berardinelli, R.E. Short. 1987. Pregnancy rates of beef heifers bred either on pubertal or third estrus. *Journal of Animal Science* 65:646–650.

Canfield, R.W., C.J. Sniffen, W.R. Butler. 1990. Effects of excess degradable protein on postpartum reproduction and energy balance in dairy cattle. *Journal of Dairy Science* 73:2342–2349.

Carpenter, B.B., L.R. Sprott. 1991. Synovex-C in replacement heifers: Effects on pelvic dimensions, hip height, body weight and reproduction. *Journal of Animal Science* 69(Suppl. 1):464.

Clark, B.L., J.H. Dufty, M.J. Monsbourgh. 1968. Vaccination of bulls against bovine vibriosis. *Australian Veterinary Journal* 44:530.

Corbeil, L.B., G.G. Schurig, J.R. Duncan, B.N. Wilkie, A.J. Winter. 1980. Immunity in the female bovine reproductive tract based on response to *Campylobacter fetus*. In *The Ruminant Immune System: Advances in Experimental Medicine and Biology*, Vol. 137, edited by J.E. Butler, pp. 729–743. New York: Plenum.

Cundiff, L.V. 1986. The effect of future demand on production programs—biological versus product antagonisms. *Proceeding of the Beef Improvement Federation Annual Meeting* May 7–9, 1986, Lexington, KY, pp. 110–127.

Cundiff, L.V., K.E. Gregory, R.M. Koch. 1974. Effects of heterosis on reproduction in Hereford, Angus, and Shorthorn cattle. *Journal of Animal Science* 38:711–727.

Cundiff, L.V., R. Nuñez-Dominguez, G.E. Dickerson, K.E. Gregory, R.M. Koch. 1992. Heterosis for lifetime production in Hereford, Angus, Shorthorn, and crossbred cows. *Journal of Animal Science* 70:2397–2410.

DeRouen, S.M., D.E. Franke, D.G. Morrison, et al. 1994. Prepartum body condition and weight influences on reproductive performance of first-calf beef cows. *Journal of Animal Science* 72:1119–1125.

Deutscher, G.H. 1985. Using pelvic measurements to reduce dystocia in heifers. *Modern Veterinary Practice* 66:751–755.

Deutscher, G.H. 1991. Growth promoting implants on replacement heifers—a review. *Proceedings, the Range Beef Cow Symposium* 12:169.

Dodson, S.E., B.J. McLeod, W. Haresign, A.R. Peters, G.E. Lamming. 1988. Endocrine changes from birth to puberty in the heifer. *Journal of Reproduction and Fertility* 82:527–538.

Doyle, S.P., B.L. Golden, R.D. Green, J.S. Brinks. 2000. Additive genetic parameter estimates for heifer pregnancy and subsequent reproduction in Angus females. *Journal of Animal Science* 78:2091–2098.

Duffy, J., J. Vaughan. 1993. Bovine venereal campylobacteriosis. In *Current Veterinary Therapy 3: Food Animal Practice*, edited by L.J. Howard, pp. 510–513. Philadelphia, PA: W.B. Saunders.

Dunn, T.G., J.E. Ingalls, D.R. Zimmerman, J.N. Wiltbank. 1969. Reproductive performance of 2-year-old Hereford and Angus heifers as influenced by pre- and post-calving energy intake. *Journal of Animal Science* 29:719–726.

Dziuk, P.J., R.A. Bellows. 1983. Management of reproduction of beef cattle, sheep and pigs. *Journal of Animal Science* 57(Suppl. 2):355.

Elrod, C.C., W.R. Butler. 1993. Reduction of fertility and alteration of uterine pH of heifers fed excess ruminally degradable protein. *Journal of Animal Science* 71:694–701.

Evans, J.L., B.L. Golden, R.M. Bourdon, K.L. Long. 1999. Additive genetic relationships between heifer pregnancy and scrotal circumference in Hereford cattle. *Journal of Animal Science* 77: 2621–2628.

Foster, D.L., S.M. Yellon, D.H. Olster. 1985. Internal and external determinants of the timing of puberty in the female. *Journal of Reproduction and Fertility* 75:327–344.

Fox, D.G., C.J. Sniffen, J.D. O'Connor, J.B. Russell, P.J. Van Soest. 1992. A net carbohydrate and protein system for evaluating cattle diets. III. Cattle requirements and diet adequacy. *Journal of Animal Science* 70:3578–3596.

Gonzalez-Padilla, E., R. Ruiz, D. LeFever, A. Denham, J.N. Wiltbank. 1975a. Puberty in beef heifers. III. Induction of fertile estrus. *Journal of Animal Science* 40:1110–1118.

Gonzalez-Padilla, E., J.N. Wiltbank, G.D. Niswender. 1975b. Puberty in beef heifers. I. The interrelationship between pituitary, hypothalamic and ovarian hormones. *Journal of Animal Science* 40:1091–1104.

Gregory, K.E., D.B. Laster, L.V. Cundiff, G.M. Smith, R.M. Koch. 1979. Characterization of biological types of cattle-cycle III: II. Growth rate and puberty in females. *Journal of Animal Science* 49:461–471.

Gregory, K.E., D.D. Lunstra, L.V. Cundiff, R.M. Koch. 1991. Breed effects and heterosis in advanced generations of composite populations for puberty and scrotal traits in beef cattle. *Journal of Animal Science* 69:2795–2807.

Hancock, R., G. Deutscher, M. Nielson, D. Colburn. 1994. Synovex-C affects growth, reproduction, and calving performance of replacement heifers. *Journal of Animal Science* 72:292–299.

Hansel, W., P.W. Concannon, J.H. Lukaszewka. 1973. Corpora lutea of the large domestic animal. *Biology of Reproduction* 8:222–245.

Holtzer, A.L.J., W. Schlote. 1984. Investigations on interior pelvic size of Simmental heifers. *Journal of Animal Science* 174(Suppl. 1): (abstract 72).

Izard, M.K., J.G. Vandenbergh. 1982. The effects of bull urine on puberty and calving date in crossbred beef heifers. *Journal of Animal Science* 55:1160–1168.

Jaeger, J.R., J.C. Whittier, L.R. Corah, J.C. Meiske, K.C. Olson, D.J. Patterson. 1992. Reproductive response of yearling beef heifers to a melengestrol acetate-prostaglandin $F_{2\alpha}$ estrus synchronization system. *Journal of Animal Science* 70:2622–2627.

Keeton, L.L., R.D. Green, B.L. Gonden, K.J. Anderson. 1996. Estimation of variance components and prediction of breeding values for scrotal circumference and weaning weight in Limousin cattle. *Journal of Animal Science* 74:31–36.

Kelton, D.F., K.E. Leslie, W.G. Etherington, B.N. Bonnett, J.S. Walton. 1991. Accuracy of rectal palpation and of a rapid milk progesterone enzymeimmunoassay for determining the presence of a functional corpus luteum in subestrous dairy cows. *Canadian Veterinary Journal* 32:286–291.

Kimsey, P.B. 1986. Bovine trichomoniasis. In *Current Therapy in Theriogenology 2*, edited by D.A. Morrow, pp. 275–279. Philadelphia, PA: W.B. Saunders.

King, B.D., G.A. Bo, C. Lulai, R.N. Kirkwood, F.D.H. Cohen, R.J. Mapletoft. 1995. Effect of zeranol implants on age at onset of puberty, fertility and embryo fetal mortality in beef heifers. *Canadian Journal of Animal Science* 75:225–230.

Kriese, L.A., J.K. Bertrand, L.L. Benyshek. 1991. Age adjustment factors, heritabilities and genetic correlations for scrotal circumference and related growth traits in Hereford and Brangus bulls. *Journal of Animal Science* 69:478–489.

Kvasnicka, W.B., R.E.L. Taylor, J.C. Huang, et al. 1989. Investigation of the incidence of bovine trichomoniasis in Nevada and the efficacy of immunizing cattle with vaccines containing *Tritrichomonas foetus*. *Theriogenology* 31:963–971.

Lacau-Mengido, I.M., M.E. Mejia, G.S. Diaz-Torga, et al. 2000. Endocrine studies in ivermectin-treated heifers from birth to puberty. *Journal of Animal Science* 78:817–824.

Lalman, D.L., J.E. Williams, B.W. Hess, M.G. Thomas, D.H. Keisler. 2000. Effect of dietary energy on milk production and metabolic hormones in thin, primiparous beef heifers. *Journal of Animal Science* 78:530–538.

Larson, R.L., L.R. Corah. 1995. Effects of being dewormed with oxfendazole and implanted with Synovex-C as young beef calves on subsequent reproductive performance of heifers. *Professional Animal Scientist* 11:106–109.

Larson, R.L., G.H. Kiracofe. 1995. Estrus after treatment with Syncro-Mate B in ovariectomized heifers is dependent on the injected estradiol valerate. *Theriogenology* 44:177–187.

Larson, R.L., L.R. Corah, M.F. Spire, R.C. Cochran. 1995. Effect of treatment with ivermectin on reproductive performance of yearling beef heifers. *Theriogenology* 44:189–197.

Laster, D.B. 1974. Factors affecting pelvic size and dystocia in beef cattle. *Journal of Animal Science* 38:496–503.

Laster, D.B., H.A. Glimp, K.E. Gregory. 1972. Age and weight at puberty and conception in different breed and breed crosses of beef heifers. *Journal of Animal Science* 34:1031–1036.

Laster D.B., G.M. Smith, L.V. Cundiff, K.E. Gregory. 1979. Characterization of biological types of cattle (Cycle II). II. Postweaning growth and puberty of heifers. *Journal of Animal Science* 48:500–508.

Laster, D.B., G.M. Smith, K.E. Gregory. 1976. Characteristics of biological types of cattle. IV. Postweaning growth and puberty of heifers. *Journal of Animal Science* 43:63–70.

Lawrence, J.R., J.D. Allen, R.C. Herschler, T.A. Miller. 1985. Synovex implants: Reimplantation and fertility in beef heifers. *Agri-Practice* 6:13–18.

Lesmeister, J.L., P.J. Burfening, R.L. Blackwell. 1973. Date of first calving in beef cows and subsequent calf production. *Journal of Animal Science* 36:1–6.

Little, P.B. 1993. The *Haemophilus somnus* complex. In *Current Veterinary Therapy 3: Food Animal Practice*, edited by J.L. Howard, pp. 546–549. Philadelphia, PA: W.B. Saunders.

Lucy, M.C., H.J. Billings, W.R. Butler, et al. 2001. Efficiency of an intravaginal progesterone insert and an injection of $PGF_{2\alpha}$ for synchronizing estrus and shortening the interval to pregnancy in postpartum beef cows, peripubertal beef heifers, and dairy heifers. *Journal of Animal Science* 79:982–995.

Marston, T.T., K.S. Lusby. 1995. Effects of energy or protein supplementation and stage of production on intake and digestibility of hay by beef cows. *Journal of Animal Science* 73:651–656.

Marston, T.T., K.S. Lusby, R.P. Wettemann, H.T. Purvis. 1995. Effects of feeding energy or protein supplements before or after calving on performance of spring-calving cows grazing native range. *Journal of Animal Science* 73:657–664.

Martin, G.B., R.J. Scaramuzzi, C.M. Oldham, D.R. Lindsay. 1983. Effects of progesterone on the responses of Merino ewes to the introduction of rams during anoestrus. *Australian Journal of Biological Sciences* 36:369–378.

Martin, L.C., J.S. Brinks, R.M. Bourdon, L.V. Cundiff. 1992. Genetic effects on beef heifer puberty and subsequent reproduction. *Journal of Animal Science* 70:4006–4017.

McGuire, W.J., R.L. Larson, G.H. Kiracofe. 1990. Syncro-Mate B induces estrus in ovariectomized cows and heifers. *Theriogenology* 34:33–37.

Morrison, D.G., W.D. Williamson, P.E. Humes. 1984. Heritabilities and correlations of traits associated with pelvic area in beef cattle. *Journal of Animal Science* 59(Suppl. 1):160.

Mortimer, R.G., J.D. Olson, E.M. Huffman, P.W. Farin, L. Ball. 1983. Serum progesterone concentration in pyometritic and normal postpartum dairy cows. *Theriogenology* 19(5):647–653.

Moseley, W.M., T.G. Dunn, C.C. Kaltenbach, R.E. Short, R.B. Staigmiller. 1982. Relationship of growth and puberty in beef heifers fed monensin. *Journal of Animal Science* 55:357–362.

Moseley, W.M., M.M. McCartor, R.D. Randel. 1977. Effects of monensin on growth and reproductive performance of beef heifers. *Journal of Animal Science* 45:961–968.

Moser, D.W., J.K. Bertrand, L.L. Benyshek, M.A. McCann, T.E. Kiser. 1996. Effects of selection for scrotal circumference in Limousin bulls on reproductive and growth traits of progeny. *Journal of Animal Science* 74:2052–2057.

Naazie, A., M.M. Makarechian, R.T. Berg. 1989. Factors influencing calving difficulty in beef heifers. *Journal of Animal Science* 67: 3243–3249.

National Research Council. 1996. *Nutrient Requirements of Beef Cattle*, 7th ed. Washington, DC: National Academy of Sciences.

Nelsen, T.C., C.R. Long, T.C. Cartwright. 1982. Postinfection growth in straightbred and crossbred cattle. II. Relationships among weight, height and pubertal characters. *Journal of Animal Science* 55:293–304.

Nelsen, T.C., R.E. Short, D.A. Phelps, R.B. Staigmiller. 1985. Nonpubertal estrus and mature cow influences on growth and puberty in heifers. *Journal of Animal Science* 61:470–473.

Neville, W.E., B.G. Mullinix, J.B. Smith, W.C. McCormick. 1978. Growth patterns for pelvic dimensions and other body measurements of beef females. *Journal of Animal Science* 47:1080–1088.

Niswender, G.D., C.E. Farin, T.D. Braden. 1984. Reproductive physiology of domestic ruminants. *Proceedings, Society for Theriogenology*, September 26–28, 1984, Denver, CO, pp. 116-136.

Núñez-Dominguez, R., L.V. Cundiff, G.E. Dickerson, K.E. Gregory, R.M. Koch. 1991. Heterosis for survival and dentition in Hereford, Angus, Shorthorn, and crossbred cows. *Journal of Animal Science* 69:1885–1989.

Oeydipe, E.O., D.I.K. Osori, O. Aderejola, D. Saror. 1982. Effect of level of nutrition on onset of puberty and conception rates of Zebu heifers. *Theriogenology* 18:525–539.

Oliver, W.M. 1975. Effect of monensin on gains of steers grazed on Coastal bermudagrass. *Journal of Animal Science* 41:999–1001.

Ott, R.S., K.N. Bretzlaff, J.E. Hixon. 1986. Comparison of palpable corpora lutea with serum progesterone concentrations in cows. *Journal of the American Veterinary Medical Association* 188(12):1417–1419.

Patterson, D.J., K.D. Bullock. 1995. Using prebreeding weight, reproductive tract score and pelvic area to evaluate prebreeding development of replacement beef heifers. *Proceedings, Beef Improvement Federation Annual Meeting*, May 31–June 3, 1995, Sheridan, WY, pp. 174–177.

Patterson, D.J., R.C. Perry, G.H. Kiracofe, R.A. Bellows, R.B. Staigmiller, L.R. Corah. 1992. Management considerations in heifer development and puberty. *Journal of Animal Science* 70:4018–4035.

Pennel, P.L., D.D. Zalesky, M.L. Day, et al. 1986. Influence of bull exposure on initiation of estrous cycles in prepubertal beef heifers. *Journal of Animal Science* 63(Suppl. 1):129.

Perry, R.C., L.R. Corah, R.C. Cochran, J.R. Brethour, K.C. Olson, J.J. Higgins. 1991. Effects of hay quality, breed and ovarian development on onset of puberty and reproductive performance of beef heifers. *Journal of Production Agriculture* 4: 13–18.

Poindron, P., Y. Cognie, F. Gayerie, P. Orgeur, C.M. Oldham, J.P. Ravault. 1980. Changes in gonadotrophins and prolactin levels in isolated (seasonally or lactationally) anovular ewes associated with ovulation caused by the introduction of rams. *Physiology and Behavior* 25:227–236.

Pollack, E.J. 1992. Expected progeny differences (within breed comparisons). In *Proceedings of Symposium on Application of Expected Progeny Differences to Livestock Improvement*, American Society of Animal Science, February 2–5, 1992, Lexington, KY, pp. 17–23.

Potter, E.L., C.O. Cooley, L.F. Richardson, A.P. Raun, R.P. Rathmacher. 1976. Effect of monensin on performance of cattle fed forage. *Journal of Animal Science* 43:665–669.

Prescott, J.F. 1993. Leptospirosis. In *Current Veterinary Therapy 3: Food Animal Practice*, edited by J.L. Howard, pp. 541–546. Philadelphia, PA: W.B. Saunders.

Price, T.D., J.N. Wiltbank. 1978. Predicting dystocia in heifers. *Theriogenology* 9:221–249.

Purvis, H.T., J.C. Whittier. 1996. Effects of ionophore feeding and anthelmintic administration on age and weight at puberty in spring-born beef heifers. *Journal of Animal Science* 74:736–744.

Rakestraw, J., K.S. Lusby, R.P. Wettemann, J.J. Wagner. 1986. Postpartum weight and body condition loss and performance of fall-calving cows. *Theriogenology* 26:461–473.

Raun, A.P., C.O. Cooley, E.L. Potter, R.P. Rathmacher, L.F. Richardson. 1976. Effect of monensin on feed efficiency of feedlot cattle. *Journal of Animal Science* 43:665–669.

Reynolds, W.L., T.M. DeRouen, J.W. High, Jr. 1963. The age and weight of Angus, Brahman, and Zebu cross heifers. *Journal of Animal Science* 22:243.

Roberson, M.S., R.P. Ansotegui, J.G. Berardinelli, R.W. Whitman, M.J. McInerney. 1987. Influence of biostimulation by mature bulls on occurrence of puberty in beef heifers. *Journal of Animal Science* 64:1601–1605.

Roberson, M.S., J.G. Berardinelli, R.P. Ansotegui, R.W. Whitman. 1983. Influence of prostaglandin $F_{2\alpha}$ and social interaction on induction of first estrus in beef heifers. *Journal of Animal Science* 57(Suppl. 1):407.

Roberson, M.S., M.W. Wolf, T.T. Stumpf, et al. 1991. Influence of growth rate and exposure to bulls on age at puberty in beef heifers. *Journal of Animal Science* 69:2092–2098.

Rogers, L.F. 1972. Economics of replacement rates in commercial beef herds. *Journal of Animal Science* 34:921–925.

Rohrer, G.A., J.F. Baker, C.R. Long, T.C. Cartwright. 1988. Productive longevity of first-cross cows produced in a five breed diallel: II. Heterosis and general combining ability. *Journal of Animal Science* 66:2836–2841.

Rosenkrans, K.S., D.K. Hardin. 2003. Repeatability and accuracy of reproductive tract scoring to determine pubertal status in beef heifers. *Theriogenology* 59:1087–1092.

Rusk, C.P., N.C. Speer, D.W. Schafer, L.S. Brinks, K.G. Odde, D.G. Lefever. 1992. Effect of Synovex-C implants on growth, pelvic measurements and reproduction in Angus heifers. *Journal of Animal Science* 70(Suppl. 1):126.

Sacco, R.E., J.F. Baker, T.C. Cartwright. 1987. Production characteristics of primiparous females of a five-breed diallel. *Journal of Animal Science* 64:1612–1618.

Sackett, D.L. 1992. A primer on the precision and accuracy of the clinical examination. *Journal of the American Medical Association* 267:2638–2644.

Schillo, K.K., P.J. Hansen, L.A. Kamwanja, D.J. Dierschke, E.R. Hauser. 1983. Influence of season on sexual development in heifers: Age at puberty as related to growth and serum concentrations of gonadotropins, prolactin, thyroxine and progesterone. *Biology of Reproduction* 28:329–341.

Schurig, G.G.D., C.E. Hall, L.B. Corbeil, J.R. Duncan, A.J. Winter. 1975. Bovine venereal vibriosis: Cure of genital infection in females by systemic immunization. *Infection and Immunity* 11:245–251.

Selk, G.E., R.P. Wetteman, K.S. Lusby, J.W. Oltjen, S.L. Mobley, R.J. Rasby, J.C. Garmendia. 1988. Relationships among weight change, body condition and reproductive performance of range beef cows. *Journal of Animal Science* 66:3153–3159.

Sherbeck, J.A., J.D. Tatum, T.G. Field, J.B. Morgan, G.C. Smith. 1995. Feedlot performance, carcass traits, and palatability traits of Hereford and Hereford Brahman steers. *Journal of Animal Science* 73:3613–3620.

Short, R.E., R.A. Bellows, J.B. Carr, R.B. Staigmiller, R.D. Randel. 1976. Induced or synchronized puberty in beef heifers. *Journal of Animal Science* 43:1254–1258.

Short, R.E., R.A. Bellows, R.B. Staigmiller, J.G. Berardinelli, E.E. Custer. 1990. Physiological mechanisms controlling anestrus and infertility in postpartum beef cattle. *Journal of Animal Science* 68:799–816.

Smith, B.A., J.S. Brinks, G.V. Richardson. 1989. Estimation of genetic parameters among breeding soundness examination components and growth traits in yearling bulls. *Journal of Animal Science* 67:2892–2896.

Spitzer, J.C. 1982. Pregnancy rate in peripubertal beef heifers following treatment with Syncro-Mate B and GnRH. *Theriogenology* 17:373–381.

Spitzer, J.C., D.G. Morrison, R.P. Wettemann, L.C. Faulkner. 1995. Reproductive responses and calf birth and weaning weights as affected by body condition at parturition and postpartum weight gain in primiparous beef cows. *Journal of Animal Science* 73:1251–1257.

Sprott, L.R., T.B. Goehring, J.R. Beverly, L.R. Corah. 1988. Effects of ionophores on cow herd production: A review. *Journal of Animal Science* 66:1340–1346.

Steffan, C.A., D.D. Kress, D.E. Doornbos, D.C. Anderson. 1985. Performance of crosses among Hereford, Angus, and Simmental cattle with different levels of Simmental breeding. III. Heifer postweaning growth and early reproductive traits. *Journal of Animal Science* 61:1111–1120.

Stewart, T.S., C.R. Long, T.C. Cartwright. 1980. Characterization of cattle of a five breed diallel. III. Puberty in bulls and heifers. *Journal of Animal Science* 50:808–820.

Stumpf, T.T., M.L. Day, P.L. Wolf, et al. 1988. Feedback of 17 beta-estradiol on secretion of luteinizing hormone during different seasons of the year. *Journal of Animal Science* 66:447–451.

Stumpf, T.T., M.W. Wolf, M.S. Roberson, R.J. Kittok, J.E. Kinder. 1993. Season of the year influences concentration and pattern of gonadotropins and testosterone in circulation of the bovine male. *Biology of Reproduction* 49:1089–1095.

Tanida, H., W.D. Hohenboken, S.K. DeNise. 1988. Genetic aspects of longevity in Angus and Hereford cows. *Journal of Animal Science* 66:640–647.

U.S. Department of Agriculture. 1996. Management practices associated with profitable cow-calf herds. *USDA-APHIS-VS Information Sheet*. Available at http://www.aphis.usda.gov/vs/cattle.htm. Accessed July 12, 2004.

Van Donkersgoed, J., C.S. Ribble, H.G.G. Townsend, E.D. Janzen. 1990. The usefulness of pelvic measurements as an on-farm test for predicting calving difficulty in beef heifers. *Canadian Veterinary Journal* 31:190–193.

Watson, E.D., C.D. Munro. 1980. A reassessment of the technique of rectal palpation of corpora lutea in cows. *British Veterinary Journal* 136:555–560.

Wheeler, T.L., L.V. Cundiff, R.M. Koch. 1994. Effect of marbling degree on beef palatability in *Bos taurus* and *Bos indicus* cattle. *Journal of Animal Science* 72:3145–3151.

Whittier, J.C., J.W. Massey, G.R. Varner, T.B. Erickson, D.G. Watson, D.S. McAtee. 1991. Effect of a single calfhood growth-promoting implant on reproductive performance of replacement beef heifers. *Journal of Animal Science* 69(Suppl. 1):464.

Whittier, W.D., A.L. Eller, W.E. Beal. 1994. Management changes to reduce dystocia in virgin beef heifers. *Agri-Practice* 15(1):26–32.

Williams, J.C., J.W. Knox, K.S. Marbury, M.E.D. Kimball, R.E. Willis. 1989. Effects of ivermectin on control of gastrointestinal nematodes and weight gain in weaner-yearling beef cattle. *American Journal of Veterinary Research* 50:2108–2116.

Wiltbank, J.N., K.E. Gregory, L.A. Swiger, J.E. Ingalls, J.A. Rothlisberger, R.M. Koch. 1966. Effects of heterosis on age and weight at puberty in beef heifers. *Journal of Animal Science* 25:744–751.

Wiltbank, J.N., C.W. Kasson, J.E. Ingalls. 1969. Puberty in crossbred and straightbred beef heifers on two levels of feed. *Journal of Animal Science* 29:602–605.

Wiltbank, J.N., S. Roberts, J. Nix, L. Rowden. 1985. Reproductive performance and profitability of heifers fed to weigh 272 or 318 kg at the start of the first breeding season. *Journal of Animal Science* 60:25–34.

Wiltbank, J.N., W.W. Rowden, J.E. Ingalls, K.E. Gregory, R.M. Koch. 1962. Effect of energy level on reproductive phenomena of mature Hereford cows. *Journal of Animal Science* 21:219–225.

Winter, A.J., B.L. Clark, I.M. Parsonson, J.R. Duncan, P.J. Bier. 1980. Nature of immunity in the male bovine reproductive tract based on response to *Campylobacter fetus* and *Trichomonas fetus*. In *Advances in Experimental Medicine and Biology: The Ruminant Immune System*, Vol. 137, edited by J.E. Butler, pp. 745–752.

Wohlgemuth, K., J.J. Melanconn, H. Hughes, M. Biondini. 1989. Treatment of North Dakota beef cows and calves with ivermectin: Some economic considerations. *Bovine Practitioner* 24:61, 64–66.

Wright, I.A., S.M. Rhind, T.K. Whyte, A.J. Smith. 1992. Effects of body condition at calving and feeding levels after calving on LH profiles and the duration of the post-partum anoestrous period in beef cows. *Animal Production* 55:41–46.

9
Breeding Bull Selection, Assessment, and Management

Peter J. Chenoweth

INTRODUCTION

The ability of bulls to transfer desired genetics to their progeny is dependent upon their fertility whether or not they are used for natural breeding or artificial insemination. In turn, herd reproductive performance has a greater effect on economic returns than either growth rate or product quality (Wiltbank 1994). Here, chute-side calculation indicates that it takes approximately five generations (15 years) of using bulls of 10% superior weaning-weight expected progeny differences (EPDs) to have the same effect on total herd kilograms weaned as a 5% increase in weaning rate, a factor that is greatly influenced by bull fertility.

Natural breeding is still the norm on most cattle breeding operations throughout the world. Natural-breeding bulls differ markedly in their reproductive capabilities. Conservative estimates indicate that at least 20% of unselected bulls are subfertile or infertile. Procedures used to evaluate bulls have been shown to be effective in identifying many of these subfertile and infertile bulls, and the procedures involved in such evaluations are remarkably consistent throughout the world. In North America, the Society for Theriogenology recommends a standardized procedure that has been modified over time as knowledge has changed and improved (Chenoweth et al. 1992). Standards for bull evaluation have also been published separately by the Australian Association of Cattle Veterinarians (Australian Association of Cattle Veterinarians 1995; Fordyce 2002), as well as by the Western Canadian Association of Bovine Practitioners (Barth 1993).

In considering bull selection and management, it is useful to consider four periods within the working "career" of a breeding bull: selection and purchase, conditioning and preparation, breeding season, and nonbreeding season. It is also helpful to identify and understand miscellaneous disorders of the bull reproductive system and to recognize their significance as well as the appropriate response to them.

SELECTION AND PURCHASE OF BULLS

Professional input at the time of bull purchase can help producers avoid a number of problems. However, such input appears to be underutilized. It is disappointing, for example, to find that, in the recent past, only 15% of U.S. producers considered that bull scrotal circumference (SC) was of great importance in bull selection, whereas 19% regarded it as being of no importance at all (National Animal Health Monitoring System [NAHMS] 1994). In this report, the major factors examined in bull selection were (in decreasing order from extremely important): breed, appearance/soundness, and temperament. Paradoxically, in this same report, the most important bull culling criteria were infertility and physical unsoundness; factors that were not included among the top selection criteria. Thus, a major opportunity exists to assist cow/calf producers by identifying subfertile/unsound bulls before their purchase or use. A proven, useful tool in achieving this objective is the bull Breeding Soundness Evaluation (BSE).

Bull BSE

The BSE represents a relatively quick and economical procedure for screening bulls before sale or use. Its objective is to establish baselines, or thresholds, above which bulls would be regarded as satisfactory potential breeders. As it is intended for wide application in a variety of breeds in different environments, it needs to be simple, repeatable, and unambiguous. Even if it fulfills all of these criteria, it should be emphasized that the BSE should not substitute for either professional judgment or common sense.

In the U.S. system, bulls are placed into categories of "satisfactory," "unsatisfactory," and "classification deferred." The system is most effective in identifying low-fertility or sterile bulls. It is relatively less effective in predicting individual bull performance at the upper end of the fertility spectrum. Reasons for this include:

- The BSE is a relatively quick and simple procedure that does not attempt to comprehensively assess all aspects of male fertility
- Fertility is a complex trait that is influenced by both male and female traits as well as by extraneous factors
- The BSE aims to identify bulls that are satisfactory (not necessarily those that are superior)
- Knowledge and understanding of male fertility keep increasing and changing.

This last factor, in particular, motivates the Society for Theriogenology to periodically revise their recommended BSE procedures. The most recent revision (Chenoweth et al. 1992) is described below. Forms and background information for this procedure may be obtained from the Society for Theriogenology (see http://www.therio.org).

BSE Procedures

Minimal requirements for an adequate BSE include key elements of history and a physical examination (Gardiner and Fordyce 2002). The following steps are usually included: physical examination, reproductive examination (including measurement of SC), collection and examination of semen, and a report.

In addition, libido/serving capacity testing may be included, as may special tests for appropriate disease entities (e.g., vibriosis, trichomoniasis). These procedures add predictive value to the assessment process and may be specifically indicated in some situations, but they are not part of the routine BSE. It is important to inform producers not only of the advantages of the BSE but also of its limitations. In addition to those described above, other limitations of the BSE exist. For example, results are most valid at the time of examination, and the BSE does not attempt to accurately predict fertility.

Timing of the BSE

The best time to conduct a BSE varies with the reason for the test. For young bulls, the BSE is particularly advantageous before sale and first breeding. Here, the optimum time for evaluation would be as close to sale, or use, as possible, allowing adequate time for either retesting or replacement before breeding. To allow for both considerations, it is common to schedule the initial BSE within 1 month of sale or introduction to the breeding herd. An important consideration is that young bulls must be pubertal for BSE classification.

For infertility investigations and insurance exams, the test should be conducted as close as possible to the events in question. The time to allow before bulls are retested is also a consideration, as there are often time constraints. Here, professional judgment is necessary. For example, low sperm motility may be caused by poor semen handling, excessive numbers of accumulated sperm in the extragonadal system, or more serious problems. In the first two cases, improvement is often seen when a second ejaculate is collected within a few minutes of the first. If a bull is young and has increased "secondary" sperm defects (e.g., droplets and bent tails), evident improvement may occur within weeks, especially once bulls are taken off test. Overconditioned bulls (particularly those that achieved rapid weight gains on a high-grain diet) are often represented in this group. However, not all such bulls improve. More serious semen "quality" problems may require 1–2 months for improvement, or they may never improve. A good policy is to schedule retests for 6–8 weeks later, so that possible differentiation may be made between temporary and more permanent problems.

Some knowledge of the time frames involved with spermatogenesis and sperm transport in the bull can assist in decision making and prognosis. It takes a total of 60 days for a spermatogonium to evolve into a spermatozoa, and another 9–14 days for it to traverse the extragonadal duct system and appear in the ejaculate. Damage may occur at the level of spermatocytes (e.g., reduced sperm numbers), round spermatids (e.g., head and acrosome morphology problems), and elongated spermatids (e.g., midpiece and tail problems), and in the extragonadal ducts (e.g., lowered motility, loose heads, increased droplets). Thus, lowered sperm motility may appear within 1–2 weeks of an insult, in conjunction with "secondary" sperm defects such as loose heads and increased droplets. Soon after this (at 15 days or more), one may observe a preponderance of diadem defects of sperm followed by crater defects, and then acrosome abnormalities (Vogler et al. 1993, Saacke et al. 2000)

Equipment for the BSE

A list of basic equipment and supplies for bull examination and semen collection and examination should include the following items (after Perry et al. 2002):

General—Rope(s), plastic obstetric sleeves, examination gloves, paper towels, waterproof marker, pen, bull forms, notepaper, scissors, antiseptic, bucket(s), extension cord(s), multielectrical outlet, thermometer, stethoscope, hoof knife, surgical pack, antibiotic(s), Vacutainers, syringes (20 mL), needles (18 g × 1.5 in), gauze squares (for restraining penis).

Semen collection—Electroejaculator(s) and leads, electrical outlet tester, probes(s), semen collection device (e.g., funnel and holder), collection tubes, insulated jacket (if cold).

Semen examination—Microscope and leads, microscope slides, coverslips, warming stage (if cold), semen extender (e.g., physiological sterile saline or 2.9% sodium citrate), pipettes, semen stain (e.g., nigrosin-eosin), other stains (e.g., new methylene blue), sperm fixative (e.g., isotonic formal saline [Campbell et al. 1960] or phosphate-buffered-saline-glutaraldehyde).

Restraint of the Bull

A safe, secure bull restraint system is a most important prerequisite for a BSE. Appropriate animal handling facilities are described here (see chapter 7) as well as previously (Grandin 2000, Jayawardhana and Thomson 2002). The examiner should have access to both the rear end of the bull and his underbelly without undue risk of injury. Injury to the bull during the evaluation process should also be avoided as much as possible. For example, if head catches are used, they should be designed so that bulls do not injure themselves if they collapse in the chute or crush. Some bulls (e.g., many *Bos indicus* types) are more tractable if their head is not restrained during the BSE. Squeeze chutes (especially hydraulic ones) are effective in constraining excessive bull movements. Use of a bar or chain behind the bull is recommended, as is providing secure footing for the bull in the chute. A backstop for the hind feet can be useful, especially for bulls undergoing electroejaculation (EE).

Identification and Ownership of Bulls

Permanent and unique identification should be established for each bull and for all associated records including samples, slides, and worksheets.

Body Condition and Other Body Measures

Body condition (or degree of fatness) of bulls can influence reproductive performance. Excessively fat or thin bulls may have problems with semen quality, libido, or mating ability as well as fertility in general (Coulter and Kozub 1989). Poor body condition may also reflect health problems apart from poor nutrition. Bull pelvic size is related to pelvic size in their daughters, which may be useful in dystocia management (Brinks 1994).

Structure, Conformation, and Movement

The bull should be observed moving freely, to assess movement and gait. He should exhibit good, symmetrical musculoskeletal formation and walk smoothly and freely. Once in the chute or crush, close attention should be paid to limbs, joints, and claws. Asymmetrical or overgrown claws and swellings over lower limb joints are common abnormalities and may reflect poor conformation, pathological conditions, or both. Conditions such as chronic lameness, sole abscesses, arthritis, severe quarter cracks, interdigital fibromas, and foot rot can adversely affect mobility, mating ability, and libido, as well as contribute to testicular degeneration if the bull is semirecumbent. High body temperatures associated with foot rot (or pyrexia from any other cause) can also lead to spermatogenic dysfunction. Because the hind limbs are most important in supporting bulls during service, hind limb and joint structure should be examined most critically. Poor conformation (e.g., "post legs," "cow hocks") is highly heritable in bulls (McNitt 1965) and contributes to bull attrition. Spondylosis or spondyloarthrosis of the lumbosacral joint is not uncommonly encountered in bulls, even relatively young ones, and this condition may contribute to service disability and related loss of libido (Chenoweth 1983).

Physical and Reproductive Examination

The primary mission of the natural-breeding bull is to efficiently impregnate all available females as early in the breeding period as possible. To achieve this, he must possess the physical and reproductive necessities to perform this task. These include good eyesight and musculoskeletal conformation as well as the necessary reproductive equipment and sex drive to produce and appropriately deliver sufficient numbers of fertile spermatozoa.

Once in the chute or crush, the bull's eyes should be examined for any indication of impaired vision. This is of particular importance, as vision is the most important special sense used by bulls in detecting estrous females. Examination of the bull's teeth allows verification of age and helps to ensure that dentition is adequate for foraging.

The sheath should be inspected and palpated. The penis is palpable within the sheath, although it is often most convenient to examine it when extended (e.g., during massage or EE). On palpation, it should be symmetrical, freely moveable, and contain no abnormal masses.

Scrotal conformation should be evaluated, as it can influence testicular development and function (Coulter 1994). Optimum spermatogenesis occurs at 4°–5°C below internal body temperature. This temperature gradient is maintained through a complex system that raises and lowers both the scrotum and testicles and transfers heat from the pampiniform plexus to the external environment through the scrotal neck. Any factor interfering with any of these mechanisms can cause defective scrotal thermoregulation, with resultant deleterious effects on semen quality and sperm production. Studies using thermography indicate that abnormal scrotal temperature gradients in bulls can provide indirect evidence of potential fertility problems (Kastelic et al. 2000). Optimum thermoregulation of the testicle occurs when the scrotum hangs vertically and the scrotal neck is distinct (Coulter 1994).

Although some variation in testicle size commonly occurs in bulls, significant differences in the size and shape of either testicle should prompt further evaluation. A disparity of more than 25% in the size of either testicle should be regarded with suspicion. The most common cause of marked scrotal asymmetry is unilateral testicular hypoplasia (often misdiagnosed as an "undescended" testicle). Definitive diagnosis of testicular hypoplasia can only be obtained via testicular histology, however, and caution should be exercised when using this term without such evidence. A variety of congenital, traumatic, or infectious conditions may also lead to reduced testicular size, with testicular degeneration and aplasia being most common.

Testicles should be palpated for both hardness and resiliency. A "normal" testicle should be firm and resilient (similar to a new tennis ball). "Soft" testicles often reflect testicular degeneration and are found more often in older bulls and in bulls that have undergone thermal stress. This latter group often includes bulls that are mostly recumbent. "Hard" testicles often reflect fibrosis, a common sequel to testicular degeneration. However, "hardness" may also occur with acute orchitis, usually in association with increased heat and other signs of inflammation. Recent developments in diagnostic ultrasound and thermography hold promise for improved diagnostic capabilities in relation to dysfunctions of scrotal contents and the testicular parenchyma (Coulter and Kastelic 1999).

The epididymides (head, body, and tail) should be palpated, with detectable problems including inflammation, fibrosis, abscessation, hypoplasia, or aplasia noted. Examination of the bull's internal reproductive organs via transrectal palpation is essential. This examination normally includes identification and palpation of the pelvic urethra, prostate gland, vesicular glands, ampullae, vasa deferentia, and internal inguinal rings. The most common abnormality encountered is accessory genital disease, most readily recognized as vesicular gland adenitis (seminal vesiculitis), as discussed later in this chapter. However, evidence of aplasia/hypoplasia of the embryonic Wolffian duct system may also be detected (see following).

Scrotal Circumference (SC)

Testicular size in beef bulls is positively associated with sperm production and semen quality, as well as with the age at puberty of related females (Brinks et al. 1978). In turn, age at puberty in heifers is favorably associated with lifetime reproduction and production traits (Brinks 1994). Thus, SC measurement represents an important component of the BSE. Adjustments for yearling bull SC based on dam age have been suggested, as shown in Table 9.1.

SC measurement is both simple and repeatable and does not require expensive, high-technology equipment. It involves manipulating the testicles to ensure that they are as close to the bottom of the scrotum as possible, with skin folds removed, and snugly applying a measuring tape around the

Table 9.1. Age-of-dam adjustment factors for scrotal circumference in yearling beef bulls.

Age of Dam (years)	Adjustment Factor (cm)
≥ 5	+ 0.0
4	+ 0.4
3	+ 0.8
2	+ 1.3

Adapted from Lunstra et al. 1988.

Table 9.2. Adjustment factors for bull scrotal circumference at weaning.

Breed	Age Adjusted (cm/day)	Regression Coefficient
Angus	0.856	1.54
Brangus	0.861	1.60
Charolais	0.077	1.54
Hereford	0.042	1.41
Simmental	0.085	1.59

Source: Geske et al. 1995.
The adjusted 205-day scrotal circumference multiplied by the regression coefficient gives the expected 365-day scrotal circumference.

greatest circumference of both testicles. It is best to take two consecutive measurements, with them not varying by more than 1.0 cm. Variations in results between evaluators led to development of a precalibrated, spring-loaded tape to standardize results (Coulter and Kastelic 1999).

The prospect of using SC as a predictive selection or screening criterion in prepubertal bulls is attractive, as there is an economic advantage in early prediction. Some studies have examined relationships with subsequent, pubertal, SC (Pratt et al. 1991, Geske et al. 1995). Both studies confirmed that bulls with small prepubertal testes were unlikely to develop large testes postpubertally. In the latter study, SC was measured on 329 bulls, representing different beef breeds, at the start of a 140-day gain test and at 365 days of age. For young Angus, Simmental, and Zebu-derived breeds, the threshold to achieve 30 cm by 365 days of age was 23 cm, whereas it was 26 cm for other continental breeds and Polled Shorthorn bulls. In Kansas, Geske et al. (1995) determined that a SC of approximately 21 cm was necessary at 205 days of age for British- and European-breed bulls to achieve 32 cm at 1 year of age, with adjustment factors being calculated for the different breeds studied (Table 9.2). SC thresholds recommended by the Society for Theriogenology for a "satisfactory" breeding bull are shown in Table 9.3.

These thresholds represent minimum standards that can be applied to different genotypes in a variety of environments. Breed associations, breeders, and sales organizations are encouraged to elevate these standards where data are supportive. Here, the system can be "fine-tuned" to reflect different biological types of cattle, especially where nutrition is predictable and consistent. For example, minimum standards have been established for different genotypes raised under similar conditions (grain fed in pens) in western Canada (Table 9.4). In addition, higher

Table 9.3. Minimum thresholds for bull scrotal circumference.

Age of Bull (months)	Minimum Scrotal Circumference (cm)
≤ 15	30
$>15 \leq 18$	31
$>18 \leq 21$	32
$>21 \leq 24$	33
≥ 24	34

Source: Chenoweth et al. 1992.

standards may be used where the objective of the evaluation is to identify "superior," rather than "satisfactory," prospective breeders.

Semen Collection

Semen can be collected from bulls by a variety of means including trans-rectal massage, use of an artificial vagina, and EE. The last method is the one most commonly used with range-type bulls in a number of countries, and it is highly efficient at obtaining a representative semen sample from pubertal bulls (Perry et al. 2002). Such samples are comparable with those obtained with an artificial vagina in terms of semen fertility and freezability (Singleton 1970). Although modern machines and probes cause much less physical reaction in bulls than did earlier machines (Ball 1986), welfare concerns have been raised concerning this procedure, as discussed in chapter 14.

Electroejaculators and Probes

Commercially available electroejaculators are powered by alternating current, internal rechargeable batteries, or by 12-volt automobile batteries. Effective EE necessitates stimulation of those pelvic nerves

Table 9.4. Recommended minimum scrotal circumferences (cm) for grain-fed bulls in western Canada.

Age of Bull (months)	Simmental	Angus, Charolais, Maine Anjou, Gelbvieh	Hereford, Shorthorn, Galloway	Limousin, Salers, Blonde d'Aquitaine	Texas Longhorn
12	32	31	30	29	28
13	33	32	31	30	29
14	34	33	32	31	30
15–20	35	34	33	32	31
21–30	36	35	34	33	32
>30	37	36	35	34	33

Source: Barth 1993.

controlling semen emission, as well as those controlling erection and ejaculation. Newer probe designs have full-length longitudinal electrodes that stimulate all functions simultaneously while causing considerably less muscle contraction than did earlier probe designs, such as those using bipolar ring electrodes (Ball 1986). Available machines vary in control mechanisms and features, with several being capable of preprogrammed stimulation regimens. Variations of stimulus patterns occur with different machines and operators. However, an overriding consideration should always be the welfare of the animal. Stimulation should be discontinued if undue stress is being caused or physical injury to the bull appears likely.

Preparation and Stimulation of the Bull

The rectum of the bull should be emptied of feces before the lubricated probe is inserted so that the anal sphincter contracts behind the main body of the unit. Procedures for EE of bulls are described elsewhere (Chenoweth et al. 1992, Perry et al. 2002). An important consideration is ensuring that a representative semen sample is collected with this technique. Lack of attention here can lead to errors in semen evaluation, as the initial portion of the ejaculate may contain large numbers of degenerating accumulated sperm in those bulls that have not been recently sexually active. Similarly, if an ejaculate shows substandard motility in the absence of an obvious physical cause, a second sample collected within a short period of time (e.g., 5–10 minutes) can often show improvement in this trait.

Another important consideration is that unless the penis is exteriorized for examination, it is difficult to ensure that the member is without problems. Penile exteriorization may be achieved with EE, although in young bulls, this may be facilitated by applying pressure on the sigmoid flexure from the rear. Exteriorization may also be achieved secondarily to massage of the internal genitalia. External manipulation through the prepuce may also be successful in exteriorizing the penis. Detectable penile conditions include trauma and adhesions within the prepuce, penile hematoma, phimosis, paraphimosis, fibropapilloma, persistent frenulum, hair rings, and circulatory disorders of the corpus cavernosum. One caveat applies when diagnosing deviations of the bull penis: Although deviations of the penis may occasionally be observed during EE, definitive diagnosis depends on observation during actual breeding attempts.

Semen Collection Devices

Semen should be collected into a prewarmed, insulated or jacketed tube. All surfaces coming into contact with semen should be clean, warm, dry, and free of spermatoxic agents. Semen should be maintained at above 30°C (preferably 35°–37°C) until on-site evaluation procedures are complete.

SEMEN EVALUATION

Initial Impressions of Semen

Volume, density, and gross characteristics of the ejaculate are not "frontline" BSE assessments, as they do not appear to be directly related to fertility. However, it is often useful to record this information. Likewise, spermatozoal concentration assessment is not mandatory, as SC measurement is considered to provide a better estimate of sperm production in range-type bulls that are infrequently examined. However, such information may help to evaluate the success of semen collection.

Other recorded observations should include evidence of contamination, hemorrhage, or inflammatory material. If the ejaculate contains sufficient purulent material for it to be obvious to the naked eye, then the bull should not be classified as satisfactory, at least until a benign cause is identified. Debris or contamination from the sheath is regarded less seriously unless it is indicative of active infection.

Sperm Motility

Sperm motility is assessed microscopically, with two methods being commonly employed: gross motility (or mass activity) and individual motility (or percentage progressive motility). It is worthwhile to use both methods where practicable, as they differ somewhat in interpretation and precision. With all motility estimations, it is important to protect semen against adverse effects (e.g., cold shock) and to do the estimation as soon as possible after semen collection.

Gross motility, or the degree of swirling (or wave motion) present in an undiluted semen sample, is a function of both sperm concentration and individual motility. Under field conditions, motility is typically assessed by placing a drop of raw semen on a warm slide and observing it at a magnification of 100 (10× eyepiece and 10× objective). With the condenser properly adjusted, mass action or "swirl" can be observed in samples that have a sufficient concentration of motile spermatozoa to produce this effect. The categories for gross motility assessment are shown in Table 9.5.

Individual progressive sperm motility is typically assessed using either a brightfield or phase-contrast microscope, preferably one equipped with a warm stage or similar means of preventing cold shock of the spermatozoa. Cover-slipped specimens are usually examined at a total magnification of 400×. In dense samples (milky or creamy), the sample should be diluted for proper observation of individual spermatozoa. Suitable extenders include 2.9% sodium citrate and physiological sterile saline, although readings should not be delayed when using the latter. The percentage of active, progressively motile cells is estimated. This procedure requires more experience to achieve competence than does gross motility estimation, but it is potentially more accurate. Individual motility ratings range from poor to very good. Greater than 70% motility achieves a rating of very good; 50%–69% is rated good, 30%–49% motility is fair, and less than 30% motility earns a poor rating for the sample. In the current SFT system, 30% motility represents the threshold for acceptability.

Table 9.5. Categories employed for gross motility assessment.

Gross Motility (Mass Activity)	Rating
Rapid swirl	Very good
Slower swirl	Good
General oscillation	Fair
Sporadic oscillation	Poor

Microscopic observation of semen samples at 400× can also help identify the presence of other cells (e.g., squamous epithelial, inflammatory, or spermatogenic cells) within the sample, with specific stains and preparations being used for more precise identification. These can include the staining of semen smears with Dif-Quik, new methylene blue, or other differential blood cell stains, whereas bacteria may be categorized using a Gram stain.

Morphology of Spermatozoa

Morphology of spermatozoa (differential counts of normal and abnormal cells) is assessed either by phase microscopy with "fixed" semen preparations (using formol-buffered saline or phosphate-buffered-saline-gluteraldehyde, for example) or by using brightfield microscopy of stained smears. Common stains used for this purpose include nigrosin-eosin, William's stain, and modified Giemsa. The Society for Theriogenology recommends use of the nigrosin-eosin stain for its combination of ease and utility. Although this stain may also be used as a "supravital" stain (i.e., sperm that are "alive" at staining will not absorb stain, whereas those that are "dead" will partially or completely absorb the red eosin color), here it is primarily used for its ability to depict sperm morphology.

Brightfield microscopy of stained smears is best done at 1000 × plus, using an oil-immersion objective. At least 100 spermatozoa should be observed in different fields and classified for normality or abnormality. To achieve a "satisfactory" BSE category in the current SFT system, bulls should have at least 70% normal sperm. This threshold applies regardless of the types of sperm abnormalities involved.

Sperm abnormalities may be placed into categories, such as "primary" or "secondary" (Blom 1950). Primary abnormalities are considered to be caused during spermatogenesis, with secondary abnormalities occurring subsequent to sperm release. Despite the fact that this is an oversimplification, this system is useful for reference. Although bulls are categorized based on total sperm abnormalities, primary and secondary abnormalities can be collated to arrive at this number. The recording of such information can also be useful for decisions on prognoses and for monitoring progress in problem bulls.

Primary sperm abnormalities include head and midpiece abnormalities, as well as proximal droplets, strongly coiled tails, double heads or midpieces, and acrosome abnormalities. Secondary sperm abnormalities include loose normal heads, loose and folded acrosomes, abaxial tails, simple bent tails, and distal droplets. Other cells in semen that may be observed under routine microscopy include epithelial cells, Medusa cells, and erythrocytes. Sperm precursors and white blood cells may also be visible.

Assessment of Frozen Semen

Assessment of frozen semen is often indicated when things go wrong, such as when artificial insemination pregnancy rates are disappointing, or when there are doubts concerning adequate liquid nitrogen levels. It is also good operating procedure to routinely perform a check on a representative sample of frozen semen before start of breeding. Assessment of frozen semen requires appropriate equipment and the use of established protocols, such as those recommended by the Society for Theriogenology as well as by most large artificial insemination organizations.

The final step of the BSE is to place bulls into categories, such as those following, recommended by the Society for Theriogenology.

Satisfactory

Satisfactory bulls equal or surpass the minimum thresholds for SC, sperm motility, and sperm morphology and have no evident genetic, infectious, or other problems or faults that could compromise breeding or fertility.

Unsatisfactory

Unsatisfactory bulls are below one or more thresholds and are highly unlikely to ever improve their status. Also included are bulls that show genetic faults or irrevocable physical problems (including infectious disease) that would compromise breeding or fertility.

Classification Deferred

Bulls that have their classification deferred are those that do not fit into the above categories and that could benefit from a retest. This category would include bulls with an "immature" semen profile as well as any bulls whose semen is substandard but considered capable of improvement. (It is not uncommon for yearling bulls to exhibit higher levels of certain types of spermatozoal abnormalities that are associated with immaturity. Such bulls will usually require a second examination before being classified as a satisfactory potential breeder.) Also in the category of "classification deferred" are bulls from which a satisfactory ejaculate could not be obtained for reasons unknown, as well as bulls with treatable problems such as vesicular adenitis (seminal vesiculitis) or foot rot. In general, if any doubt exists about a bull fitting into either the satisfactory or unsatisfactory categories, he should be considered a candidate for a retest and placed into the "classification deferred" category.

Other suggested assessment categories for bulls include "fail," "retest," "suitable for natural mating only," and "pass" (Gardiner and Fordyce 2002).

"Red Flags"

Bulls should not pass the BSE if any of the following are present:

- A genetic fault (e.g., umbilical hernia)
- One testicle (even if very large)
- Frank pus in semen
- Active accessory genital disease (e.g., seminal vesiculitis)
- Significant loss of vision
- Positive vibriosis or trichomoniasis test
- Lameness.

BSE Utilization

The BSE, in various forms, has been in use for at least 50 years. However, surveys in the United States indicate that its use with sale bulls is static and prone to go down when cattle prices drop. Particularly disappointing is the low number of herd bulls tested before each breeding season (NAHMS 1998a). More encouraging is the trend for larger operations and those that derive most or all income from cattle to have relatively high adoption rates (NAHMS 1998b). For greater acceptance of procedures such as the BSE, producers should be assured that BSE procedures are cost-effective, that they are not too disruptive or difficult to implement, and that the results are credible and unambiguous.

Cost-Benefits of Bull Evaluation

Bulls vary in their reproductive capabilities. Although it may be relatively simple, at times, to recognize a sterile bull, identifying subfertile bulls is more difficult, especially in multisire breeding herds. Use of the BSE allows bulls to be categorized reasonably accurately. For example, several trials have used bulls with groups of estrus-synchronized heifers to increase breeding pressures. In one such study, bulls classified as satisfactory using the BSE obtained 10.5% more pregnancies than those categorized as questionable. In a similar study, satisfactory bulls obtained a pregnancy rate advantage of 6% over the synchronized breeding period, and 9% ($P < .10$) for the entire breeding season, when compared with questionable bulls.

Another approach has determined bull fertility from calf parentage in multisire herds. In one such study, the most important bull factors for herd fertility included large SC and low numbers of sperm defects. In another, prior BSE screening of bulls in a large herd increased pregnancies 3.5% while using 40% fewer bulls over a shortened breeding period. A large Australian study (Table 9.6) employing DNA parentage (Holroyd et al. 1998, p. 420) concluded that "semen quality, particularly percent normal spermatozoa, was consistently related to calf output." The summation from this study was that, "(T)hese results confirm that semen examination, including sperm morphology, should be standard procedure when assessing bulls for reproductive soundness" (Fitzpatrick et al. 1998, p. 407).

From these studies, producers can conclude that the BSE is an effective tool for reducing the uncertainty associated with bull purchase and usage. A conservative estimate, based on research (Wiltbank and Parish 1986) is that bulls passing a BSE (or semen quality test) should have a 6% or greater fertility advantage over unevaluated bulls. Larger differences are, of course, possible when satisfactory bulls are compared with those that fail the BSE.

In addition to increased calf crop, benefits accrue through increased weaning weights of older calves at weaning because females become pregnant earlier in the breeding season (Figure 9.1). In this example, two bulls ("Fast Track" and "Slow Track") are compared, both of which eventually achieve the same percentage of pregnancies (94%) in their respective herds. However, the "Fast Track" bull achieves this in three cycles, whereas the "Slow Track" bull takes six cycles to do the same job. For a hypothetical U.S. herd of 100 females, this resulted in a difference of $7000 in calf sales (based on $1/lb). Missing one breeding cycle costs approximately 20 kg in individual calf weight at weaning time. A Texas study showed that cows placed with satisfactory BSE bulls weaned 7.4% more calves than did those bred to untested bulls (Wiltbank and Parish 1986). Based on current U.S. prices, a conservative 6% increase in calf crop at weaning would represent an approximate return of $20–$25 for each $1 invested in the BSE.

The last two items on the producer acceptance list relate to aspects of equipment, technique, and interpretation. Examples of poor procedures that can reduce the validity of results, as well as producer acceptance, include

- Poor microscope quality and maintenance
- Poor preparation of semen stains/smears
- Use of 400 × for sperm morphology assessment (instead of 1000 ×)
- "Rough" and inappropriate EE technique
- Poor SC measuring technique.

Behavioral Aspects of Bull Fertility

Bull Sex Drive

The ability of a bull to produce an adequate amount of good-quality semen does not necessarily guarantee superior reproductive performance, which can be compromised by other considerations such as bull sexual behavior (or that behavior associated with the detection and service of estrous females). Bull sexual behavior includes libido, or sex drive, which is the willingness or eagerness of a bull to attempt mounting and service. "Mating ability" refers to the ability of the bull to complete service. "Serving capacity" is a measure of the number of services achieved by a bull under stipulated conditions, and it thus includes aspects of both libido and mating ability.

Table 9.6. Fertility relationships in range-bred bulls in Queensland.

Site (n)	A (10)	B (8)	C (12)	D (10)	E (9)	F (23)
Domin.	–	NS	* * *	* * *	–	–
Scrot. circ.	**	NS	NS	NS	NS	NS
% motile	NS	NS	NS	–	NS	NS
% normal	NS	**	NS	**	*	* * *
Sex drive	NS	**	NS	–	NS	NS

Adapted from Holroyd et al. 2002.

*, **, * * * = relative significance of relationship, NS = not significant.

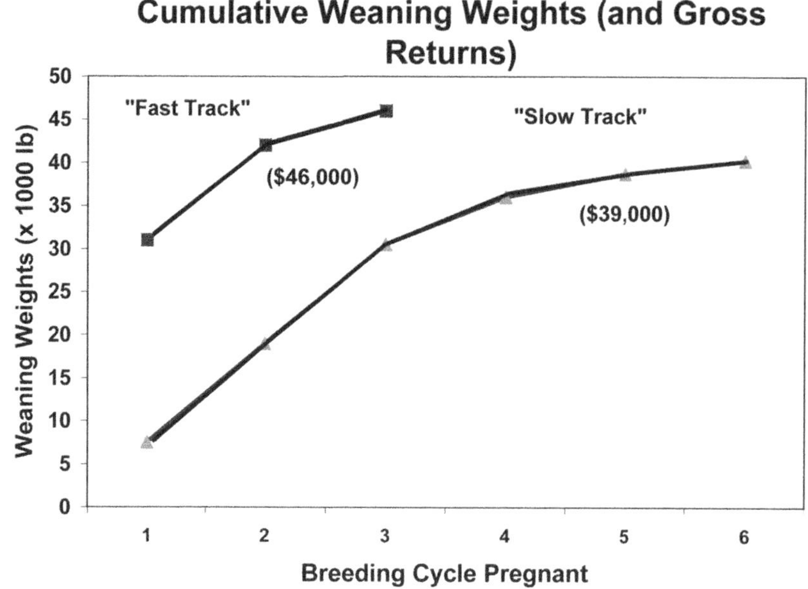

Figure 9.1. Cumulative weaning weights and estimated economic returns for "Fast track" and "Slow track" bulls.

Producers who use testing procedures developed for bulls benefit from knowing that bull libido has a significant genetic component, bulls are polygynous and tend to distribute their services among receptive females, the greatest single stimulus for bulls to mount and service is an immobile object that resembles the rear end of a female, prestimulation of bulls increases their sexual response, and competition among bulls can increase their sexual response (Chenoweth 1986).

Methods to Assess Bull Sex Drive

A number of procedures have been used to assess sex drive in bulls. Although it would appear simple to observe bull behavior in the breeding pasture or paddock, this approach has proven disappointing, at least in terms of repeatable, quantitative results. At present, the best way to generate such information is via a structured, pen-type testing regimen. With one of these procedures, for example, the libido score test, prestimulated bulls are introduced to the pen containing mildly sedated, restrained females at a bull-to-female ratio (BFR) of 2:2 and observed for 10 minutes. Eleven score categories are described in Figure 9.2.

Fertility Relationships

There is convincing evidence that bull libido influences fertility in natural breeding herds. For example, in one study, better first-cycle pregnancy rates were obtained in heifers mated with higher-serving-capacity bulls than in those mated with bulls of low serving capacity (Blockey 1978). More recently, differences in pregnancy rates were demonstrated between high-, medium-, and low-serving-capacity Hereford bulls (Blockey 1989). Other studies have shown advantages in herd fertility for bulls with higher sex drive (Chenoweth 1997). Close relationships were reported between bull rankings for fertility, libido score, and testosterone response to gonadotropin-releasing hormone (Post et al. 1987). In Texas, one study showed that bull libido assessment provided greater prediction of bull fertility than did semen assessment alone (Smith et al. 1981). Several studies employing multisire mating and progeny identification have shown positive relationships between measures of bull sex drive and their calf-getting (Coulter and Kozub 1989, Holroyd et al. 2002a).

However, other studies have revealed poor or inconclusive relationships between bull libido/serving capacity assessment and herd fertility or pasture performance (Chenoweth 1997), even though higher-libido bulls serviced more often and serviced more females than did lower-libido bulls in the breeding pasture (Boyd et al. 1989, 1990, Farin et al. 1989, Godfrey and Lunstra, 1989).

Such apparently contradictory findings may occur as a result of differing approaches and

Table 9.7. Least-square means of mating performance within breeding soundness examination and libido classifications.

	Exam Classifications		Libido	
	Satisfactory	Questionable	High	Medium
No. bulls	80	12	69	23
No. mounts	146.3	120.7	112[a]	155[b]
No. services	47.8	42.4	52.8[c]	37.5[d]
Mounts: services	5.8	4.8	3.1[c]	7.5[d]
Serviced/estrus (%)	73.5	71.4	81.3[c]	63.5[d]
Pregnant/serviced (%)	56.1	50.8	51.8	56.1
Pregnant/estrus (%)	44.8	36.7	43.7	37.8
Total pregnancy rate (%)	45.6[e]	36.5[f]	41.5	40.6

Source: Farin et al. 1989.
[a,b]Means differ ($P < .05$).
[c,d]Means differ ($P < .01$).
[e,f]Means differ ($P < .10$).

Libido score	Sexual behavior
0	No sexual interest
1	Sexual interest shown once only
2	Sexual interest shown more than once
3	Active pursuit–persistent interest
4	One mount or attempt; no service
5	Two mounts or attempts; no service
6	Three mounts or attempts; no service
7	One service; no further sexual interest
8	On service; continued interest
9	Two services; no further sexual interest
10	Two services followed by further sexual interest, including service

Figure 9.2. Libido score categories. (Chenoweth 1983).

methodologies. In some trials, bulls were not placed under sufficient breeding stress to demonstrate differences, which may have been more apparent with use of more demanding BFRs or shorter breeding periods. In addition, bulls of potentially low fertility were excluded from cooperative breeding trials.

Even though bulls may be superior in one trait, their fertility can be compromised by a deficiency in another. This was particularly well illustrated in one study (Farin et al. 1989), in which 92 beef bulls were placed both into "satisfactory" and "questionable" BSE categories, as well as into high-libido (score 9–10) and medium-libido (score 7–8) categories before single-sire mating with groups of estrus-synchronized heifers (Table 9.7). Here, the pregnancy rate was 9.1% higher for satisfactory BSE bulls compared with those in the questionable BSE category, but it did not differ between bulls of high and medium libido scores, even though high-libido bulls serviced more females and served more times than did medium-libido bulls. This paradox apparently occurred because a lower percentage of serviced females in the high-libido group became pregnant than did those in the medium-libido group. Here, differences between bulls in sex drive were clearly masked by differences in BSE components. Factors associated with BSE (SC, sperm motility, and sperm morphology) can separately influence fertility and are apparently not linked with sex drive.

Test repeatability was addressed by Landaeta-Hernandez and Chenoweth (2001) with 26 *Bos taurus* bulls, approximately 2 years of age, that were subjected to eight tests for libido score, serving capacity, and reaction time over a 2-month period. Although overall repeatabilities for all tests were low, those bulls that performed well ($n = 8$) in the initial test were consistent throughout, whereas another group of bulls ($n = 4$) consistently performed poorly. The remainder showed varying degrees of improvement over the course of testing, which was attributed to increased experience and confidence.

The ability of bulls to service females is related not only to their inherent sex drive but also to their mating, or breeding, ability. Problems in mating ability may be caused by a number of physical and pathological causes, including skeletal and penile abnormalities (Chenoweth 1983). Identification of such problems may be difficult when animals are at pasture, being more easily observed during the course of a libido/serving capacity test procedure.

Other factors can influence the development and expression of bull sex drive and competent mating behavior. These include not only genetic aspects, including breed differences (Chenoweth 1991), but also age and rearing effects (including all-male raising and previous heterosexual experience) and social effects in the breeding pasture or paddock (Chenoweth 1986). The effects of environmental factors, such as heat stress, on bull reproductive behavior are not well established, although there is good evidence that they can adversely affect estrous manifestation in females. Other stressors (e.g., restraint, EE) can lower gonadotropin levels, and thus libido, via physiological mechanisms mediated by glucocorticoids, cytokines, and other factors.

Alternative Bull Libido Assessment Procedures

The indirect determination of bull libido (e.g., through blood hormone levels) is an attractive proposition, as it could reduce or eliminate the time, labor, and aesthetic or welfare concerns that might occur with current methods of assessing sex drive in bulls, and it might perhaps increase predictability. It would also allow assessment of bulls that react poorly to the handling aspects of such tests. However, attempts to associate luteinizing hormone (LH) or testosterone (T) levels with bull sex drive have been generally disappointing (Chenoweth 1997), probably because of the episodic nature of hormone release and the inhibiting effects of handling or restraint on the animal. By inducing LH or T release with parenteral administration of gonadotropin-releasing hormone, there is potential to avoid some of these difficulties. In one study (Post et al. 1987), a significant relationship occurred between induced T levels and bull fertility; in another (Perry et al. 1990), positive relationships were obtained with induced LH levels. However, other studies have obtained disappointing results when attempting to relate gonadotropin-releasing-hormone-induced levels of testosterone or LH with bull sex drive (Byerley et al. 1990, Chase et al. 1997).

Complexities of Bull Fertility Assessment

The expression of bull sexual behavior is strongly influenced by genetics, although environment and bull age and experience can also be influential. Interactions occur between genetics and environmental factors, and these are currently poorly understood.

Serving capacity tests and libido scores are useful for categorizing bulls in terms of their mating activity and ability, particularly for top-performing bulls. However, such tests vary in their ability to predict bull fertility per se. This is because there are other influences on fertility, such as BSE factors, and these are not necessarily positively associated with libido. In addition, social relationships and interactions can influence bull sexual behavior, especially when bulls are placed together in multisire mating groups. Even when multiple assessment factors are included in models used to predict fertility, these models account for only a relatively small proportion of the variability in bull fertility in multisire breeding herds. This indicates that research has yet to reveal some highly important factors relating to bull performance, fertility, and interactions with environmental factors, as well as genetic relationships with other traits. Current work with paternity testing using DNAfp, both in Australia and the United States, and the potential offered by the use of such markers, will help to resolve some of these issues.

Pitfalls in Libido/Service Capacity Testing

Successful testing of bulls for libido and mating ability requires careful planning. The following points represent some recommendations for testing procedures that employ restrained, nonestrus females.

- Testing of bulls that are excessively apprehensive or agitated is not recommended. Precautions should be taken to handle cattle quietly and calmly and to avoid unnecessary distractions during the test.

- Bulls should not be tested immediately following their subjection to other procedures such as EE, vaccination, or parasite control measures.
- Testing should not occur under adverse weather conditions, such as in extreme heat, cold, or inclement weather.
- Testing of bulls in groups in which one or more bulls are markedly dominant, such as with mixed-age groups of bulls, should be avoided. The exposure of only two bulls to a test pen at any one time, and subsequent retesting with different combinations of bulls, helps to minimize this problem. It should be noted, however, that a dominant bull can exert an inhibitory effect from a distance (e.g., from an adjacent pen). It is also recommended that restrained females be located at an appropriate distance from each other (5 meters or more). Females used for testing should be docile and tractable. Other considerations are discussed in Chapter 14.
- Use of inadequate stimuli will affect results. Restrained females should be incapable of excessive movement, or some bulls may be deterred. Overly restless or agitated females should be replaced. The service crates used should not impede mounting and service.
- Spreading of venereal diseases can occur during testing. Precautions should be taken to ensure that diseases such as vibriosis and trichomoniasis are not transmitted via the testing procedure.

It is important to acknowledge that animal welfare concerns have been expressed concerning the conduct of such tests, as discussed in chapter 14.

CONDITIONING AND PREPARATION OF BULLS BEFORE BREEDING

Acclimation and quarantine represent an important period in the breeding career of a bull, for several reasons. First, it enables the bull to adapt to his new environment (temperature, feed, water, "germs and bugs"). Second, this period enables the owner to quarantine new bulls and observe them for any signs of disease (as well as other problems). Third, it provides an opportunity to introduce the new bull to the same health program (vaccinations, antiparasiticides, trace elements, and so forth) that is adopted for the rest of the herd.

Bulls are often relocated for various reasons. The distances involved may be great, and the "new" location may be markedly different, both environmentally and socially, from the previous one. Problems associated with such relocation and attempted adaptation are often not diagnosed until bulls have been long resident on the new property, by which time the problems may be evident in reduced pregnancy rates or delayed and strung-out calving (McCool and Holroyd 1993). This topic is complex and poorly researched. In the meantime, although acknowledging that multiple causal factors exist and interact, recommendations to minimize risks associated with bull relocation include

- Purchasing young bulls because they have more capacity to adapt, both environmentally and socially
- Introducing bulls when conditions are optimal at the new site
- Introducing bulls to their new environment gradually by holding and feeding them in yards initially, then releasing them into small paddocks under supervision.

BREEDING SEASON

USE OF YOUNG (1- TO 2-YEAR-OLD) BULLS

If genetic progress in a population is a function of both heritability of favorable traits and generation interval, then young bulls should help maximize genetic progress in the herd (especially if mated with heifers). It is not uncommon for young bulls to be placed into service soon after purchase, with variable results. This may occur simply because the "normal" bull spermiogram is usually not attained until several months postpuberty (Coulter 1994). In addition, rate of testicular development, puberty onset, and development of a stable spermiogram are all largely dictated by genotype–environment interactions (Fields et al. 1979, Perry et al. 1991). Despite this, scrotal size at puberty is remarkably consistent (26–27 cm) across different breeds and environments (Coulter 1994).

Surveys indicate that young bulls are generally given lower mating loads than are older bulls (Chenoweth 2000), and research tends to vindicate this approach. For example, studies on sexual development in early postpubertal bulls show that there is a variable period during which they exhibit elevated levels of secondary sperm abnormalities (Coulter 1994, Soderquist 2000). Overconditioned bulls often require several months of "let-down" before they achieve acceptable semen quality. In breeding trials with synchronized females in Colorado, yearling bulls achieved lower fertility than did older bulls, despite having passed a BSE and exhibiting similar sexual activity (number of estrous females detected and serviced) to that of the older bulls (Farin et al. 1989). Similarly, in tropical Australia, a comparison of the pregnancy patterns achieved by yearling and 2-year-old bulls indicated significant differences in favor of the older bulls in terms of pregnancy rate (5.2 pregnancies/week vs. 4.3 pregnancies/week, respectively). At any given time, the yearling bulls were only 59% as successful as the older bulls in achieving pregnancies (G. Fordyce, unpublished data). Thus, young bulls appear to have reduced reproductive capability compared with older bulls, and this is most likely associated with aspects of sperm production, both quantitative and qualitative.

Multisire Breeding and BFRs

Traditional recommendations for BFRs of 1:20–1:30 for herd bulls considerably underestimate the capabilities of competent bulls. For example, one study compared single-and multisire systems with Hereford bulls at BFRs of 1:25, 1:44, and 1:60 (Rupp et al. 1977). Here, the conclusion was that the reproductive capability (fertility, libido, and mating ability) of individual sires was more important than either BFR or use of single-sire versus multisire breeding systems.

In Colorado, yearling Hereford bulls that were preassessed for breeding soundness and libido were compared at BFRs of 1:20 and 2:40 with estrus-synchronized crossbred heifers (Farin et al. 1982). Overall, bull mating performance and pregnancy rates did not differ between BFRs. Comparison of a variety of single-sire BFRs (1:7 to 1:51), also with estrus-synchronized females, found that BFR was not a limiting factor to fertility, even at the lowest BFRs (Pexton et al. 1990). Two studies conducted in vastly different environments showed that BSE-screened bulls increased herd pregnancy rates at reduced BFRs (Prince et al. 1987, McCosker et al. 1989).

Thus, it is apparent that sound bulls that have passed a BSE can handle considerably more females during a generic breeding season than traditional recommendations would indicate. It is also evident that most producers have yet to take full advantage of these findings. For example, U.S. surveys indicate that overall, cattle-breeding operations use yearling bulls at a BFR of 1:17.5 and mature bulls at one of 1:25—figures that have changed little in recent years (NAHMS 1998a). Cattle breeders in the Rocky Mountain West used a mean BFR of 1:21, with 25% of herds employing a BFR of less than 1:18 (Sanderson and Gay 1996). Belated recognition of the capabilities of competent bulls has come with more recent U.S. industry recommendations for bull ratios of 1:40 (mature) and 1:15–20 (yearling) (Ewbank 1996).

Influence of Breeding Paddock Size and Terrain

Female cattle tend to form a sexually active group when more than several are in estrus or proestrus at the same time. This group is very mobile but tends to seek out the bull or bull group and keep it within eyeshot. In a smaller, easily traversed paddock, this is a relatively simple system. However, as distances get greater, especially when there are physical obstacles (e.g., creeks, hills, and so forth), then several groups may form, each with their own bull or bulls. Under such conditions, it is difficult to make inflexible recommendations about optimal BFRs.

Managing Bull Social Behavior

In multisire breeding herds, it is recommended that the bull group be relatively homogenous (age, size, and breed type) and relatively young (<3.5–4 years old) and that they have been allowed to establish their social ranking before (at least 1–2 weeks) commencement of the breeding season.

Bull Rotation

Although supportive data are lacking, it is often recommended that producers rotate bulls during the breeding season to optimize performance and reduce risk of disastrous results (Ewbank 1996). One approach is to use more mature bulls at commencement of the breeding season, and younger bulls later in the season. Another approach is to rotate mature bulls with younger bulls at 2-week intervals. Either approach is believed to limit the harmful effects of social dominance between bulls, while maximizing the possible effects of biostimulation.

Monitoring of the Breeding Herd

It is not uncommon for bulls to suffer an injury or disease during the breeding season that adversely affects performance. Common injuries include preputial prolapse (and laceration), "broken penis," and feet and leg injuries. Disease entities include bovine ephemeral fever (not found in the United States), which can adversely affect bull semen for 3–4 months or more, pinkeye, cancer eye, and so forth. Monitoring the breeding herd at regular intervals will help to detect problems early, and thus help to minimize costly losses.

"Feed to Breed"

A natural-breeding bull should be a "sexual athlete." This means that he should be structurally sound, have good sperm production and delivery systems, and be physically fit. The importance of these aspects is reinforced by knowing that it is not unusual for normal bulls to service 20–30 times per day.

As active breeding bulls can lose 50–100 kg during the breeding season (Ellis 2003), they need to be carrying some condition (but not too much) at the start of the breeding season. For this reason, it is recommended that the body condition scores of bulls at the start of breeding be greater than 4 and less than 7.

Nutrient restriction during the prepubertal period will retard the onset of puberty, reduce testicular growth at any given age or stage, and possibly reduce future sperm production potential (Entwistle 2003). However, overfeeding of bulls can also be problematic. Although this practice is now less common than before, young bulls are often fed high-energy diets as part of their development and assessment process. If this continues until bulls are overconditioned, a number of adverse effects may be seen:

- They may no longer be "athletes," a loss of status that may be associated with feet and leg problems.
- Overconditioning has been shown to adversely affect bull sex drive.
- Overconditioning can adversely affect semen quality. This is particularly true when bulls have poor scrotal conformation or presence of fat in the scrotal neck and when the temperature–humidity index is high.

High-energy diets are, in general, not beneficial for reproductive traits in postpubertal bulls (Coulter 1994) and may actually adversely affect reproductive traits. Feeding high-energy diets to bulls for a prolonged period can compromise their testicular thermoregulatory capacity, leading to greater sensitivity to the spermatoxic effects of high ambient temperatures (Entwistle 2003).

Micronutrients (e.g., zinc, selenium, and vitamins A and E), as well as spermatoxic agents, can also influence bull reproductive performance. An example of the latter is gossypol (derived from cotton seed), which, under the right conditions, can adversely affect semen quality and sperm production in bulls (Chenoweth et al. 1994).

NONBREEDING SEASON

Bulls should not be ignored during the period that they are removed from females. For example, it can be useful to conduct a postbreeding bull appraisal. This allows for bulls that have severe problems to be identified and culled and for those with fixable problems to be attended to. During the nonbreeding season, bulls should be managed to ensure optimal performance in their next breeding season and to prolong their reproductive lifespan. Thus, emphasis should be placed on maintaining bull condition, attending to preventive health measures (vaccination, parasites, hoof treatment), minimizing opportunities for bull injury and adverse environmental effects, and promoting growth and development of young bulls.

Attention to nutrition over this period is important. Mature and young bulls should be fed different diets according to their different needs. This may involve separation based on age. It is also useful to keep bulls in homogenous groups in paddocks that are large enough to allow for adequate exercise and that provide protection (e.g., shade trees) from extreme weather conditions. In more extensive regions, such as Australia, positive nutritional intervention in bulls during the "dry season" has been shown to have beneficial effects on sperm production during the ensuing breeding season (Entwistle 2003).

MISCELLANEOUS CONCERNS AND DISORDERS

Heat Stress

Heat stress is caused primarily by elevated environmental temperature, and it can be influenced by high humidity, thermal radiation, and low air movement. Zones of thermal comfort for cattle are available whereby a temperature–humidity index is often employed to describe the combined effects of heat and humidity. Breed types vary greatly in their adaptability to heat stress. For example, British-breed beef bulls have shown a summer depression of semen quality in the Gulf States of the United States, whereas Zebu bulls showed little or no adverse effects. Both scrotal and testicular configuration are important for maintaining appropriate testicular thermoregulation, with poor conformation or fat in the scrotal neck acting to compromise this ability.

In cattle, common problems associated with heat stress include depressed voluntary intake, decreased growth, spermatogenic dysfunction (including lowered sperm production), depressed fertilization, and early pregnancy rates (particularly the first 15 days).

The spermatogenic epithelium generally reacts to stressors, including heat, in a stereotypic fashion (Kenney 1970, Saacke et al. 2000). Even under "normal" conditions, spermatogenesis occurs in an environment that is oxygen poor and very susceptible to the adverse actions of reactive oxygen species. Even a relatively mild testicular insult such as scrotal insulation for 48 hours will cause a predictable cascade of effects that are represented by an abnormal spermiogram, commencing within days of the insult and continuing for over 1 month (Vogler et al. 1993). Major spermatogenic damage from heat stress may not be evident for a month or two after the event, and recovery, if it occurs, can take as long.

The similarity of the spermatogenic response to a wide variety of stressors indicates that a combination of stressors could have a compounding effect on the degree of spermatogenic damage. Thus, a combination of relocation, acclimatization, and heat stress could combine to produce long-term severe endocrinological and spermatogenic disruption in bulls (Welsh and Johnson 1981).

To minimize heat stress problems, producers should use environmentally adapted cattle and ensure adequate shade in pastures and paddocks. Producers should rely on bulls with good scrotal conformation for effective thermoregulation and should avoid combining stressors such as transport, acclimation, and heat. In addition, producers should avoid breeding at times of the year when the temperature–humidity index is highest.

Vesicular Adenitis (Seminal Vesiculitis)

Seminal vesiculitis is often diagnosed during BSEs of bulls (Larson 1997). Although it is most often encountered in young, peripubertal bulls housed together and fed high-energy diets, seminal vesiculitis is also diagnosed in aged bulls. Prevalence rates in different bull groups have ranged from 0.9% to 49% (Linhart and Parker 1988). Initial detection generally occurs when internal reproductive organs are examined transrectally. The pathogenesis and etiology are still unclear, with ascending (from the urethra), descending (from other sex organs), and hematogenous routes of infection all being incriminated. Some evidence indicates that reflux of semen and urine into the vesicular glands may be a contributory factor (perhaps in association with maldevelopment) or that dysfunction of the duct systems involved could be etiological factors (Linhart and Parker 1988). Grain-fed young bulls raised together in pens tend to ride each other, a behavior that may cause ascending infections via the urethra. Recommended control measures have included lowering both the energy level of feed and the population density of pens, as well as feeding the bulls erythromycin or tetracycline.

Physical findings often include enlargement and thickening of the gland in conjunction with loss of lobulation, increased firmness, heat, and pain on palpation. Bulls with active seminal vesiculitis often exhibit pain on EE. Two distinct syndromes have been described based on palpation findings (Larson 1997). The first is characterized by irregular enlargement of (usually) only one gland, peritoneal adhesions, loss of lobulation, and fluctuation, as well as by formation of abscesses. Bacteria, particularly *Actinomyces pyogenes*, are often associated with this condition. The second syndrome is characterized by bilateral inflammation and loss of lobulation, with no evidence of bacterial involvement. Diagnostically, any of these signs in conjunction with purulent material (often in clumps) or blood in the ejaculate can indicate active seminal vesiculitis. Sequelae can include fibrosis, adhesions, and fistulation as well as progression of infection to the epididymis and testicle.

Active seminal vesiculitis is often associated with depressed semen quality. The condition can resolve spontaneously in young bulls, especially when they are dispersed and placed on a lower-energy diet. However, it is currently not possible to identify those individuals that will undergo spontaneous recovery. A culture sample, obtained by passing a catheter up the urethra and massaging the glands, may be useful in deciding treatment options (Larson 1997). Causal organisms, either proven or suggested, have included *Aeromonas hydrophila*, *Actinomyces bovis*, *Actinomyces pyogenes*, *Brucella abortus*, *Chlamydia psittaci*, *Mycoplasma bovigenitalium*, *Mycoplasma mycoides subsp mycoides SC*, *Haemophilus somnus* (Groteluesehen et al. 1994), streptococci, staphylococci, *Escherichia coli*, *Ureaplasma diversum*, and infectious bovine rhinotracheitis–like viruses. However, organisms are not isolated in a number of cases, even when appropriate sampling techniques are employed.

Clinicians can find treatment options frustratingly inconsistent. The treatment of choice would be a long-term antibiotic to which the causative organism is susceptible. To determine sensitivity, however, requires collecting an uncontaminated sample from the vesicular glands, using a procedure such as that described by Parsonson et al. (1971). Nontargeted options have included 1–2-week regimes of parenterally administered penicillin or tetracycline, with a number of more recent reports employing Micotil or Naxcel. The latter products may have some advantage here, as they are macrolides with a weak organic base, and it is possible that they could trap ions within the accessory genitalia. However, this action has not been firmly established. More adventurous treatments have included intraglandular injection of antibiotics or sclerosing agents, surgical removal of the glands, and laser stimulation.

Segmental Aplasia/Hypoplasia of the Wolffian Duct

Segmental aplasia of the mesonephric or Wolffian duct system is a congenital condition in bulls, with strong indications that it may be hereditary in nature. In most cases, the problem is unilateral, often being presented on the right side. In these cases, the body, tail, entire epididymis, or even a part or all of the vas deferens may be missing or hypoplastic. The vesicular gland on the affected side may also be hypoplastic or missing, as may the ampullae. The reported incidence varied from 0.59%–1.18% in bulls in Denmark to 5% in infertile bulls in the Netherlands. In unilateral cases, the bull is fertile, although only one testicle is producing semen.

In severe cases, often the epididymis on the affected side is small. Palpation of a very small or missing epididymal tail in pubertal bulls can be indicative of segmental aplasia. However, variations occur depending on the site of the problem. For example, if the problem is located in the vas deferens, the epididymal tail may be enlarged. Blockage in the region of the epididymal head can cause testicular degeneration secondary to fluid accumulation, which generally does not occur if the blockage is in the epididymal body or tail.

Persistent Penile Frenulum ("Tied Penis," "Persistent Fibrous Raphe")

At birth, the penis of the male calf is short, slender, and fused to the internal lining of the prepuce. Under androgenic influence, the penile and preputial epithelia progressively keratinize and separate, commencing at the anterior end and proceeding posteriorly. At puberty, this separation is mostly complete, although it is within normality for the process to continue for several months following the attainment of puberty. On occasion, normal separation does not occur, leading to the condition known as persistent frenulum (or "tied penis"). Here, a band of tissue remains, connecting the anterior end of the penis to the internal preputial lining. This connection may be represented by a thin band of tissue ($\leq$0.5 cm) that is easily severed (either at natural service or by intervention), or the attachment may be much more extensive and require more serious surgical intervention to separate it. When present, a persistent penile frenulum generally prevents normal service because the affected bull is unable to fully protrude the penis. A genetic basis exists for this condition (McNitt 1965), with certain breeds and lines of bulls being more affected than others. This indicates that some caution and counseling are indicated when human intervention is invoked.

Penile Hematoma ("Broken Penis")

Penile hematomas are formed at the time of service by the rapid escape of blood from the corpus cavernosum penis through a ruptured tunica albuginea. The presenting signs usually include a noticeable swelling of the sheath immediately cranial to the scrotum, often accompanied by some degree of preputial prolapse. The everted preputial tissue is often bluish in color because of the accumulation of venous blood. Predisposing causes include high blood pressure within the corpus cavernosum penis at the time of service ($>14,000$ mm Hg), a relative lack of reinforcement of the tunica albuginea at the rupture site, and severe downward depression of the erect penis (Morgan 1997). The rupture, usually 2–7 cm in length and transverse to the long axis of the penis, almost always occurs on the dorsal aspect, opposite the site of initial attachment of the retractor penis muscles.

The treatment objective is to restore the integrity of the tunica albuginea and to avoid infection, which easily occurs in the environment created by the hematoma. Avoiding infection is most important, as sequelae such as abscessation and adhesion formation can often have more serious consequences than does the original injury. Thus, an appropriate antibiotic coverage should accompany any other form of treatment. Apart from fixation of the penis resulting from adhesions, other sequelae can include the desensitization of those neural connections just behind the glans penis that induce the ejaculatory thrust and the development of vascular "shunts" between the corpus cavernosum penis and the surrounding tissues. In the former case, bulls cannot complete service even with a full erection, and in the latter case, a full erection cannot be achieved.

Treatments have included benign neglect (approximately 50% successful with 90 or more days of sexual rest), the use of therapeutic ultrasound, and several surgical approaches that are well covered in other texts (e.g., Wolfe 1986, Morgan 1997).

Preputial Prolapse

Preputial prolapse (eversion of the preputial epithelium) of bulls can occur for a number of reasons. A common sequence is for protrusion of the parietal preputial lining to be followed by trauma that induces inflammation and subsequent inability to retract the prepuce. If this persists, then dehydration, denaturation, and permanent tissue injury can result. A number of bulls will habitually and temporarily evert part of their preputial lining, with no apparent harm. Some breeds (e.g., Angus, Polled Hereford, Brahman, Santa Gertrudis, and Beefmaster) evert more than do others. Several predisposing factors have been described (Larsen and Bellenger 1971):

- Absence of the retractor prepuce muscle (genetically linked with the polled gene)
- Pendulous sheath
- Excessive parietal preputial membrane.

Correction of the problem has varied from conservative measures, such as suspension of the damaged sheath in conjunction with regular hydrotherapy, to surgical approaches, such as circumcision or resection. Although high success rates are possible, untoward sequelae include infection, fibrosis and constriction of the preputial orifice, and phimosis.

Testicular Hypoplasia

True testicular hypoplasia becomes a recognizable clinical entity only at, or after, puberty, although definitive diagnosis is often not possible until 18–24 months of age, when it may be differentiated from other conditions such as undescended testis, small testis, and testicular atrophy. Definitive diagnosis must be based on histological (as below) and, perhaps, historical, information (Buergelt 1997). Testicular hypoplasia in bulls may present either unilaterally or bilaterally. It is considered to be a congenital condition and possibly to be genetic. This condition was shown to have a genetic basis in Swedish Highland cattle, where it was termed "gonadal hypoplasia" because of its genetic relationship with ovarian hypoplasia in related females. Estimations of the overall incidence of testicular hypoplasia are generally low (<0.5%), although higher levels (12%–25%) have been reported in certain populations (Roberts 1986).

Testicular hypoplasia implies incomplete or inadequate development of the seminiferous epithelium, with a lack of germinal cells being the prime etiological factor (Buergelt 1997). Underlying chromosome anomalies ("sticky chromosomes" and multipolar spindle cell hypoplasia) have been described. Marked testicular hypoplasia may occur in bulls that have the rare XXY (Kleinfelter's) syndrome (Roberts 1986).

A spectrum of manifestations falls under the banner of testicular hypoplasia in the bull. Usually, however, the affected testicle or testicles are considerably smaller than normal, with the corresponding scrotum and epididymis also being undeveloped. Bull sex drive is usually unaffected. Varying degrees of fertility may occur, as well as complete sterility. It is not uncommon for owners to confuse testicular hypoplasia with imperfect testicular descent, which is a rare condition in bulls. Palpation of such a "retained" testicle will usually reveal that the problem is one of relative testicular size and not one of imperfect descent.

Testicular Degeneration

Testicular degeneration is relatively common, especially in older bulls. Apart from normal age-related degeneration, it often occurs subsequent to testicular injury or insult, within a response spectrum that is often preceded by factors such as testicular retention (cryptorchidism), inflammation (orchitis), protracted elevated testicular temperature (e.g., with pyrexia, inguinal hernia, poor scrotal thermoregulation, hydrocele, scrotal dermatitis, protracted recumbency), or hypoplasia, and is followed by testicular fibrosis (Lagerlof 1934). Testicular degeneration has also been observed subsequent to scrotal frostbite. Nutritional factors associated with this condition in bulls, apart from spermatoxic agents, are rare, although testicular atrophy and degeneration have been linked with severe underfeeding or starvation, as well as with deficiencies of zinc, manganese, vitamin A, and vitamin E. Depression of libido, or sex drive, has not been associated with either inanition or starvation in bulls, although debility may adversely affect its expression.

A number of spermatoxic agents have been identified in bulls. These include arsenic, gossypol, ethylene dibromide, fescue, cadmium, and other heavy metals. These cause a variety of effects depending on action and duration of exposure, with testicular degeneration being just one possible outcome. Finally, some procedures employed to neuter male cattle nonsurgically, such as the use of intratesticular sclerosing agents, will also cause testicular degeneration as an intended effect. Table 9.8 describes the histological differences between testicular hypoplasia and degeneration.

Table 9.8. Histological differentiation between testicular hypoplasia and degeneration.

Hypoplasia	Degeneration
Normal seminiferous tubule circumference	Irregular seminiferous tubule circumference
Regular Sertoli cell lining	Loss of tubular lining and collapse
Thickened, even, basal lamina	Thickened, wavy, basal lamina
Absence of inflammatory cells	Secondary inflammation of the interstitium
Absence of lipofuscin	Presence of lipofuscin in tubular cells

Source: Buergelt 1997.

Squamous Cell Carcinoma

The occurrence of squamous cell carcinoma (SCC) in cattle is related to the effects of cumulative damage of ultraviolet radiation on unpigmented areas of epithelium. The eye region is particularly susceptible to the development of SCC, particularly when areas such as the eyelids, third eyelid, and sclera lack pigment. The most common site for development of SCC is the lateral limbus of the eyeball, with the third eyelid being the next most common site. SCC is most commonly seen in older animals (those 4–6 years old or older) although precursor lesions ("plaques") may be observed in bulls less than 2 years of age. Although SCC is not regarded as heritable per se, there is a definite breed-type predilection related to pigmentation around the eye, as well as to conformation of the eye and its placement. The breed types most often affected by SCC include Horned and Polled Herefords, Holsteins/Friesians, and Simmentals. As eyelid and scleral pigmentation is moderately heritable, selection for improved eye pigmentation can reduce SCC in herds. Culling of affected animals is also useful in reducing the incidence of SCC.

Testing Bulls for Trichomoniasis

Trichomoniasis, caused by *Tritrichomonas foetus*, is an insidious venereal disease of cattle that is prevalent in most regions where range-type cattle are run. Although it can cause significant adverse effects on the breeding program of naive herds, it tends to cause steady, if not remarkable, loss in herds once it is endemic. Further description of this disease entity may be found in chapter 8. Sampling is best performed on bulls, especially those that have achieved sexual maturity (3–4 years of age or older). Subsequent diagnosis is dependent on a number of factors, as discussed below.

Collecting preputial samples from bulls for detection of trichomoniasis has long been a routine diagnostic procedure in a number of countries. In the United States, this procedure generally entails using a plastic pipette to obtain aspirated material from the region of the fornix. This material is then placed into an appropriate transport/culture media for subsequent culture and examination. The inoculated media is preferably stored at 35°–37°C until microscopic examination for *T. foetus* is performed. This examination may be conducted by a practitioner, or it may be sent to a diagnostic laboratory, where it is commonly examined after incubation at both 2 and 5 days.

A transport/culture system that has gained wide acceptance within the United States over the last 12 years is the In-pouch system. This consists of a self-contained, microaerophilic culture system containing a buffered proprietary media that is inhibitory to bacterial and fungal growth. Under good conditions, its sensitivity in detecting *T. foetus* from cultured bovine preputial smegma is approximately 90%. However, success depends on a number of factors, and sensitivity as low as 70% has been reported. One complication that can compromise the test is contamination and overgrowth of the sample. (This is not an uncommon complication for samples submitted to the Kansas State University Diagnostic Laboratory.) To minimize this problem, the following guidelines are suggested.

Storage of Testing Equipment for *T. foetus* before Use

- Store at room temp (15°–25°C), vertically, and in the dark
- Respect recommended shelf life (1 year from date of manufacture)
- Do not freeze
- Do not use if medium is cloudy or contains precipitate.

Collection Regime for *T. foetus*

- Minimize contamination (from both bull and environment)

- Insert a small sample (0.5–1.00 cc; avoid rinse fluids and urine, as maintenance of pH is important
- Delete air from upper pouch before closure
- Store upright (to concentrate cellular material).

Selected Relatively Common Genetic Faults and Diseases in Bulls

Most bovine genetic defects of economic significance are inherited as autosomal recessives. The following compilation is based on those by Leipold and Dennis (1986), Barth and Oko (1989), Holroyd et al. (2002b) and Perry et al. (2002). D = dominant; R = recessive; P = polygenic; ? = probably genetic, etiology not clear; * = incomplete penetrance.

General Genetic Defects

- α Mannosidosis: Angus, Galloway, Murray Grey (R)
- β Mannosidosis: Saler (R)
- Bovine Leukocyte Adhesion Deficiency: Holstein/Friesian (R)
- Cataract: (?)
- Coloboma: Charolais (D)
- Cryptorchidism: Hereford (R; sex-limited recessive, probably involving at least two sets of genes)
- Dermoid, eye: (P)
- Double-muscling ("muscular hypertrophy"): All major beef breeds in United States and Europe (R)
- Epitheliogenesis imperfecta (R)
- Factor XI deficiency: Holstein/Friesian (R)
- Generalized glycogenesis (Pompes disease): Shorthorn, Brahman (R)
- Haemophilia A: Herefords (sex-linked recessive)
- Hip dysplasia: Hereford and others (?)
- Hypospadia (possible recessive)
- Hypotrichosis: Poll Hereford (R)
- Inferior brachygnathia: Various breeds (?)
- Inguinal hernia (?)
- Inherited congenital myoclonus: Poll Hereford (R)
- Knee hyperflexion: Gyr (R)
- Maple Syrup Urine Disease: Polled Hereford, Polled Shorthorn (R)
- Osteitis: Hereford (?)
- Osteogenesis imperfecta: Charolais (R)
- Osteopetrosis: Various beef breeds (?)
- Penile deviation: (?)
- Persistent penile frenulum: Shorthorn, Angus (?)
- Protoporphyria: Limousin, Blonde d'Aquitaine (R)
- Segmental aplasia of the Wolffian duct: Red-Brown Danish, Hereford (?)
- Short penis: (possibly recessive)
- Short retractor penis muscle: Dutch Friesian (R)
- Spastic syndrome ("stretches," "crampy"): (R)
- Spastic paresis ("Elso heel"): Angus, Beef Shorthorn, Holstein/Friesian, and others (P, ?)
- Spiral deviation of the penis: Poll Hereford (Australia) (?)
- Syndactyly: Holstein/Friesian, Chianina, Angus, Simmental (?)
- Testicular hypoplasia: Swedish Highland, possibly British breeds (R, *)
- Tibial hemimelia: Galloway (R)
- Umbilical hernia: Friesian (P)
- Ventricular septal defects: Limousin, Hereford (?)

Defects of Semen or Sperm

- Decapitated sperm defect: Hereford, Guernsey, Holstein, Swedish Red and White, and others (Hancock and Rollinson 1949) (sex-limited recessive)
- Knobbed acrosome defect: Holstein/Friesian, Angus (Teunissen 1946) (R)
- "Dag" defect: Jersey et al. (Blom 1966) (R)
- "Tail stump" defect: (Hancock 1959) (?)
- "Pseudo-droplet" defect: Friesian (Blom 1968) (?)

REFERENCES

Australian Association of Cattle Veterinarians. 1995. *The Veterinary Examination of Bulls.* Indooroopilly: Australian Association of Cattle Veterinarians.

Australian Association of Cattle Veterinarians. 2002. *Bull Fertility: Selection and Management in Australia.* Indooroopilly: Australian Association of Cattle Veterinarians.

Ball L. 1986. Electroejaculation. In *Applied Electronics for Veterinary Medicine and Animal Physiology*, edited by W.R. Klemm, pp. 395–441. Springfield, IL: CC Thomas.

Barth, A.D. 1993. *Bull Breeding Soundness Evaluation.* Saskatchewan: Western Canadian Association of Bovine Practitioners.

Barth, A.D., R.J. Oko. 1989. *Abnormal Morphology of Bovine Spermatozoa*. Des Moines: Iowa State Press.

Blockey, M.A. de B. 1978. The influence of serving capacity of bulls on herd fertility. *Journal of Animal Science* 46:589–595.

Blockey, M.A. de B. 1989. Relationship between serving capacity of beef bulls as predicted by the yard test and their fertility during paddock mating. *Australian Veterinary Journal* 66:348.

Blom, E. 1950. Interpreting spermatic cytology in bulls. *Fertility and Sterility* 1:223–238.

Blom, E. 1966. A new sterilizing and hereditary defect (the "Dag" defect) located in the bull sperm tail. *Nature* 209:739–740.

Blom, E. 1968. A new sperm defect—"pseudodroplets"—in the middle piece of the bull sperm. *Nordisk Veterinaermedicin* 20:279–283.

Boyd, G.W., D.D. Lunstra, L.R. Corah. 1989. Serving capacity of crossbred yearling bulls. I. Single-sire mating behaviour and fertility during average and heavy mating loads at pasture. *Journal of Animal Science* 67:60–71.

Boyd, G.W., V.M. Healy, R.G. Mortimer, J.R. Piotrowski, K.G. Odde. 1990. Serving capacity tests unable to predict fertility of yearling bulls. *Theriogenology* 36:1015–1025.

Brinks, J.S. 1994. Relationships of scrotal circumference to puberty and subsequent reproductive performance in male and female offspring. In *Factors Affecting Calf Crop*, edited by M.J. Fields and R.S. Sand, pp. 363–370. Boca Raton, FL: CRC.

Brinks, J.S., M.J. McInerney, P.J. Chenoweth. 1978. Relationship of age at puberty in heifers to reproductive traits in young bulls. *Journal of Animal Science* 47(1):135–136.

Buergelt, C.D. 1997. *Color Atlas of Reproductive Pathology of Domestic Animals*. St Louis, MO: Mosby.

Byerley, D.J., J.K. Bertrand, J.G. Berardinelli, T.I. Kiser. 1990. Testosterone and luteinizing hormone response to GnRH in yearling bulls of different libido. *Theriogenology* 34:1041–1049.

Campbell, R.C., J.L. Hancock, I.G. Shaw. 1960. Cytological characteristics and fertilising capacity of bull spermatozoa. *Journal of Agricultural Science* 55:91–99.

Chase, C.C. Jr., P.J. Chenoweth, R. Larsen, et al. 1997. Growth and reproductive development from weaning through 18 months of age among breeds in sub-tropical Florida. *Theriogenology* 47:723–745.

Chenoweth, P.J. 1981. Libido and mating behavior in bulls, boars and rams. A review. *Theriogenology* 16(2):155–177.

Chenoweth, P.J. 1983. Examination of bulls for libido and breeding ability. *Veterinary Clinics of North America: Large Animal Practice* 5:59–74.

Chenoweth, P.J. 1986. Reproductive behavior of bulls. In *Current Therapy in Theriogenology*, edited by D. Morrow. Philadelphia, PA: W.B. Saunders.

Chenoweth, P.J. 1991. *Bos indicus* bulls—how different are they? *Proceedings of the Society for Theriogenology*, Montgomery, AL: Society for Theriogenology, pp. 117–122.

Chenoweth, P.J. 1997. Bull libido/serving capacity. *Veterinary clinics of North America: Food Animal Practice* 13:331–344.

Chenoweth, P.J. 2000. Rationale for using bull breeding soundness evaluations. *Compendium on Continuing Education for the Practicing Veterinarian*. 22:S48–S55.

Chenoweth, P.J., J.C. Spitzer, F.M. Hopkins. 1992. A new bull breeding soundness evaluation form. *Proceedings of the Society for Theriogenology*, Montgomery, AL: Society for Theriogenology, pp. 63–70.

Chenoweth, P.J., C.A. Risco, R.E. Larsen, J. Velez, C.C. Chase Jr., T. Tran. 1994. Effect of dietary gossypol on aspects of semen quality, sperm morphology and sperm production in young Brahman bulls. *Theriogenology* 40:629–642.

Coulter, G.H. 1994. Beef bull fertility: Factors affecting seminal quality. In *Factors Affecting Calf Crop*, edited by M.J. Fields and R.S. Sand. Boca Raton, FL: CRC.

Coulter, G.H., J.P. Kastelic. 1999. Management programs for developing beef bulls. In *Current Veterinary Therapy 4: Food Animal Practice*, edited by J.L. Howard and R.A. Smith, pp. 127–136. Philadelphia, PA: W.B. Saunders.

Coulter, G.H., G.C. Kozub. 1989. Efficacy of methods to test fertility of beef bulls used for multiple-sire breeding under range conditions. *Journal of Animal Science* 67:1757–1766.

Ellis, R.W. 2003. Factors influencing reproductive performance of yearling beef bulls. *Proceedings of the Nebraska Veterinary Medical Association Summer Meeting*, pp. 97–117.

Entwistle, K. 2003. Bull reproductive anatomy and physiology. In *Bull Fertility: Selection and Management in Australia*, edited by G. Fordyce, pp. 2.1–2.17. Indooroopilly: Australian Association of Cattle Veterinarians.

Ewbank, K. 1996. Beyond breeding. *Drovers Journal* September: 66–70.

Farin, P.W., P.J. Chenoweth, E.R. Mateos, J.E. Pexton. 1982. Beef bulls mated to estrus synchronized heifers: Single vs multi-sire breeding groups. *Theriogenology* 17:365–372.

Farin, P.W., P.J. Chenoweth, D.F. Tomky, et al. 1989. Breeding soundness, libido and performance of beef bulls mated to estrus-synchronized heifers. *Theriogenology* 32:717–725.

Fields, M.J., W.C. Burns, A.C. Warnick. 1979. Age, season and breed effects on testicular volume and semen traits in young beef bulls. *Journal of Animal Science* 48:1299–1304.

Fitzpatrick, L.A., G. Fordyce, M. McGowan, et al. 1998. Bull selection and use in northern Australia. 4. Semen traits. In *Proceedings of the 20th World Buiatrics Congress*, Sydney Australia, 1:407–412.

Fordyce, G., editor. 2002. *Bull Fertility: Selection and Management in Australia*. Indooroopilly: Australian Association of Cattle Veterinarians.

Gardiner, B., G. Fordyce. 2002. Evaluation and certification of bull fertility. In *Bull Fertility: Selection and Management in Australia*, edited by G. Fordyce, pp. 111–114. Indooroopilly: Australian Association of Cattle Veterinarians.

Geske, G.M., R.R. Schalles, K.O. Zollner. 1995. Yearling scrotal circumference prediction equation and age adjustment factors for various breeds of beef bulls. *Proceedings of the Cattleman's Day Report*. Manhattan: Kansas State University.

Godfrey, R.W., D.D. Lunstra. 1989. Influence of single or multiple sires and serving capacity on mating behavior of beef bulls. *Journal of Animal Science* 67:2897.

Grandin, T. 2000. *Livestock Handling and Transport*. Wallingford: CAB International.

Grotelueschen, D.M., R.G. Mortimer, R.P. Ellis. 1994. Vesicular adenitis syndrome in beef bulls. *Journal of the American Veterinary Medical Association* 205:874–877.

Hancock, J.L. 1959. The morphological characteristics of spermatozoa and fertility. *International Journal of Fertility* 4:347–359.

Hancock, J.L., D.H.L. Rollinson. 1949. A seminal defect associated with sterility of Guernsey bulls. *Veterinary Record* 61:742–743.

Holroyd, R.G., V.J. Doogan, J. De Faveri, et al. 1998. Bull selection and use in northern Australia. 6. Calf output and predictors of fertility of bulls in multiple-size herds. In *Proceedings of the 20th World Buiatrics Congress*, Sydney Australia, 1:417–422.

Holroyd, R.G., V.J. Doogan, J. De Faveri, et al. 2002a. Bull selection in northern Australia. 4. Calf output and predictors of fertility of bulls in multiple-sire herds. *Animal Reproduction Science* 71:67–79.

Holroyd, R.G., E. Taylor, D. Galloway. 2002b. Physical examination of bulls. In *Bull Fertility: Selection and Management in Australia*, edited by G. Fordyce, pp. 1–3, 19. Indooroopilly: Australian Association of Cattle Veterinarians.

Jayawardhana, G.A., D. Thomson. 2002. Handling bulls. In *Bull Fertility: Selection and Management in Australia*, edited by G. Fordyce, pp. 12.1–12.6. Indooroopilly: Australian Association of Cattle Veterinarians.

Kastelic, J.P., R.B. Cook, G.H. Coulter. 2000. Scrotal/Testicular thermoregulation in bulls. In *Topics in Bull Fertility*, edited by P.J. Chenoweth. Ithaca, NY: International Veterinary Information Service.

Kenney, R.M. 1970. Selected diseases of the testicle. *Proceedings of the 6th International Conference on Animal Diseases,* Philadelphia, PA. Pp. 295–314.

Lagerlof, N. 1934. Changes in the spermatozoa and in the testes of bulls with impaired or enhanced fertility. *Acat. Path. Et Micobiol. Scand., Suppl.* 19:1–254.

Landaeta-Hernandez, A.J., P.J. Chenoweth. 2001. Assessing sex-drive in young *Bos taurus* bulls. *Animal Reproduction Science* 66:151–160.

Larsen, L.H., C.R. Bellenger. 1971. Surgery of the prolapsed prepuce in the bull; its complications and dangers. *Australian Veterinary Journal* 47:349–357.

Larson, R.L. 1997. Diagnosing and controlling seminal vesiculitis in bulls. *Veterinary Medicine* 92:1073–1078.

Leipold, H.W., S.M. Dennis. 1986. Congenital defects affecting bovine reproduction. In *Current Therapy in Theriogenology*, 2nd ed., edited by D.A. Morrow, pp. 177–199. Philadelphia, PA: W.B. Saunders.

Linhart, R.D., W.G. Parker. 1988. Seminal vesiculitis in bulls. *Compendium in Continuing Education for Veterinarians: Food Animal Practice* 10(12): 1428–1433.

Lunstra, D.D., K.E. Gregory, L.V. Cundiff. 1988. Heritability estimates and adjustment factors for the effects of bull age and age of dam on yearling testicular size in breeds of beef bulls. *Theriogenology* 30:127.

McCool, C., R.G. Holroyd. 1993. Mating practices. In *Bull Fertility, Proceedings of Conference QC93008*, edited by R. Holroyd, pp. 50–51. Brisbane: Queensland Department of Primary Industries.

McCosker, T.H., A.F. Turner, C.J. McCool, et al. 1989. Brahman bull fertility in a North Australian rangeland herd. *Theriogenology* 32:285–300.

McNitt, J.I. 1965. *Genetic aspects of estimated breeding soundness of bulls.* M.Sc thesis, Department of Animal Sciences, Colorado State University.

Morgan, G. 1997. Surgical correction of abnormalities of the reproductive organs of bulls and preparation of teaser animals. In *Current Therapy in Large Animal Theriogenology*, edited by R.S. Youngquist, pp. 240–251. Philadelphia, PA: W.B. Saunders.

National Animal Health Monitoring System. 1998a. *Cow/Calf Health and Productivity Audit. Part IV: Changes in the U.S. Beef Cow-Calf Industry 1993–1997.* U.S. Department of Agriculture.

National Animal Health Monitoring System. 1998b. *Importance of income in cow-calf management and productivity.* U.S. Department of Agriculture Info Sheet, September.

National Animal Health Monitoring System. 1994. *Cow/Calf Health and Productivity Audit. Part II: Beef Cow-Calf Reproductive and Nutritional Management Practices* and *Part III: Beef Cow-Calf Health and Health Management.* U.S. Department of Agriculture.

Parsonson, I.M., C.E. Hall, I. Settergren. 1971. A method for the collection of bovine seminal vesicle secretions for microbiological examination. *Journal of the American Veterinary Medical Association* 158:175–177.

Perry, V.E.A., P.J. Chenoweth, T.B. Post, R.K. Munro. 1990. Fertility indices for beef bulls. *Australian Veterinary Journal* 67:13–16.

Perry, V.E.A., P.J. Chenoweth, T.B. Post, R.K. Munro. 1991. Patterns of development of gonads, sex-drive and hormonal responses in tropical beef bulls. *Theriogenology* 35:473–486.

Perry, V., N. Phillips, G. Fordyce, et al. 2002. Semen collection and evaluation. In *Bull Fertility: Selection and Management in Australia*, edited by G. Fordyce, pp. 5,1–5,19. Indooroopilly: Australian Association of Cattle Veterinarians.

Pexton, J.E., P.W. Farin, G.W. Rupp, P.J. Chenoweth. 1990. Factors affecting mating activity and pregnancy rates with beef bulls mated to estrus synchronized females. *Theriogenology* 34:1069–1077.

Post, T.B., H.R. Christensen, G.W. Seifert. 1987. Reproductive performance and productive traits of beef bulls selected for different levels of testosterone response to GnRH. *Theriogenology* 27:317–328.

Pratt, S.L., J.C. Spitzer, H.W. Webster, et al. 1991. Comparison of methods for predicting yearling scrotal circumference and correlations of scrotal circumference to growth traits in beef bulls. *Journal of Animal Science* 69:2711–20.

Prince, D.K., W.D. Mickelsen, E.G. Prince. 1987. The economics of reproductive beef management. *Bovine Practitioner*. 22:92–97.

Roberts, S.J. 1986. *Veterinary Obstetrics and Genital Diseases (Theriogenology)*, 3rd ed. Ithaca, NY: S.J. Roberts.

Rupp, G.P., L. Ball, M.C. Shoop, et al. 1977. Reproductive efficiency of bulls in natural service: Effects of bull to female ratio and single vs. multiple sire breeding groups. *Journal of the American Veterinary Medical Association* 171:639–642.

Saacke, R.G., J.C. Dalton, S. Nadir, R.L. Nebel, J.H. Bame. 2000. Relationship of seminal traits and insemination time to fertilization rate and embryo quality. *Animal Reproduction Science* 60–61:663–677.

Sanderson, M.W., J.M. Gay. 1996. Veterinary involvement in management practices of beef cow-calf producers. *Journal of the American Veterinary Medical Association* 208:488–491.

Singleton, E.F. 1970. Field collection and preservation of bovine semen for A.I. *Australian Veterinary Journal* 46:160–163.

Smith, M.F., D.L. Morris, M.S. Amoss, N.R. Parish, J.D. Williams, J.N. Wiltbank. 1981. Relationships among fertility, scrotal circumference, seminal quality and libido in Santa Gertrudis bulls. *Theriogenology* 16:379.

Soderquist, L. 2000. Are Swedish yearling beef bulls mature enough for breeding? *Proceedings of the 14th International Congress on Animal Reproduction (Stockholm, Sweden)* 3:12.

Teunissen, G.H.B. 1946. Een afwijkning van het acrosom (kopkap) bij de spermatozoiden van een stier. *Tijdschr Diergeneeskd* 71:292.

Vogler, C.J., J.H. Bame, J.M. DeJarnette, M.L. McGuiliard, R.G. Saacke. 1993. Effects of elevated testicular temperature on morphology characteristics of ejaculated spermatozoa in the bovine. *Theriogenology* 40:1207–1219.

Welsh, T.H. Jr., B.H. Johnson. 1981. Stress-induced alterations in secretion of corticosteroids, progesterone, luteinizing hormone, and testosterone in bulls. *Endocrinology* 109:185–190.

Wiltbank, J.N. 1994. Challenges for improving the calf crop. In *Factors Affecting Calf Crop*, edited by

M.J. Fields and R.S. Sand, pp. 1–22. Boca Raton, FL: CRC.

Wiltbank, J.N., N.R. Parish. 1986. Pregnancy rate in cows and heifers bred to bulls selected for semen quality. *Theriogenology* 25:779–783.

Wolfe, D.F. 1986. Surgical procedures of the reproductive system of the bull. In *Current Therapy in Theriogenology,* 2nd ed., edited by D.A. Morrow, pp. 353–379. Philadelphia, PA: W.B. Saunders.

10
Assisted Reproduction

Peter J. Chenoweth

INTRODUCTION

Accelerating biotechnological developments hold exciting promise for providing cow/calf producers with the tools necessary to improve the quality and productivity of their cattle. Today, we have an impressive list of available techniques for assisted reproduction, ranging from artificial insemination (AI) and embryo transfer (ET) to cloning and in vitro fertilization. Despite these advances, overall usage of reproductive technologies in beef cow/calf herds in the United States is still relatively low, with larger operations making more use of available technologies than do smaller ones (National Animal Health Monitoring System 1998). Not all technologies are cost-effective in all situations, and results can vary widely. Knowledge of the physiological makeup of cattle, as well as of the various benefits and drawbacks of different artificial breeding methods, is the key to making assisted reproductive technologies work most effectively.

AI AND ESTRUS SYNCHRONIZATION

AI is well proven as an effective tool for promoting genetic progress in livestock. Although the dairy industry has adopted AI technology wholeheartedly, the beef industry has tended to be less receptive, despite evidence that many economically important traits in beef cattle are relatively highly heritable and that AI is the best means to rapidly exploit genetically superior sires (Patterson 1999). This reluctance also exists despite indications that an effective genetics program, managed with AI, is cost-effective (Toombs 1998).

Associated benefits of using AI include disease control, particularly of those diseases known to be venereally transmitted. Other benefits include the use of genetically proven bulls, herd improvement through increase in desired traits, better quality of replacement heifers, and the capability of using proven easy-calving bulls with heifers. The producer's risks and costs of bull handling are reduced. When used in conjunction with estrus synchronization, AI allows for calves to be born earlier, during a shorter calving season.

In the United States, the most recent estimate of the usage of AI in beef cow/calf operations was 13.3% (National Animal Health Monitoring System 1998), with larger operations (>300 head) using AI most commonly (37.1%). Only 11.9% of operations used estrous synchronization, with larger operations again having the greatest representation (31.8%). In addition, the proportion of herds using AI was twice as great for operations in which cattle were the primary source of income, in contrast to those herds for which this was not true.

The reasons for these relatively low adoption rates vary. Contributing factors include unwillingness to change management practices, fear of the unknown, and the perception that an AI program, with or without synchronization, will involve an unacceptable amount of extra effort with no guarantee of success.

A certain amount of effort is necessary to ensure success with artificial breeding methods. This includes attention to details such as identification and management of "eligible" females, animal confinement, estrus detection, breeding facilities, time constraints, and AI technique. Time/labor concerns were the most common constraints listed by cow/calf producers for both AI (38.8%) and estrus synchronization (36%) in the National Animal Health Monitoring System 1998 study, whereas nearly 20% of producers listed "complicated" as a constraint for both procedures. Lack of appropriate facilities was also mentioned as a constraint (7%–8%).

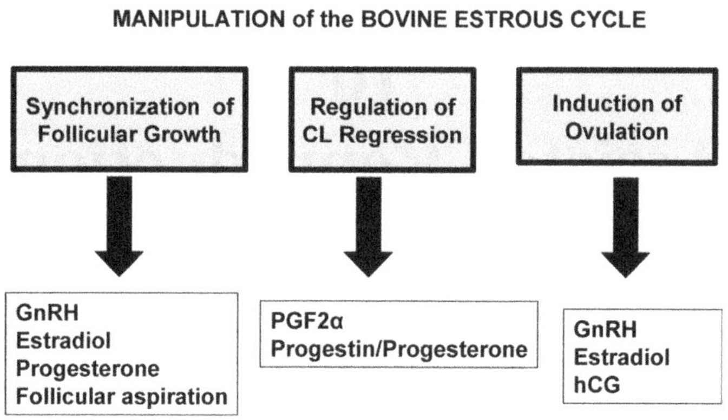

MANIPULATION of the BOVINE ESTROUS CYCLE

Synchronization of Follicular Growth → GnRH, Estradiol, Progesterone, Follicular aspiration

Regulation of CL Regression → PGF2α, Progestin/Progesterone

Induction of Ovulation → GnRH, Estradiol, hCG

Figure 10.1. Manipulation of the bovine estrous cycle.

Cow/calf producers have reservations concerning the cost and effectiveness of AI/synchronization. One of the biggest deterrents to date has been a relative lack of success in appointment or timed AI (TAI), although related concerns have included overall duration of programs, number of cattle handling opportunities, and the cost of both synchronization products and semen. Many of the technical problems have been overcome in recent years. Despite this, the average producer still does not realize that there are systems available today that can achieve a 50% or better pregnancy rate with just three animal handlings, including TAI at the last handling (Johnson 2003).

Synchronization allows more cows to be available for AI, shortens the period of AI, reduces or eliminates the need for estrus detection, allows more efficient scheduling of breeding and calving, and can allow females to get pregnant earlier than otherwise. Although hormonal means to effectively group estrus in beef females have been available for some time, beef producers have not moved rapidly to adopt them (Patterson 1999). Today we have a number of products available for manipulation of the bovine estrous cycle, including those that synchronize follicular growth, cause regression of the corpus luteum (CL), or induce ovulation (Figure 10.1).

The following is a list of products currently available for bovine estrus synchronization programs.

- Prostaglandins
  - Lutalyse (Pfizer) 5 mL; intramuscularly in female beef and dairy cattle
  - Estrumate (Schering-Plough) 2 mL; intramuscularly in beef heifers and cows
  - In-Synch (Agrilabs) 5 mL; intramuscularly in beef heifers and cows
  - ProstaMate (Phoenix Scientific) 5 mL; intramuscularly in beef females and dairy heifers
- Progestins
  - Syncro-Mate B implant 6 mg, N (Norgestomet) injection, 3 mg N and 5 mg EV (estradiol valerate); intramuscularly in beef heifers and cows
  - MGA (melengestrol acetate) 0.5 mg/head per day; orally in beef heifers (estrus suppression only)
  - EAZI-BREED CIDR (controlled internal drug release; Pfizer) 1.38 g progesterone; implant in female beef and dairy cattle
- GnRH (gonadotropin-releasing hormone)
  - Cystorelin (Merial) 2 mL; intramuscularly or intravenously in bovine females
  - Factrel (Fort Dodge) 2 mL; intramuscularly in dairy females
  - Fertagyl (Intervet) 2 mL; intramuscularly or intravenously in bovine females
  - OvaCyst (RXV) 2 mL; intramuscularly or in dairy females.

When using any of these products, observe strict adherence to label warnings and precautions. Syncro-Mate B (known as Crestar in Australia) is not currently approved for use in the United States. Estradiol valerate is not currently approved for food animal use in the United States. MGA is approved in the United States for suppressing estrus in heifers. It does not carry Food and Drug Administration approval as a product to synchronize estrus. Cystorelin and Fertagyl are labeled in the United States to treat follicular cysts in beef and dairy females.

Combinations of products are being increasingly employed in AI/synchronization programs. For example, although prostaglandin causes regression of the CL, the time of resultant ovulation is dependent on the stage of the dominant follicle at the time of prostaglandin administration. This leads to inadequate synchrony for TAI. This problem can be solved by using GnRH, which causes ovulation of the dominant follicle at certain stages of the cycle, as well as recruitment of a new dominant follicle. An injection of PGF 7 days later causes CL regression when a new dominant follicle is present, thus synchronizing CL regression with follicular development. Further synchronization of ovulation can be achieved with a second injection of GnRH 48 hours following the PGF injection.

Planning and Implementing AI/Synchronization Programs

There are a number of options available for synchronization programs in beef cattle, and producers must also take into account relevant managerial considerations to ensure a successful program. Regardless of the product and program employed, a number of factors are common in successful programs. For instance, a good year-round nutrition program is crucial, producing mature cows with a body condition score (BCS) ≥ 5 at calving and primiparous cows with a BCS ≥ 6 at calving, with a 2–3-week longer postpartum interval (PPI) than that of mature cows. Also important are the maintenance of adequate PPIs and an effective general health and biosecurity program. A short breeding season ($\leq$60 days), appropriate facilities, and knowledgeable, competent help also are all important for the success of an AI/synchronization program.

Advance planning for AI programs, especially those involving synchronization, should commence well ahead of breeding (Johnson and Stevenson 2003b). For larger beef operations, this may include prior education and training of personnel as well as ensuring that facilities and personnel can handle the estimated numbers of synchronized females. Semen should be ordered early, as should liquid nitrogen equipment, synchronization products, and AI materials. If necessary, frozen semen should be checked by a competent laboratory before the commencement of breeding. Procedures and standards for frozen semen assessment are available from the American Society for Theriogenology, as well as from most large AI organizations (see Resources).

Selection of a particular synchronization system includes a number of considerations, of which cost is but one. Different systems, as described below, have specific advantages, as well as disadvantages. Complex models will not work well on extensive range-type operations. If anestrus is a potential problem, consideration should be given to programs that are likely to induce estrus in noncycling females, such as those that incorporate MGA, GnRH, or CIDRs or calf removal (Johnson and Stevenson 2003b). The duration of the treatment protocol may be a limiting factor, as may the ability to handle cattle on multiple occasions. Other considerations include the ability, equipment, and facilities to detect estrus and to administer treatments (e.g., feeding MGA). See Resources for materials that provide a more complete discussion of different systems and their relative strengths and weaknesses.

Scheduling of the different tasks associated with planning and conducting an effective AI/synchronization program can be facilitated by using computer planning programs such as the Estrus Synchronization Planner produced by the Iowa Beef Center (http://www.iowabeefcenter.org/synchplanner/synchplanner.asp). The synchronization regimes described later in this chapter are based on considerable research. Changing the timing or intervals between injections will often result in a major loss in synchrony. When there is some flexibility in timing, it will be indicated in the notes. It is advisable to be skeptical of other systems that involve an additional cost or steps before evaluating the controlled research to support their adoption.

Selection of Female Candidates

As stated previously, one of the biggest deterrents to greater acceptance of assisted reproductive technologies in beef cattle, such as AI/synchronization programs, is poor results. In turn, one of the greatest causes of poor results in cow/calf enterprises is the use of females that are poor reproductive candidates. Often these are females that are anestrous or in poor condition. Animals selected for AI/synchronization programs should be those that would be good candidates for successful natural breeding. For heifers, this means that they weigh at least 55%–60% of their expected mature weight, have a BCS of 6 plus, and are cyclic (preferably at least 3 weeks before program commencement). Higher conception rates were achieved in heifers bred after their third postpubertal estrus than in those bred at the first estrus (Byerley et al. 1987).

Use of reproductive tract scoring (Bellows and Staigmiller 1994) is effective in determining heifer cyclicity and response to synchronization and AI. Multiparous cows should calve at a BCS of 5, and primiparous cows at a BCS of 6. Herds with a controlled breeding season length of 60 days or less and 80% or more calving in the first 40 days of the calving season will often undergo an adequate PPI. However, this also varies with nutritional status as well as with postpartum problems. In primiparous cows calved 2–3 weeks ahead of multiparous cows, there were still 8% fewer cycling females at the start of the breeding season than were cycling in the multiparous group (Stevenson 2001). Thus, a minimum of 50 days postcalving is recommended for commencement of synchrony in primiparous *Bos taurus* beef cows.

Breed differences do exist, with *Bos indicus*–type cattle often requiring a longer PPI than do *Bos taurus* cattle to resume fertile cyclicity. Several strategies to help reduce this interval may be considered, including early weaning, temporary calf removal, and once-a-day suckling (see chapter 11), as well as exploitation of the male effect, or biostimulation, (Chenoweth and Spitzer 1995).

Herd Health

AI programs should be attempted in only those herds that have a good herd health and biosecurity program. The best breeding programs will be unsuccessful if they must compete with infertility diseases such as vibriosis and trichomoniasis. Protection should be provided against abortion-causing diseases such as leptospirosis, infectious bovine rhinotracheitis, and brucellosis (where applicable). Herd health considerations extend to the source of semen: Semen, whether frozen, chilled, or fresh, should come from sires of known herd health and from an organization that maintains good health and sanitary requirements. In the United States, this is ensured by using frozen semen from organizations accredited by Certified Semen Services, a subsidiary of the National Association of Animal Breeders. See Resources for details of minimal health requirements for Certified Semen Services–accredited bull studs and AI centers.

In addition to disease problems, other health considerations such as respiratory and digestive problems and lameness can adversely affect programs. It is advisable to refrain from combining the administration of synchronization drugs with routine vaccinations (especially those employing modified live vaccines) or with other major managerial tasks.

Semen Quality

Semen quality can vary considerably, even when taken from the same bull. All bulls used in AI programs should be subject to breeding soundness evaluations as well as periodic assessment of semen quality, especially postfreezing. Purchase of semen from organizations that are approved by Certified Semen Services is highly recommended. Even when bulls have passed a BSE evaluation, there can be considerable variation between them in the results obtained from using their semen in AI/synchronization programs (J. Yelich unpublished data). In a number of cases, more stringent analysis of semen quality, particularly sperm morphology, can determine the reasons for such variation. In other cases, differences become apparent only after the program has been completed. Close attention to information provided by AI organizations can ensure that bulls of known high fertility are used, especially for timed insemination programs. Checking of frozen semen quality is advisable, both routinely and when problems arise. This inspection should be done using appropriate equipment and protocols (see chapter 9).

Semen Handling

One issue related to semen quality is that of semen handling. Although semen may be initially of good quality, and processed appropriately, it can be adversely affected by subsequent handling. Frozen bull semen can be stored indefinitely if constantly maintained at appropriately low temperatures (e.g., in solid carbon dioxide at −79°C, or liquid nitrogen at −196°C). Exposure of semen to higher temperatures, even transiently, can cause permanent damage. Particular attention should be paid to avoiding thermal damage while handling semen in the tank by ensuring that straws remain below the frost line of the tank until used. Thawing procedures, as recommended by the major AI organizations, should be strictly followed. New technologies and methods for use in evaluating frozen semen have been described (Rodriguez Martinez 2000).

Facilities

Effective artificial breeding programs require good facilities, including those for congregating cattle, feeding and watering them if necessary, observing estrus, sorting animals quietly, and breeding. For breeding, an enclosed (dark) breeding box has advantages in reducing animal movement and stress. (See Chapter 1). The average squeeze chute often does not provide enough shoulder room for the inseminator to work easily when squeezed. For synchronization programs, prior calculation should be made of the maximum number of animals that can be handled (i.e., bred) during the breeding window of time. Protection from the elements while handling semen and breeding is also important for good results. Footing surface can influence estrous activity, with dry, natural footing (dirt, grass) being best.

Animal Identification and Records

Implementation of AI on beef operations often mandates an upgrading of record keeping as well as improvement in animal identification systems. If heat detection is used, animals should be easily identifiable from some distance away. Record-keeping systems should keep tabs on date of estrus and insemination, bull used, and technician employed. Other information, such as estrus signs, ease of AI, potential problems, and any other relevant comments are also helpful.

Stress Reduction

Stress reduces reproductive success and probably plays a greater role in AI/synchronization programs than in natural breeding. Stress is increased by environmental factors such as excessive heat (above 30°C) or cold, and unstable weather. It is also increased when animals are handled roughly, such as when they are driven a long distance and subjected to dogs, whips, and electric prods. Stress is reduced when animals are conditioned to both facilities and handling and are handled gently and breeding is performed in an enclosed breeding box. Transport should be avoided at critical times, as this can affect results. For example, work with heifers in Colorado showed that transport either between 8 and 19 or between 29 and 33 days following AI was more detrimental to synchronized pregnancy rates than transport soon after AI (1–4 days).

Technicians-Inseminators

Technicians-inseminators used in beef AI programs may vary widely in experience and technique. Inseminators fresh out of AI school would do well to get some experience before getting into any timed insemination activities, where there is pressure to work more quickly than when inseminating a few animals. Variation in conception rates even between experienced technicians can range up to 20% or more. Evaluation of accurate records will allow problems to be recognized. Adequate training in semen handling and AI is essential to avoid potential problems. In addition, when large groups of "synchronized" females are involved, the inseminator or inseminators should have prior experience in breeding large numbers within a restricted period. A sufficient number of technicians should be on hand to complete the job, and it is good procedure to rotate jobs to minimize the effects of fatigue. A number of AI training schools exist in the United States, with minimum standards recommended by the National Association of Animal Breeders. These standards include instruction (and practical experience where appropriate) in insemination techniques, semen handling, and reproductive management (see Resources).

Estrus Detection

Beef producers are often relatively unaware of estrus detection details, in contrast to their dairy colleagues. Here, prior education could prove to be a sound investment, whether or not estrus detection aids are also employed. Personnel involved with heat detection should be both knowledgeable and motivated. Estrus detection is best carried out in smaller groups of animals in a confined area such as a feedlot, pen, or trap. It is recommended that one observer not attempt to check more than 100 females at a time in a synchronized group. Individual animals should be easily identifiable from a distance.

The best time for observation is first thing in the morning and in the late afternoon/early evening. At least 30–60 minutes of uninterrupted time should be devoted to this task on each occasion, although results improve with increased time spent on detection. For example, in synchronized estrus, detection for 2 hours both in the morning and in the evening, as well as 1 hour at noon, identified 40% more females in estrus than did twice-daily checks of 30 minutes each (Geary 1999). During periods of intense estrus activity, it is advantageous to sort off estrous females several times per day to allow better identification of females just coming into estrus.

Estrus detection aids contribute to efficiency in most types of programs, although many programs still require regular observation. Aids run the gamut from tail paint or chalk to Kamar heat mount detectors and Heat-Watch (electronic pressure-activated devices that transmit data to a recording device). Such aids may be used in conjunction with teaser bulls or, equally effectively, with androgenized cows (Burns and Spitzer 1992).

Calving

Following a successful synchronization program, calving will occur within a relatively compressed time period. If all females are bred on the same day, most calves will be born over a 2-week period, with a 4–5-day period of maximum activity. Preparations should be made for an adequate calving watch over this period, with appropriate personnel and facilities being available to handle contingencies. In this situation, it is helpful if easy-calving sires were used for heifer breeding.

Program Options

A vast—and increasing—array of options exists for the reproductive management of cattle. Choosing the right program for a particular cow/calf operation depends on many factors, some of which are discussed below. In making a choice, producers should not assume that assisted reproductive technologies will solve existing reproductive or management problems. Such basic requirements as good nutrition, herd health, and cow comfort will also help determine the success, or failure, of the most sophisticated AI/synchronization program.

Prostaglandin Programs

Single PGF Injection

Prostaglandins cause regression of the CL, with subsequent lowering of circulating progesterone levels. Females that do not have a functional CL (e.g., prepubertal heifers or acyclic cows) will not respond to PGF. Best results occur with females that were in estrus 7–17 days before injection. Peak estrous activity occurs approximately 72 hours after PGF injection. Advantages of single PGF injection include low cost, minimal handling required, and good conception rates in eligible females. However, a single PGF injection will not induce estrus in noncycling females, and even if all females are cycling, it will not synchronize 100% of the herd. Finally, this program can be compromised by low cyclicity and heat detection. It is recommended that producers breed based on observed estrus (Johnson and Stevenson 2003a). The options given here are suitable for either heifers or cows (Figure 10.2).

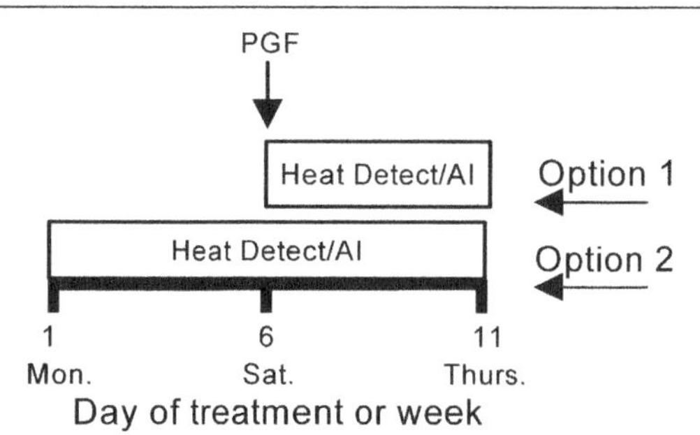

Figure 10.2. Single-prostaglandin injection options.

Option 1 (Fig. 10.2) Administer PGF to all females, perform AI for 5 days, heat check and breed according to heat. Approximately 75% of the animals will show heat in 2–5 days if all are cycling.

Option 2 (Moody System) (Fig. 10.2) Perform heat detect and AI for 5 days, inject PGF to females not detected in heat on day 6, perform heat check and AI through day 11. This option provides a check on female cyclicity before administering PGF. If all animals are cycling, approximately 25% will show estrus in the first 5 days. If the proportion cycling is too low, synchronization and insemination plans could be modified accordingly.

Option 3 (not shown) PGF is used as a follow-up treatment for females not detected to be in heat after the use of another synchronization protocol. Inject PGF to nonresponding females 10–14 days after responding females are in heat to the original protocol. This approach is useful in females that ovulated in response to the initial treatment but that were not detected to be in heat (e.g., heat detection rate was low because of extreme weather conditions).

Option 4 (not shown) PGF is used at any time after 5 or more days of a heat check/AI program to finalize the AI period. With this approach, animals with missed heats will hopefully respond, and the total AI period could be shortened by a few days.

Double PGF Injection

A double PGF-injection regime, if timed appropriately, will bring most cycling females into estrus 2–5 days after the second injection. Cattle that were not in the appropriate stage of the cycle to respond to the first injection should be in the appropriate stage by the time of the second injection. Any of the options described can be applied to replacement heifers or cows. The advantages of this program include its relatively low cost and good conception rate (in responding females). However, a double PGF-injection regime will not induce estrus in noncycling females, and some animals do not respond to the second PGF injection. This program can be compromised by low cyclicity and heat detection. It is recommended that females be bred based on observed estrus (Figure 10.3; Johnson and Stevenson 2003a).

Option 1 In option 1, the interval between injections can vary between 10 and 14 days (with best results at the longer interval).

Option 2 Estrus detection is conducted following the first injection, and those females showing estrus

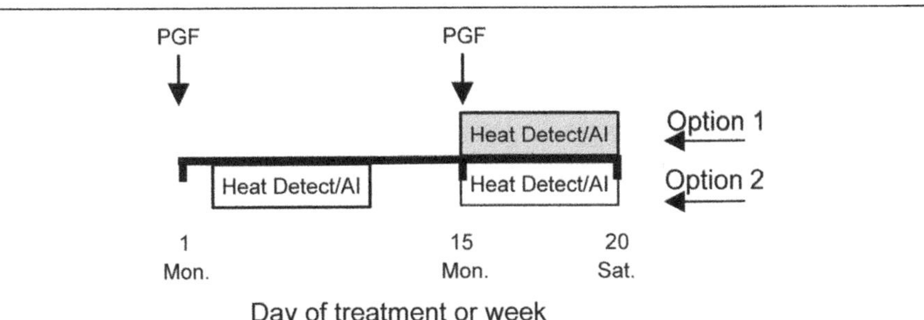

Figure 10.3. Double-prostaglandin injection options.

are bred. The second injection of PGF is given only to those females that did not show estrus after the initial injection, followed by estrus detection and AI.

Option 3 (not shown) *Bos indicus* cattle in Florida showed improved response to a double PGF injection program when the second dose was split and delivered over 2 days (Cornwell et al. 1985).

Progestin Programs

Progesterone and its derivatives were integral to the earliest attempts to synchronize estrus in cattle because of its ability to suppress GnRH pulses. Disappointing early fertility results led to the development of the progestin regimes employed today. Long-term treatment with a progestin (i.e., 14 days) allows spontaneous CL regression. Subsequent withdrawal of the progestin results in a synchronized estrus. Such programs generally provide good synchronization of estrus, including the capability of inducing estrus in some noncycling females. However, conception and pregnancy rates may suffer because aged-persistent follicles are used. One way to reduce this problem is by using progestins to presynchronize females for breeding at the subsequent estrus. Further improvement in results has come with the combined use of progestins with other products that synchronize follicular growth waves, as discussed below.

Syncro-Mate-B

Syncro-Mate-B (SMB, Crestar) consists of an ear implant containing progesterone. This is usually left in place for 9 days, with an injection of estradiol valerate given at implant insertion. SMB allows breeding either at detected estrus or TAI. This system has the advantage that it may be used with all cycling females, regardless of where they are in their estrous cycle, and it may also stimulate cyclicity in some noncycling females. If implants are inserted on day 1 and removed on day 10, most responding females will show heat between days 11 and 14. In lactating females, calf removal between implant removal and start of breeding (48 hours) can improve estrus and pregnancy responses (see below). SMB is currently not available in the United States.

Different options exist for SMB programs.

TAI All treated females are mass inseminated between 48 and 54 hours after implant removal.

Estrus Detection Animals are inseminated 12 hours after first standing heat.

Combination of the above Methods Females that show estrus before 48 hours are inseminated using the "a.m.–p.m." rule (cows observed in heat in the afternoon are inseminated the following morning, and vice versa), and nonresponding females are mass inseminated at 48–54 hours after implant removal. Mass insemination in this fashion can result in pregnancy in about 30% of nonresponding females.

Natural Breeding (Bull/Sync) Fertile, active bulls are admitted to treated groups within 24 hours of implant removal at ratios of one bull to between ten and twenty treated females. This program is best conducted within a relatively small pasture or dry lot. Most breeding occurs over the ensuing 4 days. Results are generally comparable to those obtained in a good AI program.

CIDRs and Progesterone-Releasing Intravaginal Devices

Both CIDRs and progesterone-releasing intravaginal devices (PRIDs) are devices used to synchronize estrus in cows and heifers. In the United States, the CIDR contains 1.38 g progesterone, whereas the PRID contains 1.55 g progesterone plus 10 mg of estradiol valerate. A common protocol is to insert the CIDR into the vagina, give an injection of PGF

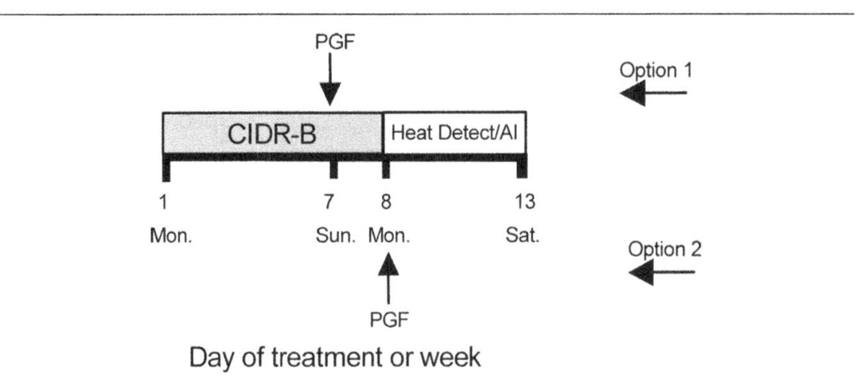

Figure 10.4. Controlled internal drug release (CIDR) options.

on day 6, and remove the device on day 7. Following removal of the CIDR, estrus should occur within 1–3 days. Breeding should be conducted with animals detected in estrus (Figure 10.4). This method will also stimulate cyclicity in some noncycling females. PRIDs are generally left in place for 12 days (although this can be shortened to 8 days with a PGF injection on day 7). CIDRs have been recently released on the U.S. market, whereas at the time of the writing of this book, PRIDs are not yet available in the United States.

Option 1 Inject PGF 1 day before insert removal. This option provides the greatest degree of estrus synchrony.

Option 2 Inject PGF on the same day as insert removal. This allows one less cattle "working," although with less synchrony of estrus.

Indications are that CIDRs also can be employed successfully in conjunction with TAI (Perich et al 1996), with a 63% pregnancy rate obtained in beef heifers inseminated 48 hours following device removal.

MGA/Prostaglandin

A relatively low-cost system for estrus synchronization employs melengestrol acetate (MGA), an oral progestin, usually in conjunction with PGF. MGA is not approved by the Food and Drug Administration to synchronize estrus, although it is approved to suppress estrus in heifers. In combination with prostaglandin, MGA is probably the most widely used system in the United States to synchronize estrus in beef heifers.

For estrus synchronization, MGA is fed at 0.5 mg/head per day for 14 days, preferably with a grain or protein supplement. To ensure females get their required daily dose, they should be conditioned to the feeding regime before the commencement of MGA feeding. Females will generally exhibit a subfertile estrus 2–5 days after withdrawal of the MGA. Breeding at this first estrus after MGA withdrawal is not recommended.

Several options exist for breeding programs subsequent to MGA administration. A simple approach is to use bulls to breed the treated group of females (as described below). The bulls can be introduced to the females as early as 10 days after MGA withdrawal.

A more sophisticated approach employs a single injection of PGF at 17–19 days after MGA feeding has ceased (with 19 days being preferred), causing regression of the ensuing CL (Figure 10.5). Most females will show estrus 48–72 hours after the PGF injection. Females may be inseminated approximately 12 hours after standing estrus is first observed.

A third approach is to employ a second PGF injection 11 days after the first. In this system, a number of females that may not have been observed in estrus after the first PGF injection (up to 20%) may respond.

Other variations include "MGA Select," where MGA is fed for 14 days, followed by an injection of GnRH on day 26 and PGF on day 33. This system, which is indicated for groups of 100 or more females, provides improved estrus synchrony. Another variant is "7–11 Synch," whereby MGA is fed for 7 days, with PGF being administered on day 7. This is followed by GnRH at day 11 and a second PGF injection on day 18. This procedure, which is suggested for groups of 100 or fewer females, allows a shorter feeding period without compromising fertility. With all MGA-based programs, uniform daily consumption of the progestin is critical. This necessitates adequate bunk space (18–24 in for heifers and

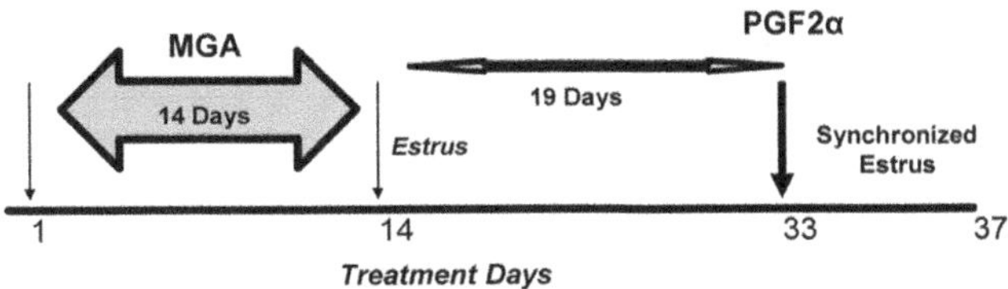

Figure 10.5. Melengestrol acetate (MGA) + prostaglandin program.

cows, respectively), as well as equal opportunity for feeding.

Select Synch, CO-Synch, and OvSynch

Programs have been devised that use various combinations of GnRH and PGF in attempts to synchronize follicular dynamics, manage time of ovulation, and facilitate TAI. Different variations on this theme have been termed Select Synch, CO-Synch, and OvSynch. For postpartum beef cows, Select Synch and CO-Synch have received the most attention, whereas OvSynch has been used mostly with dairy cows.

Select Synch targets follicular development. This program incorporates conventional estrus detection and insemination. GnRH is given on day 1 and PG on day 7, followed by estrus detection and AI. This is the only current GnRH system that should be considered for heifers.

CO-Synch aims to synchronize ovulation. GnRH is given on day 1, PGF on day 7, and GnRH on day 9 to coincide with mass insemination (TAI) at 48 hours following PGF. This could be combined with 48-hour calf removal between PGF and AI.

With OvSynch, a second injection of GnRH is given 48 hours after PGF, with mass insemination taking place 16–18 hours later. This option is used most with dairy cows. However, when CO-Synch and OvSynch have been compared in beef cows, results are similar, so here the additional trip through the chute for GnRH in OvSynch is probably not justified. OvSynch does not appear to work very successfully in heifers.

An important caveat with all three systems is that when breeding occurs subsequent to estrus detection, the time and quality of estrus detection can influence results by up to 20%. Thus, when comparing programs, duration of estrus detection at least should be stipulated (Figure 10.6).

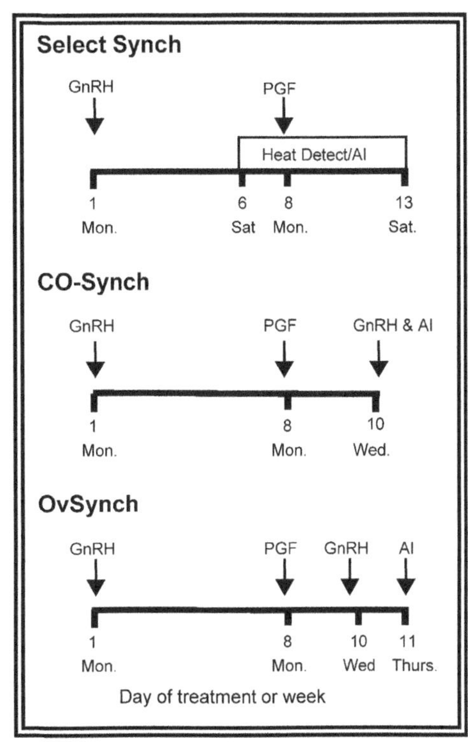

Figure 10.6. Select Synch, CO-Synch, and OvSynch. Adapted from Johnson and Stevenson 2003a.

Other Options

In an attempt to prevent inconvenient early estrous expression, researchers have experimented with administration of different progestins between injections of GnRH and PGF. For example, MGA has been fed during this interval. Alternatively, a CIDR has been inserted on day 1 and removed on day 8. Pregnancy rates have improved 0%–15% by

Table 10.1. Comparison of pregnancy rates (%) to a 5-day artificial insemination period or to a single timed insemination.

	Heifers		Cows	
Program	Average	Range	Average	Range
MGA + PGF	60	40–70	55	40–60
MGA Select	60	40–65	60	40–65
MGA-CO-Synch*	—	—	60	45–65
Select Synch	50	40–65	45	25–55
CO-Synch	—	—	50	30–55
CO-Synch + CIDR*	—	—	—	above + 15
OvSynch	—	—	50	45–57
CIDR + PGF	45	35–60	45	35–60
7–11 Synch	—	30–55	—	35–65
2 × PGF	50	30–65	40	20–45

*Single-timed in semination.
Source: Johnson 2003.
Dash indicates that no data are available.

including a progestin, probably because of increased female cyclicity. However, the best time for mass insemination when using such systems has still not fully been determined.

Comparisons of Pregnancy Rates across AI/Synchronization Programs

Attempts have been made to compare AI/ synchronization programs in terms of results, as shown in (Table 10.1). Here, the assumption is that we are dealing with well-managed programs under optimal conditions. Achieving common ground for purposes of comparison is essential, although difficult in effect. Results may vary even between otherwise identical programs because of differences in female selection criteria, heat detection intensity and duration, and criteria employed to categorize pregnancies.

Cost-Benefits of Different Programs

To compare the cost-benefits of different programs, producers first need to know their costs of production, with a logical starting point being those costs associated with conventional natural breeding (Johnson 2003). These costs, in turn, vary because of factors such as the purchase price of the bull, his fertility and longevity, and the fixed and variable costs associated with maintaining the herd. Although AI/synchronization programs often include extra investment in terms of labor and supplies, this investment should be balanced against the potentially superior genetic merit (and value) of calves as well as against the benefits of a shortened calving season. Using different combinations of factors, the costs per natural-bred pregnancy in the United States Midwest were shown to vary from approximately \$16 to \$90 (Johnson 2003). The cost of adding bulls varied with the one-cycle pregnancy rate (here assumed as a constant 50%) and with herd size, ranging from \$8.27 per pregnant female for herds of 100 head to \$2.61 for herds of 300 head. The authors constructed a partial budget for estrus synchronization and AI in beef herds, using assumptions as shown below (Table 10.2). Based on mid-2003 prices, costs of different components were also calculated (Table 10.3).

To directly compare cost-benefits requires standardization of terms and some assumptions. One assumption is that AI calves were more valuable based on weaning weight and genetic value (\$25) than were calves derived from natural service, with a weaned calf crop of 82%. On the basis of cost per pregnant female, this model indicated that natural service using fertile bulls was the most economical system, followed by MGA + PGF and MGA-Select, or Select Synch (depending on herd size). In this simulation, CO-Synch + CIDR was the most expensive option. On the basis of 500-lb equivalent weaned calves, several systems were approximately equal to natural service, including MGA + PGF,

Table 10.2. Partial budgeting factors for artificial insemination/synchronization programs.

Budget Effect	Source	Budget Effect	Source
Increased returns	Heavier calves (earlier average birth date) Uniformity of calf crop (use of fewer sires, shorter breeding season)	Decreased returns	Fewer cull bulls to sell
Decreased costs	Fewer bulls to buy and maintain Less labor for more concentrated calving season More predictable calving ease	Increased costs	Planning and management for estrus synchronization and artificial insemination Synchronization products and supplies Labor Improved facilities?

Source: Johnson 2003.

Table 10.3. Estimated costs of components of artificial insemination/synchronization programs.

Item	Cost per Unit ($)
Semen	13.00/straw
PGF2α	2.00/dose
Gonadotropin-releasing hormone	4.00/dose
Controlled internal drug release	8.00/dose
Supplies	0.50/insemination
Fixed costs[a]	176.30

[a]Semen tank, carrying case, pipette gun, thaw box, liquid nitrogen
Source: Johnson 2003.

MGA Select, and Select Synch, for all herd sizes. For larger herds (300 head), several systems also were competitive in cost to natural service, including 7–11 Synch, CIDR + PGF7, and CIDR + PGF8 (Johnson 2003).

SYNCHRONIZATION PROGRAMS WITH NATURAL BREEDING (BULL/SYNC)

Early work in which bulls were placed with groups of synchronized females (Bull/Sync) aimed to improve understanding of the reproductive capabilities of bulls. However, it soon became apparent that Bull/Sync represented a viable management option in itself. Although natural breeding and estrus synchronization may not appear to be natural bedfellows, there are situations in which this combination may be advantageous. For example, Bull/Sync can represent an interim step toward AI/synchronization, which is less demanding on management, facilities, labor, and expense. Here, a number of managerial aspects can be honed, and eligible female groups assembled, with less risk of disaster than might occur with the sudden imposition of a full AI/synchronization program. Such programs can be successful using "normal" bull-to-female ratios, provided the bulls employed are active and fertile (Pexton et al. 1990).

Another application of Bull/Sync is to concentrate breeding and calving periods in select groups such as heifers. Here it can be advantageous both for selection of replacement heifers and for managing the calving period. Bull/Sync has also proven to be a useful management tool for mixed enterprises, in which owners wish to confine both breeding and calving within manageable, predictable periods.

Most of the earlier studies were conducted with the use of either PGF or SMB as estrus synchronization agents. Results showed little difference between these two methods in terms of pregnancy rates achieved. These results, in turn, were generally comparable with results obtained in well-run AI programs (Pexton et al. 1989). Bull/Sync has also been employed successfully with both MGA/PGF and Select Synch protocols. However, synchronization systems that produce the tightest synchrony (as intended for

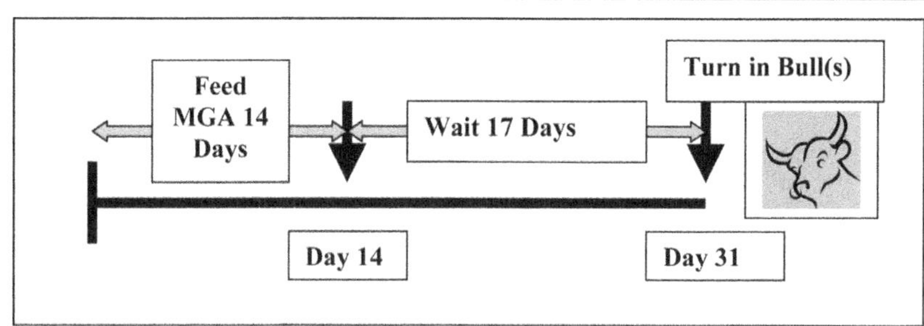

Figure 10.7. MGA Bull/Sync Program.

TAI) would add unnecessary cost to a Bull/Sync program and add extra breeding pressure on the bulls. The advantages of Bull/Sync include synchrony of breeding and calving, as well use as a possible interim step to AI, but with less effort and less need for special facilities, making it a more flexible breeding option. Bull/Sync also provides possible biostimulation effects benefits.

However, Bull/Sync also poses some disadvantages. It requires additional managerial time and effort as compared with natural breeding of nonsynchronized females. Females need to be selected and synchronized. Bulls need to have passed a BSE. Some monitoring of the intensive breeding period (2–5 days initially) is recommended. Also, Bull/Sync has less potential to create genetic progress than does AI, which can employ bulls of superior genetic merit. Natural breeding of bulls comes with inherent risks such as danger when handling, misadventure, and inconvenience. The risk of infertility and venereal disease is also greater than with an AI/synchronization program.

Despite these considerations, Bull/Sync has its adherents. If due care is taken, as described in the guidelines below, the results are comparable with well-run AI programs and are sometimes better (Farin et al. 1989).

Bull/Sync generally involves intense bull sexual activity within a contracted period. Thus, a number of considerations are pertinent to the success of such programs. Bulls must be thoroughly assessed before breeding, and young (2–4-year-old) bulls should be used. The breeding should be conducted in a small pen or yard, using normal bull-to-female ratios (1:15 to 1:25) and single sires, if possible. The breeding process must be monitored as to length (2–5 days), and adequate "rest and relaxation" (2–3 weeks or more) must be provided to the bulls if they are to be used repetitively.

Bulls can be used in concert with most synchronization programs in which overt estrus is displayed. Two common Bull/Sync programs are to give an injection of prostaglandin and then place the bulls with the cows or to feed them MGA for 14 days, wait 17 days, and then place the bulls with the cows (see Figure 10.7).

Advantages of the MGA Bull/Sync protocol include low drug cost, no heat detection, and a less concentrated breeding period so the bulls have more time to service the cows. Another advantage of feeding MGA is that it helps stimulate cyclicity in prepubertal heifers and anestrous cows.

PREGNANCY TESTS

The conventional method of determining pregnancy status in cattle for many years has been via transrectal palpation. Comprehensive descriptions of this procedure are available (Youngquist 1997, Australian Association of Cattle Veterinarians 1998) and beyond the scope of this discussion. In addition to transrectal palpation, other methods of examination are available for pregnancy diagnosis in cattle. For example, estimation of blood or milk progesterone levels at approximately 21d post-breeding can differentiate between a pregnant animal and one that is in either proestrus or estrus (Australian Association of Cattle Veterinarians 1998) if accurate breeding time is known. This method is more accurate in detection of "open" females (90%) than in correct diagnosis of pregnancies (32%–34%). As stated, this method has the best application when accurate breeding dates are available; for example, when used in conjunction with AI programs.

Ultrasonography, particularly when employing real-time, B-mode technology, has shown rapid development and acceptance for pregnancy diagnosis within the beef industry. B-mode ultrasound has advantages beyond those of transrectal palpation. These include the ability to accurately (and conclusively) diagnose early pregnancies (before day 30), to determine fetal sex (after day 55), to detect early twin pregnancies, to assess fetal viability via heartbeat (from approximately day 21), and to help diagnose problems in the female reproductive tract.

Ultrasound images of actual pregnancies are useful for legal verification of pregnancy and can be used as a marketing tool for purebred and specialty cattle operations. In addition, ultrasonography allows consistent measurement and characterization of ovarian structures (follicles and corpora lutea). More specialized uses include ultrasound-guided follicular puncture for ovum pick-up. However, ultrasound equipment is expensive (although the price is steadily decreasing) and susceptible to damage, and extra training is necessary for proficiency. In addition, early embryonic death is common in cattle, and earlier detection of pregnancy, whether by ultrasound or by other technologies, will still fail to detect or forecast some of this loss. A number of published resources are available that describe the reproductive applications of ultrasound in cattle (Stroud 1994, Ribadu and Nakao 1999, Jones and Beal 2003).

Increasing attention is being given to the identification of pregnancy-specific proteins, including a large group of pregnancy-associated-glycoproteins. One of these, the early pregnancy factor, has shown promise since it was first described over 20 years ago. Although a simple, cow-side test is still to come, indications are that the early pregnancy factor could eventually provide a very early (1–2-day) indicator of pregnancy in cattle (Youngquist 2003). At present, a similar test protein, the early conception factor, is being promoted as a test to identify nonpregnant cows 6–30 days postinsemination. A test for another protein, the bovine pregnancy specific protein, has been employed commercially to detect pregnancy in serum collected approximately 1 month (28 days for heifers, 30 days for lactating cows) after insemination in dairy cattle.

One limitation to the cow-side use of such tests has been that most of them have been based on radioimmunoassay, which necessitates appropriate laboratory equipment and training. As indicated above, however, rapid cow-side tests are being developed using different technologies. Availability of such tests, at an economical price, would add considerable value to AI programs, with and without the use of estrus synchronization, by allowing earlier reenrollment and breeding of nonpregnant animals. Because the caveat of early embryonic loss remains, early markers of pregnancy might better be considered as indicators of "openness" than of pregnancy per se (Youngquist 2003).

HEAT DETECTION AIDS

Inability to accurately detect estrus in females is one of the greatest hindrances to a wider acceptance of AI in the cattle industries. A number of aids are available to help detect estrus in cattle. However, even with use of these aids, the efficiency of heat detection varies with herds, being influenced by management and facilities (Stevenson and Phartak 1999). Some of these aids are well proven and are currently widely employed, such as the use of tail paint or chalk, chin-ball markers with teaser bulls or androgenized cows, and Kamar heat mount detectors. This discussion will dwell on newer procedures.

HeatWatch

This technology, available commercially in the United States since 1994, comprises a small transmitter attached to the cow's tail head, which sends a signal when pressure is applied. This signal is picked up by a receiver attached to a computer. The software can be programmed to suit particular programs and preferences. With the advantage of having coverage 24 hours a day, 7 days a week, this device has been shown to be 95% accurate and more than 95% effective at detecting standing heat. Comparison with other systems (Bovine Beacon and Kamar) in Kansas indicated that HeatWatch seemed most accurate (Stevenson and Phartak 1999). However, this system is somewhat expensive (approximately $50 per cow) and requires fairly close proximity between transmitters and receiver.

Estrus Alert

Estrus Alert comprises a stick-on patch placed over the tail head that is friction activated to reveal color changes. Approximately five mounts are required to reveal most of the signal layer. It is available in different colors to facilitate management groupings.

Bovine Beacon

The Bovine Beacon is a patch that is applied with glue to the cow's tail head and that glows bright pink when activated by mounting pressure. The glow will persist for 12–18 hours, which can be an advantage in low-light situations. In different trials, this patch has proven equal to or superior to Kamar devices.

ET

Embryo recovery and transfer is the third most commonly used biotechnology in cattle in the United States, after AI and estrus synchronization. Approximately 40,000–50,000 beef calves, mostly within registered cow herds, are derived annually from ET using frozen embryos in the United States and Canada (Bazer et al. 2003), each at an estimated cost of $500 to $1000 above conventional breeding means. ET will probably see increased utilization as new techniques evolve, such as inexpensive, reliable embryo sexing or cloning.

The term "MOET" (multiple ovulation and ET) is used to help emphasize the place that ET has within the context of a program of genetic improvement. The application involves large nucleus herds in which the results of genetic improvement are disseminated to a wider population via ET, AI, and natural breeding, using young bulls. The process commences with a donor cow being treated with hormones to induce maturation and release of multiple oocytes. These eggs are fertilized within the donor cow, using semen from a select bull. The eggs are allowed to develop for approximately 7 days and are then removed and implanted in recipient females for the remainder of gestation. Eggs may be transferred soon after collection or cryopreserved for later transfer. Guidelines for general procedures used in bovine ET are available from the International Embryo Transfer Society (see Resources), as well as from the Food and Agriculture Organization of the United Nations (FAO) Training manual for ET in cattle (Seidel and Seidel 1991).

Pregnancy rates using fresh embryos can approximate those from a good AI program (approximately 60%). Use of frozen embryos in otherwise well-conducted programs can lower pregnancy rates by 10% or more. Success in ET programs is attendant on good training of personnel, good management of fertile donors and recipients, and good herd health. At least 12–15 recipients are recommended for each donor per cycle. Good results are largely dependent on good management of the recipient herd, which is a major expense in ET programs. Evaluating and selecting embryos is another area requiring good technique and attention to detail. The use of single-sex sperm will greatly benefit ET in the respective livestock industries (Bazer et al. 2003).

ET has other applications of note. One is to assist in the detection of carriers of undesirable recessive traits for both cows and bulls. Here, multiple embryos from homozygous, fertile females can be used to test suspected carrier bulls, and daughters of carrier bulls can be tested to determine whether or not they carry the deleterious allele (Seidel and Seidel 1991). Other technologies that use many of the micromanipulation techniques employed in ET include in vitro fertilization, semen sexing, sexing of embryos, cloning by nuclear transplantation, and production of transgenic animals.

Semen Sexing

Today, bovine sperm can be sorted according to sex (X or Y chromosome-bearing) with 90% plus accuracy using high-speed flow cytometry in conjunction with fluorescent probes. However, the process is not yet sufficiently rapid to provide a sufficient quantity of sexed sperm for acceptable results using conventional AI methods. Research with low-dose insemination, especially when deposited deep within the ipsilateral uterine horn, has shown improved results. Cost-effective semen sexing will lead to greater utilization of twinning to increase production per beef cow and to increase heifer births in dairy cattle.

Sexing Embryos

Although a number of potential techniques exist to determine the sex of embryos, only one method is currently available commercially. Here, the DNA of several cells removed microsurgically from the embryo is probed to detect the presence or absence of the male-determining Y chromosome. Although the success rate for skilled technicians using nonfrozen embryos is over 90%, the process requires time, specialized equipment, and expertise. Development of a quick, noninvasive, accurate, and inexpensive method would allow wider acceptance of this procedure.

Cloning by Nuclear Transplantation

Cloning, or making a genetically identical copy of an individual, is not necessarily a new concept, as it has been employed in horticulture for some time.

Cloning by nuclear transplantation involves taking an in vitro–matured oocyte, or unfertilized egg, and microsurgically removing its chromosomes. Somatic cells can also be used for nuclear transfer to produce embryonic stem cells. Individual cells of separate embryos are then fused to the denuded oocytes by electrical pulses. The process results in 20–30 genetically identical, one-cell embryos that are cultured in vitro for 7 days and then transferred to recipients, as is done in a conventional ET program. The individuals produced may not be completely identical because of environmental and genetic interactions, as well as cytoplasmic components of inheritance that can be transferred with somatic cell procedures.

In general, this process is expensive and relatively inefficient and has a tendency to produce low-viability, congenitally defective calves. Welfare issues associated with this technology are mentioned in chapter 14. Despite such reservations, the research and technology associated with the pursuit of successful cloning have contributed greatly to our knowledge of genetics and developmental biology, and there is little doubt that the current shortcomings will be overcome in time.

Production of Transgenic Animals

Transgenic techniques generally involve insertion of a specific gene (or genes) via DNA transfer in anticipation that the gene will become incorporated into and replicated within the recipient's genome. The genes inserted may be derived from the same species and used to improve production traits (growth, efficiency, milk production) or confer resistance to specific pathogens or parasites. They may also be derived from other species and used as modifying genes to produce, for example, therapeutic proteins in milk, or to develop animal models for human disease. Developmental and health problems in the recipients, or their offspring, have been associated with some of these procedures. In the United States, transgenic animals are regulated by the Food and Drug Administration, U.S. Department of Agriculture, or Environmental Protection Agency, with oversight responsibilities differing with the nature and use of the intended transgenic product.

Quantitative Trait Loci

With beef cattle, in common with other livestock species, most traits of economic importance are of quantitative nature; that is, they are influenced by a number of genes and their environmental interactions. Thus, major efforts are currently underway to identify and map the individual quantitative trait loci that control important traits. By using the linked marker loci, quantitative trait loci can be used in genetic improvement programs via marker-assisted introgression (Soller 1990) to transfer useful alleles from a resource population to a breeding nucleus. In dairy cattle, such research has aimed at identifying genes affecting milk production.

In beef cattle breeding, use of such technology has great potential in areas such as establishing paternity in multisire herds and linking individual bulls with the fertility, health, and performance parameters of their calves.

RESOURCES

Biotechnology

General information: http://www.nal.usda.gov/bic/; http://biotech.icmb.utexas.edu; http://www.ncbi.nlm.nih.gov/bic/
Regulation/Policy: http://www.usda.gov/agencies/biotech
Genome information: http://gnn.tigr.org; http://www.genome.gov
Agricultural biotechnology: http://www.aphis.usda.bov/brs

AI

Bull Health/Quality: http://www.NAAB-CSS.org
Standards for instruction in insemination techniques, semen handling, and reproductive management: NAAB, P.O. Box 1033, Columbia, MO 65205

Estrus Synchronization

Programs—strengths and weaknesses: http://www.oznet.k-state.edu/nwao/livestock.htm
Planner: http://www.iowabeefcenter.org
Synchronizing estrus and ovulation in cows and heifers: http://www.oznet.ksu.edu/nwao/LIVESTOCK.htm
Applied Reproductive Strategies in Beef Cattle Workshop, 2002. CD-Rom. Kansas State University Cooperative Extension Service/College of Veterinary Medicine

CIDRs

General information: http://www.cidr.com
Applied Reproductive Strategies in Beef Cattle Workshop, 2002. CD-Rom. Kansas State University Cooperative Extension Service/College of Veterinary Medicine

Control of the Bovine Estrous Cycle. 2002. DVD. Society for Theriogenology. Available at: http://www.therio.org

ET

International Embryo Transfer Society.

Manual of the International Embryo Transfer Association, 3rd ed. Available at: http://www.iets.org.

Seidel, G.E., S.M. Seidel. 1991. *Training manual for embryo transfer in cattle.* Food and Agriculture Organization of the United Nations Animal Reproduction and Health Paper 77. Savoy, IL: International Embryo Transfer Society.

American Embryo Transfer Association: http://www.aeta.org

Canadian Embryo Transfer Association (includes details of certification of practitioners, which is a requirement of the Canadian Food Inspection Agency): http://www.ceta.ca

Semen Assessment

American Association for Theriogenology: http://www.therio.org

Heat Detection

Heatwatch: http://www.heatwatch.com

Estrus Alert: http://www.estrusalert.com

REFERENCES

Australian Association of Cattle Veterinarians. 1998. *Pregnancy testing in cattle.* Indooroopilly: Australian Association of Cattle Veterinarians.

Bazer, F.W., T.E. Spencer, E. Gootwine, A. Gertler. 2003. Emerging technologies in animal reproduction. *Proceedings of the Society for Theriogenology*, Montgomery, AL: Society for Theriogenology, pp. 9–19.

Bellows, R.A., R.B. Staigmiller. 1994. Selection for fertility. In *Factors Affecting Calf Crop*, edited by M.J. Fields, R.S. Sand. Boca Raton, FL: CRC Press, pp. 197–212.

Burns, P.D., J.C. Spitzer. 1992. Influence of biostimulation on reproduction in post-partum beef cows. *Journal of Animal Science* 70:358–362.

Byerley, D.G., R.B. Staigmiller, J.G. Berardinelli, R.E. Short. 1987. Pregnancy rates of beef heifers bred either on puberal or third estrus. *Journal of Animal Science* 65:645–650.

Chenoweth, P.J., J.C. Spitzer. 1995. Biostimulation in livestock with particular reference to cattle. *Assisted Reproductive Technology/Andrology* 7:271–278.

Cornwell, D.G., J.F. Hentges Jr., M.F. Fields. 1985. Lutalyse as a synchronizer of estrus in Brahman heifers. *Journal of Animal Science* 61(Suppl. 1): 416.

Farin, P.W., P.J. Chenoweth, D.F. Tomky, et al. 1989. Breeding soundness, libido and performance of beef bulls mated to estrus-synchronized females. *Theriogenology* 32:717–725.

Geary, T. 1999. Heat detection systems and estrus management. *National Association of Animal Breeders (NAAB) Symposium: Improving Reproductive Performance, Proceedings.* Columbus, MO: National Association of Animal Breeders.

Johnson, S. 2003. Impact of estrous synchronization on the economics of reproductive management of beef cattle and factors to consider in protocol selection. *Proceedings of the Society for Theriogenology, Annual General Meeting.* Montgomery, AL: Society for Theriogenology, pp. 432–443.

Johnson, S., J. Stevenson. 2003a. *Synchronizing Estrus and Ovulation in Cows and Heifers.* MF2573. Manhattan: Kansas State University Agricultural Experiment Station and Cooperative Extension Service. Available at http://www.oznet.ksu.edu. Accessed July 19, 2004.

Johnson, S., J. Stevenson. 2003b. *Tips for a Successful Synchronization Program.* MF2574. Manhattan: Kansas State University Agricultural Experiment Station and Cooperative Extension Service. Available at http://www.oznet.ksu.edu. Accessed July 19, 2004.

Jones, A.L., W.E. Beal. 2003. Reproductive applications of ultrasound in the cow. *Bovine Practitioner* 37:1–9.

National Animal Health Monitoring System. 1998. Part III. Reference of 1997 Beef Cow-Calf Production Management and Disease Control. USDA:APHIS:VS. Available at http://www.aphis.usda.gov/vs/ceah/cahm. Accessed July 18, 2004.

Patterson, D.J. 1999. Estrous synchronization programs for replacement heifers and post-partum cows. *Proceedings of the Symposium on Reproductive Management Tools and Techniques, University of Missouri.*

Perich, M.R., R.C. Mulley, I.J. Lean, R. Reinberger. 1996. A comparison of three methods of oestrus synchronization for fixed time insemination in dairy heifers. *Proceedings of the Australian Society of Animal Production* 21:384–385.

Pexton, J., P.W. Farin, R.A. Gerlach, et al. 1989. Efficiency of single-sire mating programs with beef bulls mated to estrus-synchronized females. *Theriogenology* 32:705–716.

Pexton, J.E., P.W. Farin, G.W. Rupp, P.J. Chenoweth. 1990. Factors affecting mating activity and pregnancy rates with beef bulls mated to estruc-synchronized females. *Theriogenology* 34:1069–1080.

Ribadu, A.Y., T. Nakao. 1999. Bovine reproductive ultrasonography: A review. *Journal of Reproduction and Development* 45:13–28.

Rodriguez-Martinez, H. 2000. Evaluation of frozen semen: Traditional and new approaches. In *Topics in Bull Fertility*, edited by P.J. Chenoweth. International Veterinary Information Service. Available at: http://www.IVIS.org. Accessed July 15, 2004.

Seidel, G.E., S.M. Seidel. 1991. Training manual for embryo transfer in cattle. Food and Agriculture Organization of the United Nations Animal Reproduction and Health Paper 77.

Soller, M. 1990. Genetic mapping of the bovine genome using DNA-level markers with particular attention to loci affecting quantitative traits of economic importance. *Journal of Dairy Science* 73:2628–2646.

Stevenson, J.S. 2001. Incidence of anestrus in suckled beef and milked dairy cattle. *Journal of Animal Science* 79:(Suppl. 1):116.

Stevenson, J.S., A. Phartak. 1999. Effective use of heat detection devices. *Large Animal Practice* January/February: 29–31.

Stroud, B.K. 1994. Clinical applications of bovine reproductive ultrasound. *Compendium on Continuing Education for the Practicing Veterinarian* 16(8):1085–1097.

Toombs, R.E. 1998. National Animal Health Monitoring System: Beef '97. Fort Collins, CO.

Youngquist, R.S. 1997. Pregnancy diagnosis. In *Large Animal Theriogenology*, edited by R.S. Youngquist. Philadelphia, PA: W.B. Saunders.

Youngquist, R.S. 2003. Emerging technologies in bovine pregnancy diagnosis. *Proceedings of the Society for Theriogenology*, Montgomery, AL: Society for Theriogenology, pp. 162–171.

11
Calving and Calf Management in Beef Herds

Michael W. Sanderson

INTRODUCTION

The goal of a cow/calf enterprise is the efficient, profitable production of calves: producing and maintaining a live calf to weaning. Management of the calving season is critical to this goal and must begin long before calving season. Research indicates that 50%–60% of calf mortality from birth to weaning occurs within 24 hours of birth. Furthermore, 40%–50% of calf morbidity and 70%–80% of calf mortality occur in the first 21 days of life (Patterson et al. 1987, National Animal Health Monitoring System [NAHMS] 1997b). Because the majority of calf morbidity and mortality occurs around the time of calving, it seems clear that proper management of the calving season is critical for control of neonatal morbidity and mortality in beef herds and for optimization of the weaned calf crop.

Many factors affect neonatal morbidity and mortality and proper management of the calving season, including those factors that occur long before calving season begins. These are precalving, calving time, and postpartum calving aspects that are important for optimal management of the calving season.

Calf management continues to be important following calving. Although the majority of calf morbidity and mortality occur in the first 21 days of life, the majority of growth occurs after these 21 days. Optimizing the pounds of calf weaned by the herd necessitates careful management of calves to promote growth during the summer grazing period. Calf morbidity continues to be of concern, but according to producer survey data, respiratory disease and infectious bovine keratoconjunctivitis (IBK) are more common in older calves than is diarrhea (NAHMS 1997b). Numerous factors affect calf growth and health postcalving, including proper management and timing of castration, dehorning, vaccination, parasite control, nutrition, and weaning.

PRECALVING MANAGEMENT

Many risk factors for early neonatal morbidity and mortality must be managed in the precalving period. Management of the calving season to control calf morbidity and mortality must begin long before calving season. Precalving management includes proper replacement heifer development and sire selection for control of dystocia, good herd nutrition for good calf vigor, a proper vaccination program, and management of the prepartum cow environment to control environmental contamination.

DYSTOCIA MANAGEMENT

Dystocia, or calving difficulty, is a major factor affecting beef calf morbidity and mortality risk, and preventive management must begin early. In one study, over 50% of all calf mortality was attributed to dystocia (Bellows et al. 1987). Calves that experienced dystocia were more likely to die within 12 hours of birth (odds ratio [OR] = 13) and were more likely to experience a morbid incident by 45 days of age (OR = 4) than were calves that did not experience dystocia (Wittum et al. 1994). At the herd level, herds with high levels of dystocia were five times more likely to experience high calf morbidity than were herds with a low level of dystocia (Sanderson and Dargatz 2000). There are many factors that influence dystocia risk, and birth weight and pelvic dimension are two important ones. Management to control their effects must begin long before calving season begins. Our prepartum efforts to manage the dystocia level and moderate its effects must center on replacement heifer development and sire selection.

Development of Replacement Heifers for Dystocia Management

Proper replacement heifer development is necessary to good calving management and has been more completely addressed in this book (see chapter 8) and elsewhere (Spire 1997, Larson and Moser 1998). Comments in this chapter will be confined to those that are particularly related to calving and dystocia. Heifers need to be placed in a nutritional development program, starting at weaning, to allow for good growth and development of an adequate pelvic area. Measurement of the pelvic area of the dam to predict dystocia has been used as a criterion for the selection of replacement heifers, under the assumption that a larger pelvic area allows for a calf of higher birth weight to be passed without assistance. However, available research indicates that pelvic area alone explains only a small proportion of the variability in dystocia risk (Johnson et al. 1988).

Pelvic area measurements taken before the breeding season or at the time of pregnancy examination are used to estimate the pelvic area before calving. Those heifers with a small pelvic area before the breeding season can be culled or selectively mated to easy-calving bulls, and those with a small pelvic area at the time of pregnancy examination may be aborted, culled, or identified for careful observation at calving (Deutscher 1985). Some on-farm studies have reported that pelvic measurements taken a few weeks before the breeding season and again at the time of pregnancy examination have poor positive predictive values and sensitivity for predicting dystocia (Van Donkersgoed et al. 1990). Although the mean pelvic area in those heifers with dystocia was slightly smaller than in those without calving difficulty, there was a large overlap in the distribution of pelvic area size between the two groups, and the measurements did not discriminate well between heifers that would experience dystocia and those that would not.

Reasons for the inconsistent results among different observations may include the definition of dystocia, the accuracy and precision of the measurement procedure, and the fact that calf birth weight in relation to the dam's pelvic area determines the degree of calving difficulty (Van Donkersgoed et al. 1990). Limited evidence indicates that identifying heifers with the narrowest pelvic width for culling may be more effective than culling based on pelvic area (Anderson et al. 1993b). Despite difficulties with the predictability of pelvic measurement, identifying heifers with small and abnormally shaped pelvises for culling may be a useful technique, though it should be used as part of a comprehensive management plan for dystocia.

Nutritional development of replacement heifers needs to begin at weaning and continue through calving time. Heifers should be fed to allow modest rates of body weight gain (0.5 kg/day) during late pregnancy. Restriction of nutritional intake during late pregnancy does not decrease dystocia rates in beef heifers (Kroker and Cummins 1979). In feeding trials, restriction of energy intake to a level that maintains maternal body weight has little influence on the development of the fetus, but maternal metabolism may be altered considerably (Prior et al. 1979). Within reasonable limits, the level of maternal energy intake does not markedly influence fetal weight or composition (Prior and Laster 1979). Birth weight may be slightly decreased, and fat stores may be decreased. However, the loss of 0.5 kg/day during the last trimester of pregnancy in beef heifers is associated with weak labor, increased dystocia rates, reduced calf growth rate, prolonged postpartum anestrus, reduced pregnancy rate, and increased morbidity and mortality as compared with heifers that are maintaining or gaining weight at a moderate rate during late pregnancy (Corah et al. 1975, Kroker and Cummins 1979).

Selection of Bulls for Dystocia Management

The birth weight of the calf is the most important factor in determining the likelihood of dystocia (Naazie et al. 1989) and accounts for 30%–50% of the variability in dystocia rates (Meijering 1984). Minimizing birth weight may not be an economically viable alternative because calves with a genetic ability for rapid growth exhibit that potential in utero. Rapid in utero growth may produce birth weights capable of causing dystocia, but it also results in superior weaning weights. Dystocia is likely a threshold trait, and once a certain threshold birth weight is reached in a herd, dystocia rates may increase dramatically. Conversely, after the birth weight decreases to a certain level, further drops have only a minimal effect on dystocia rates (Anderson et al. 1993a).

Efforts to decrease the incidence of dystocia without using some method of controlling or predicting calf size will likely be ineffective. However, a combination of culling heifers with small pelvic areas and using bulls that sire calves with acceptable birth weights may reduce dystocia significantly (Makarechian and Berg 1983). Using only the sire's birth weight as the criterion for selecting which bulls to use in a breeding program to control calf birth weight and dystocia is of limited effectiveness. A number of nongenetic influences affect the birth weight of a sire, such as age of the dam, environment, nutritional level, and twinning. This confounds our

BIRTH WEIGHT EPD

Breed associations report EPDs in the same units as the trait they reflect; for example, birth weight EPDs are reported in pounds. Their intent is to estimate the genetic effect of a bull on the average cow in the breed in the base year, which varies between breeds. EPDs are useful for comparing bulls but do not identify the specific effect a bull will have on a herd. For example, a bull with a birth weight EPD of +3.0 would be expected to sire calves 4 lb heavier on average than a bull with a birth weight EPD of −1.0 when bred to the same set of heifers. EPD distributions and breed mean values can be obtained from the respective breed associations to assess the ranking of bulls within the breed. They are a useful guide to estimating the bulls' effect on the average cow of today.

An accuracy value, ranging from 0 to 1, is reported for each EPD. The accuracy is a measure of how reliable the EPD value is. It provides a statistical confidence in how accurately the EPD reflects the true effect of the bull on the herd. High accuracies indicate a high level of confidence that the stated EPD truly reflects the bull's effect. EPDs can be calculated on yearling bulls with no progeny, but the accuracy is low. High-accuracy EPDs (>0.8 to 0.9) are usually only obtained in AI bulls with a large number of progeny. As illustrated in Figure 11.1, a low-accuracy birth weight EPD is consistent with a much wider range of true birth weights. The high-accuracy birth-weight distribution from a bull is much narrower, resulting in much higher confidence in the EPD estimate. Also included in Figure 11.1 are approximate standard deviations and confidence intervals for birth weight EPD at different accuracies.

ability to identify bulls that have a genetically low birth weight. Compared to using sire birth weight only, expected progeny differences (EPDs) for birth weight are a more effective way to identify bulls that will sire acceptable birth weight calves (see box; Larson and Herring 1998). To maintain good preweaning gains, an attempt should be made to identify low–birth weight EPD bulls for use on heifers while maintaining at least moderate weaning and yearling weight EPDs. The use of AI sires with high-accuracy EPDs is the most effective way to accomplish this. EPDs can be calculated on yearling bulls with no progeny based on the bulls' performance as well as on the performance of other animals in their pedigree, but the accuracy is low. Still, EPDs provide better information than relying on the sire's birth weight alone. EPDs were only useful for comparisons within breeds until recently. Methods have been developed for comparison of EPDs across breeds to allow estimation of the effect of bulls on different breeds. These comparisons are of particular use in selecting bulls to control dystocia in crossbreeding programs. At present, adjustment factors between breeds are available to assess the effect of a bull in a crossbreeding system (VanVleck and Cundiff 1998). Two recent innovations in the use of EPDs for management of dystocia are the calving-ease EPD and the maternal calving-ease EPD. Calving-ease EPD is related to birth weight EPD, but it may more effectively predict calving ease. Maternal calving ease is a measure of the effect of the maternal grandsire, and the ease with which a bull's daughters will give birth.

Herd Nutrition in Calving Management

The role prepartum nutrition of dams plays in the incidence of failure of passive transfer (FPT) is not firmly established. In one study of 2-year-old heifers, body condition scores at calving did affect calf serum IgG1 levels. Calves from heifers with a body condition score of less than 5 had lower IgG1 levels than did calves from heifers with body condition scores greater than or equal to 5 (Odde 1988). Other work, however, has not supported a relationship between restricted nutrition in the final trimester and lowered calf immunoglobulin levels (Hough et al. 1990) Heifers that are fed a restricted diet produce decreased colostral volume, but the IgG1 concentration of the colostrum is increased (Odde 1988). The number of animals included in most studies has been small, perhaps limiting the power available to detect the effect of prepartum nutrition on FPT.

A well-balanced prepartum ration is a prudent management factor despite the lack of clarity in the relationship between prepartum nutrition and FPT. Proper prepartum nutrition supports improved reproductive performance and milk production in cows. Some evidence also supports an association between calf health and vitamin and trace mineral nutrition of dams. Increasing levels of vitamin E given to cows may increase the level of calf passive immunity (Corah et al. 1998). The ration should be balanced for trace minerals and vitamins to try to ensure optimal immune response in both the dam and the calf.

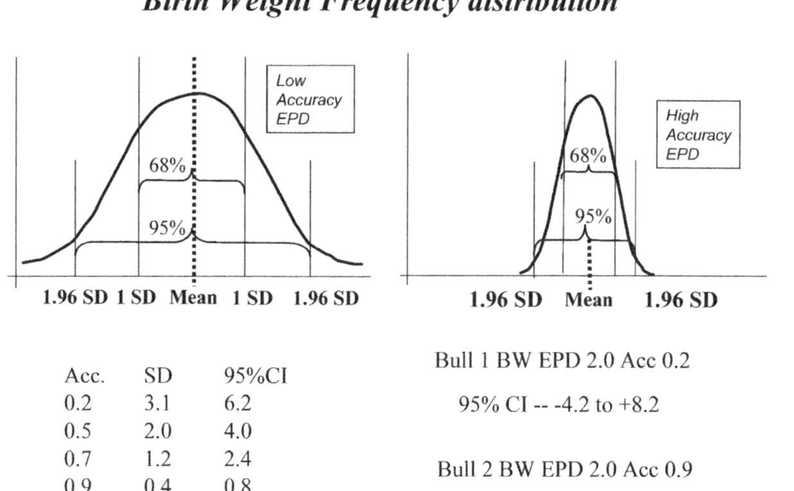

Acc.	SD	95%CI
0.2	3.1	6.2
0.5	2.0	4.0
0.7	1.2	2.4
0.9	0.4	0.8

Bull 1 BW EPD 2.0 Acc 0.2

95% CI -- -4.2 to +8.2

Bull 2 BW EPD 2.0 Acc 0.9

95% CI - +1.2 to +2.8

Figure 11.1. Birth weight EPD frequency distribution.

SD = Standard deviation; CI = Confidence interval; Acc = Accuracy

Inadequate protein nutrition in late pregnancy has been associated with weak calf syndrome at birth. Calves that are weak and slow to rise and suckle are at increased risk for FPT. As such, it may be a factor contributing to neonatal morbidity and mortality in beef calves. In one study, beef cows fed a low-protein ration during the last 4 months of pregnancy had a decreased gestation period and decreased weight gains compared to controls fed a protein-adequate ration (Waldhalm et al. 1979). Calf mortality in the low-protein group was associated with dystocia and prematurity. Calves born to cows on restricted-protein or restricted-energy diets have a decreased thermogenic response and are more susceptible to cold stress (Carstens 1994).

Proper prepartum nutrition is critical for optimal neonatal calf health, as well as for subsequent cow reproductive performance. Cow weight and weight loss are functions of interactions within a herd as well as of the diet presented to the cows. Some of the factors affecting cow weight loss include breed of cow, general management, amount and quality of feed presented, cow age and dominance, and degree of shelter provided during cold and windy weather. Cows should be sorted and managed to provide adequate feeding space so that dominant cows do not overeat at the expense of timid or young cows. A ration should then be formulated to meet the needs of the cows on the basis of their breed, size, environmental conditions, and stage of gestation.

Vaccination Program

The cow and heifer vaccination program must account for disease agents that may cause in utero infections resulting in late-term abortions or weak-born calves. A number of infectious agents may be associated with late-term abortions or weak-born calves with subsequent high mortality rates. These include infectious bovine rhinotracheitis (IBR), bovine viral diarrhea (BVD), *Leptospira* ssp., and *Brucella abortus*. Vaccination programs should be designed to provide protection to the cow during gestation. Options for vaccination timing include use of a modified live vaccination between calving and breeding, and a killed vaccination at pregnancy examination time. Recently, one modified live vaccination four-way viral vaccine with or without *Leptospira* has been cleared for use in pregnant cows, provided they have been previously vaccinated according to the label directions. A new *Leptospira* vaccine has also recently become available that may provide improved immunity. Anecdotal information from dairy herds indicates an improvement in reproductive performance, but no clinical trial data to evaluate the efficacy of the vaccine are currently available. As with all vaccination programs, the cost of the vaccination must be weighed against the value of lost production and vaccine efficacy. Even under the best circumstances, vaccination will not provide complete protection to a herd, and additional biosecurity and biocontainment practices should be implemented to control disease. The cost and efficacy of biosecurity and biocontainment programs should be weighed against the cost of disease in the herd. Possible strategies include limiting imports to the herd and instituting appropriate testing and quarantine protocols (see chapter 5). For *Leptospira*, environmental control such as drainage of standing water and wildlife control is also important.

A vaccination program should also include measures against agents of disease for the neonatal calf. Vaccines are commercially available for rotavirus, coronavirus, *Escherichia coli* K99, and *Clostridium perfringens* types B and C. To be effective, these vaccines should be given to cows and heifers before calving to elevate specific immunoglobulin levels in the colostrum. An initial vaccination and booster followed by a yearly booster vaccination is required. The yearly booster vaccination is given at least 2 weeks and not more than 6 weeks before calving.

Clinical trial data for vaccines used in this manner are variable, with some trials reporting no effect and others reporting significant decreases in neonatal morbidity (Hjerpe 1990). This variability may be related to differences in the incidence of sufficient causes for diarrhea in the different study populations. Overall, the data indicate that vaccination may be a useful adjunct to proper management in controlling neonatal diarrhea (Cornaglia et al. 1992, Boland et al. 1995). Vaccination may not result in effective control when the level of management is poor, and it may not be cost-effective with exceptional management. In addition, there is an oral modified live rotavirus and coronavirus vaccine available for administration to calves at birth. Administration must occur at least 1 hour before colostrum ingestion. Clinical trial data have not supported this vaccine's effectiveness (Hjerpe 1990).

Prepartum Cow Environment Management

Management of the physical environment of cows before parturition is a valuable tool in the control of environmental contamination in the calving area and, subsequently, in the infection pressure placed on the calf. Movement of cows and heifers to the calving grounds should occur no more than 2–3 weeks before the onset of the calving season to minimize the contamination of the calving pastures (Schumann et al. 1990). Poorly drained calving pastures, resulting in standing water and mud, increase risk for elevated morbidity and mortality in calves. Excess population density will increase risk for calf disease, and inadequate observation caused by large pasture size will increase the effects of dystocia on calf disease. Calving pasture size should attempt to balance these two factors: small enough to allow for observation, but large enough to avoid excessive population density and pathogen exposure. Data on optimal population density in the calving pasture to balance these two issues are not available; however, 4000–6000 square feet of pasture per cow may be reasonable on good-quality pastures.

Heifers have elevated dystocia rates and should be calved separately from cows to allow for increased intensity of observation. Both heifers and cows should be sorted into management groups on the basis of their estimated calving date, which is calculated from estimated gestational duration collected at the time of pregnancy examination. Separating the herd into management groups based on class and due date allows concentrated observation of the smaller number of cows or heifers that are most likely to be calving and in need of intervention. In addition, both segregation and concentrated observation can be done on a small number of heifers or cows without increasing population density.

Feeding cows late in the evening may be useful in increasing the proportion of calves born during the day versus the night (Lowman et al. 1981), although not all studies have supported this effect. If this strategy is attempted, cows should be fed late in the evening (9–11 p.m.), beginning at least 2–3 weeks before the onset of calving. The likelihood of success is decreased if cows are fed earlier or if the program is not begun far enough in advance of the start of calving.

The season in which calves are born has also been associated with increased neonatal morbidity and mortality levels, most likely resulting from weather stress. One Canadian study found that calves born from December to March were four times more likely to die within 24 hours and two times more likely to have had diarrhea by 30 days of age than were calves born in April and May (Ganaba et al. 1995). Calving in the late spring or early fall may be a valuable tool that can be used to significantly decrease both environmental stress on calves and their risk for disease. The overall management, financial, marketing, and economic effects of calving at this time can also be significant and should be carefully considered and planned before a change is made to a late-spring calving time.

CALVING TIME MANAGEMENT

Calving time management requires timely intervention in dystocia cases and proper management and cleaning of the calving environment. Confinement of cows or heifers will improve early detection of dystocia but will also increase population density and concentrate pathogen levels in the calving pens. Good management of the environment through proper surface preparation, good cow and pair movement, and good drainage and cleaning practices is necessary to control pathogen exposure level.

Calving Time Dystocia Management

Calving difficulty, or dystocia, is the major cause of calf loss for cow/calf producers. In one study involving 13,296 parturitions, 51% of 798 necropsied calf mortalities were attributed to dystocia, which exceeded losses from all other causes (Bellows et al. 1987). Wittum et al. (1994) showed that dystocia increased calf risk for mortality in the first 12 hours after birth by 13 times and calf risk for morbidity by four times in the first 45 days of life, as compared to calves that did not experience dystocia. When heifers calve at 2 years of age, the prevalence of dystocia ranges from 15% to 30%, resulting in significant financial loss. In the NAHMS Beef '97 survey, producers reported a dystocia incidence in heifers of 17%.

Dystocia results in increased perinatal calf mortality, morbidity, and decreased productivity in cow/calf herds. At the herd level, an increasing incidence of dystocia has been associated with increased risk of calf morbidity (Sanderson and Dargatz 2000). When the incidence of dystocia was above 20%, herd risk for high morbidity increased fivefold. Dystocia has also been associated with an increased incidence of mothering problems in dams of all ages. Some studies have related increased morbidity (Wittum et al. 1994) and stillbirth risk (McDermott et al. 1991) to calves born to heifers; however, in one of the studies, no effect of dam age was noted when the effect of dystocia was accounted for (Wittum et al. 1994). Dystocia may also decrease pregnancy rate in the next breeding season in females that lose calves (Patterson et al. 1987), and prolonged parturition may result in a decrease in subsequent reproductive performance in beef cattle (Doornbos et al. 1984).

Early Intervention for Dystocia Management

Despite our best efforts to avoid dystocia, some illness will occur. Indeed, as discussed previously, minimizing dystocia incidence may not be an economically justifiable goal, as small–birth weight calves generally grow more slowly postpartum and have lower weaning weights. Dystocia can cause hypoxia, acid–base disturbance, and decreased vigor, leading to a delay in the ingestion of colostrum and to an increased incidence of FPT (Odde 1988, Besser et al. 1990). In one study, approximately one-third of the anatomically normal calves that died from illness associated with dystocia had gross pathologic evidence of parturition injury. This effect was evident even though the cows had not been given assistance. The authors suggested that these deaths could have been prevented if surveillance and obstetric assistance had been provided (Bellows et al. 1987). Early intervention in dystocia cases is an effective way to minimize the effects of dystocia on calves (Bellows and Lammoglia 2000). A research study indicates that stage II labor should not exceed 1 hour in heifers and 30 minutes in cows, and this same study indicated that heifers that complete stage II labor within 1 hour do not have extended postpartum intervals (Rice 1994).

Heifers need to be regularly monitored, and assistance provided promptly, if stage II labor is not to exceed 1 hour. Data from the NAHMS Beef '97 survey showed that 39% of operations allowed heifers to labor for 3 or more hours, and 72% allowed heifers to labor for 2 or more hours before assistance was given (NAHMS 1997a). If heifers cannot be regularly monitored, it may be prudent to focus bull selection on minimizing birth weight at the expense of growth to avoid the negative effects of dystocia on calf health and survivability. The individual producer will need to identify the most economical level of dystocia and growth and select bulls to match the cows and heifers.

Producers must be well trained to intervene appropriately in dystocia and to recognize when to call the veterinarian. Early intervention in dystocia cases can decrease calf stress (Bellows and Lammoglia 2000) and reproductive effects of dystocia in the cow (Doornbos et al. 1984). Because of the higher incidence of dystocia in replacement heifers, a concentrated, separate calving season may be helpful to allow labor resources to be focused on early intervention with heifers. Feeding late in the evening to encourage daytime calving may also help in the allocation of labor resources (Lowman et al. 1981). Calves that cannot be delivered vaginally should be identified early and presented for a Caesarean section to minimize calf stress (Bellows and Lammoglia 2000) and injuries from excessive traction (Schuijt 1990). A general rule of thumb for the producer is that if a cow has not made significant progress in delivering the calf within 30 minutes, it is time to seek veterinary assistance.

Functional working facilities are necessary to make early dystocia intervention practical. At minimum, an easily accessible 12 × 12-ft pen with a head catch is needed. If facilities are not accessible or functional, then labor requirements for early intervention increase dramatically, and producers are less likely to intervene in a timely manner. This delay can result in elevated morbidity and mortality in calves and in decreased reproductive performance in heifers and cows.

Management of the Calving Environment

Research data on the optimal management of the calving season at the herd level are sparse, perhaps because of the difficulty in assembling large numbers of producers who will cooperate in providing a statistical basis for inference at the herd level. One such study examined risks for high herd mortality level using 56 herds in a case–control design (Schumann et al. 1990). This study found a number of univariate associations of management practices and herd mortality incidence. In the final multiple logistic regression model, percentage of heifers in the herd, poor drainage in the calving area, and wintering cows and heifers on the same ground were associated with increased risk for high herd mortality. For every 1% increase in poorly drained calving area, the risk for high herd mortality increased 1.4 times. Calving cows and heifers together was associated with increased herd mortality in the univariate analysis. It was highly correlated with wintering cows and heifers together and was not included in the final multivariate model. Dystocia incidence was not collected in the study, and risk estimates might change significantly if the effects of dystocia were included.

Season of calving may be associated with morbidity and mortality levels as well, likely through the stress of cold temperatures and wet conditions. Cool, damp conditions are ideal for pathogen survival and likely facilitate the build-up of pathogen levels to supply an infectious dose to calves. Winter-born calves may be more likely to die and more likely to have diarrhea than are calves born in spring (Ganaba et al. 1995). In a Nebraska study, decreasing temperature and increasing precipitation on the day of calving decreased calf survival, and calves born to heifers were particularly susceptible to weather conditions (Azzam et al. 1993). A Louisiana study found that calf death loss was greater in spring-born calves compared to fall-born calves (Bagley et al. 1987)

Calving time management is focused around decreasing stress and infection pressure on the calf. Unfortunately, these two goals are somewhat antagonistic. Controlling calf stress is achieved by confinement and observation during calving to ensure prompt dystocia intervention and the provision of shelter from inclement weather for calves. Both of these practices, however, increase population density and, subsequently, infection pressure on the calf. Balancing the need for observation and timely intervention with the need to control population density to optimize the calf crop is the business of calving management. Unfortunately, hard data on what the optimal combination may be are scarce. Clearly, dystocia is an important, and perhaps the most important, risk for calf morbidity and mortality.

Calving in confined conditions appears to be associated with increased morbidity, but population-attributable fractions indicate that it is a less important determinant of morbidity than is dystocia (Sanderson and Dargatz 2000). One study from Australia indicates that closely confined and monitored heifers may have an increased risk of dystocia and stillbirth (Dufty 1981). Data on the optimal square footage necessary for each cow or heifer in the calving corral are not available. Estimates vary from 1000 to 2000 square feet per cow in the observation pen (Radostits et al. 1994). From an infection pressure standpoint, the more space provided, the lower the infection pressure; however, monitoring becomes more difficult as the area increases. Facilities designed to bring heifers or cows quietly and quickly from observation pastures into calving pens for early intervention may take advantage of the best of both worlds.

Current knowledge of proper management of the calving environment indicates a number of important management areas for morbidity and mortality control. Heifers and cows should be calved separately to provide the opportunity for increased observation of first-calf heifers and for prompt dystocia intervention. Calving observation corrals must be adequately drained to avoid standing water and mud/manure build-up. Pathogen accumulation and survival are increased by a moist environment, resulting in increased infection pressure on the calf. Calving corrals should have a slight slope to encourage runoff of water and a compacted surface to minimize the amount of water absorbed. Calving pastures should have good grass cover to minimize mud and be managed to maintain a healthy stand of grass. All calving corrals and pastures should have a southern exposure to facilitate drying and maximize solar radiation. Pens should be scraped and allowed to dry over the summer. Ideally, calving grounds should be left vacant the rest of the year to allow drying and solar radiation to decrease the pathogen levels.

Dystocia pens into which dystocias or mothering problems are brought must be managed to minimize pathogen build-up and infection pressure on calves. They should be easily cleaned and support minimal pathogen loads. Concrete floors are easily cleaned and disinfected, but they may not be economically practical in many instances. Compacted sand or gravel may be a useful alternative. Sand and gravel have low organic content, support minimal bacterial growth, and can be scraped and replaced between calving seasons. Dystocia pens should be bedded and the bedding removed between calvings. An adequate number of pens is necessary to allow for drying time after cleaning and to alternate the pens before using them again. Calving at a time of year when weather factors are less stressful on calves and the environment is easier to manage may be beneficial. This recommendation will require a comprehensive plan for marketing calves and an assessment of the economic effects of calving season change in all areas of the enterprise. Such a change should not be entered into in a casual way but may be indicated after thorough evaluation.

A veterinarian's visit to the farm approximately 3–4 weeks before the onset of the calving season will provide him or her with the opportunity to evaluate the management preparations made by the producer and to recommend any final changes. A written record of the visit and the recommendations made should be sent to the producer. Proper intervention rules and procedures for dystocia can be reviewed with the producer at this time, as should rules for handling cow/calf pairs after calving. A producer calving management seminar each year can serve to refresh the memories of producers and improve management performance. Ideally, records on calving ease, along with morbidity and mortality dates for individual calves, will be kept to allow analysis of risks and risk groups. If good records are kept, an increased incidence of disease will be easily detected and the management factors responsible will be more likely identified.

POSTPARTUM CALVING MANAGEMENT

Postpartum management of the calving season requires proper planning to ensure transfer of passive immunity to calves and proper cow/calf pair management. Good passive transfer requires proper nutritional management of cows and management of calves at high risk for FPT. Cow/calf pair management must be designed to segregate high-risk and morbid calves from the healthy herd to minimize disease risk.

Passive Transfer in Calving Management

Calves receive immunity passively from the dam through the ingestion of colostrum. The calf's immune system is immature at birth and is dependent on the acquisition of passive immunity for disease protection in early life. Immunoglobulin (IgG1 and IgM) and lymphocytes are absorbed directly across the intestinal wall into the calf's circulation to provide immunity. The ability of the intestinal wall to absorb these large molecules and cells is a transient phenomenon. Intestinal closure is complete by 24 hours, and absorption has decreased significantly by 6–8 hours of age (Radostits et al. 1994). Ingestion of adequate amounts of quality colostrum as early as possible after birth is important for calf survival and growth. Limited surveys of beef calves have shown a passive transfer failure rate of 11%–30% (Perino 1997). Calves with FPT are three to nine times more likely to become sick before weaning and are five times more likely to die before weaning as are calves with adequate passive transfer (Perino et al. 1993, Wittum and Perino 1995). Taken together, these data indicate that FPT in beef calves may be responsible for a significant proportion of neonatal calf morbidity and mortality.

The relationship between dystocia and FPT has not been consistent in the literature. In a set of 73 crossbred, 2-year-old heifers, increasing the calving difficulty score was associated with decreasing serum IgG1 and IgM levels in calves, even though assisted heifers were milked out and calves force-fed (Odde 1988). Increasing calving difficulty was also associated with an increased interval from birth to standing, which may delay colostrum ingestion and decrease passive transfer. Other studies have not shown significant differences in immunoglobulin levels between calves that experience dystocia and those that do not when both groups were either force-fed an equal amount of colostrum (Stott and Reinhard 1978) or allowed to suckle (Perino et al. 1995). Postnatal respiratory acidosis resulting from dystocia has been associated with decreased colostral absorption despite adequate colostral intake (Besser et al. 1990).

The discrepancy between research results may be caused by differences in the severity of dystocia experienced or in colostral quality. It seems likely that dystocia would have some negative effect on passive transfer. Therefore, reasonable additional efforts should be made to ensure sufficient colostral intake by calves experiencing dystocia.

Poor udder conformation in beef cows has been associated with increased time spent teat-seeking and with time to first suckle in calves (Ventorp and Michanek 1992), perhaps increasing the likelihood of FPT. Mastitis in beef cows has been associated with decreased serum immunoglobulin levels in their calves as well (Perino et al. 1995). Mastitis has not been extensively studied in beef cows; however, available studies indicate a wide range of prevalence from 9% to 67% (Perino et al. 1995, Simpson et al. 1995, Paape et al. 2000). The effects of mastitis on calf growth and weaning weight are not clear. Some studies have shown no effect on weaning weight (Paape et al. 2000), and others have indicated a depression in weaning weight (Newman et al. 1991). Treatment of cows with intramuscular tetracycline was not effective in reducing the prevalence of mastitis in beef cows (Duenas et al. 2001).

Efforts to minimize the incidence of FPT should focus on dystocia management, proper nutrition, and intervention with calves at high risk for FPT. Cows that experience dystocia should be milked out immediately and the calf force-fed colostrum to ensure ingestion. Cows with poor udder conformation or mastitis should be milked and the colostrum fed to the calf to ensure timely intake. Colostrum supplements have not proven useful in controlled clinical trials of elevating serum IgG levels (Mihura et al. 1997, Perino et al. 1995). Vaccination of the cow before calving with pathogens causing enteric disease in calves may be a useful adjunct to good overall management in reducing morbidity through increasing the quality of colostrum (Cornaglia et al. 1992, Boland et al. 1995). Vaccination in both studies resulted in an approximately 50% decrease in morbidity, although the studies were not blinded, so treatment bias cannot be ruled out. For optimal effect on colostrum, cows should receive their final booster vaccination 2–3 weeks before the onset of calving season.

Multiple tests are available to test the blood of calves for adequate transfer of immunity. The incidence of FPT can be monitored in beef herds by quantifying serum total protein at 1–8 days of age (Tyler et al. 1996) Serum total protein correlates well with IgG levels in serum and is easily and quickly tested for. Recently, a commercially available whole-blood immunoassay has become available for assessment of passive transfer. The test is adaptable to field use, and the sensitivity and specificity are better than serum total protein in predicting FPT (Dawes et al. 2002). However, by the time the calf is 24 hours of age, should either test be performed, options for intervention to increase immunoglobulin levels are limited to intravenous administration of plasma. Monitoring calf FPT incidence may be best suited to outbreak investigations to determine whether FPT is a factor in the current disease outbreak or for client education to improve management. Some data indicate that serum TP levels may be a useful indicator of FPT as early as 10 hours after birth; however, administration of commercial colostrum supplements had no effect on 24-hour TP levels (Perino et al. 1995). When used as a diagnostic test in clinical cases, dehydration can falsely elevate TP levels, leading to false negative results.

Cow/Calf Pair Management

Physical/spatial management of the postpartum pair is an important component of calving season management. Hard data on optimal practices are scarce, but some reasonable practices can be recommended. Once cows or heifers have calved and the calf has suckled, move the pair out of the calving area to a healthy calf nursery pasture within 24 hours. Monitor cows and heifers to ensure they accept and mother their calves and the calves suckle. Heifers may be at increased risk for rejecting their calves, particularly following a dystocia, which would subject the calves to increased risk for disease. The healthy calf nursery should be a well-drained, established grass pasture. Density recommendations for a healthy nursery pasture have ranged from one (Church and Janzen 1978) up to five (Radostits et al. 1994) cows per acre.

If the pair does not bond and the calf fails to suckle within a short period of time, the animals should be brought into the calving barn and restrained to allow the calf to suckle. Pairs entering the calving barn are at increased risk for disease because of the reasons for entering the barn, such as dystocia, failure to suckle, or hypothermia, and also because of exposure to the population-dense environment of the barn. Once a pair enters the calving barn they should not go to the healthy nursery pasture but go to a separate, high-risk nursery. The high-risk nursery should be a quality pasture with a similar population density as that of the healthy nursery. The smaller number of calves in the high-risk nursery can be more intensively monitored for morbidity and treated promptly. Segregation of these high-risk calves also avoids exposure of the rest of the herd. The healthy nursery pasture provides decreased population density to minimize infection pressure.

Calves identified as morbid in the healthy nursery pasture should be removed for treatment and placed in a morbidity nursery. Clinically ill calves can serve to multiply infectious agents and result in high levels of environmental contamination. One study indicated that quarantine of calves with diarrhea, along with their dams, from the rest of the herd decreased morbidity incidence (Clement et al. 1995). The morbidity nursery may be combined with the high-risk nursery, but this will increase exposure of the high-risk calves to pathogens. Morbid calves should not be returned to the healthy nursery area. All equipment used to treat morbid calves should be thoroughly disinfected between calves to avoid transfer of infectious agents from calf to calf.

Cold temperature and precipitation can significantly increase calf stress and morbidity and mortality (Azzam et al. 1993). Providing shelter for calves has been associated with decreased mortality rates (Schumann et al. 1990). Congregation of calves in shelters also produces an area of increased population density and infection pressure. If not managed carefully, shelters may increase morbidity and mortality because of increased infection pressure. Construction of shelters that can be easily transported will allow producers to move shelters away from contamination to fresh ground.

The most common cause of calf morbidity in the neonatal period is diarrhea. According to a major survey of beef producers in the United States, 80% of morbidity from birth to 21 days of age is caused by diarrhea (NAHMS 1997b). It is generally not possible to differentiate diarrheas associated with different etiologic agents on the basis of clinical signs, but the difference in the ages of calves affected can be clinically helpful in identifying the likely agents involved. Specific identification of the etiologic agent is not always necessary because risk factors for diarrhea are similar for the common agents of disease. Cows and calves often serve as carriers of the common agents of diarrhea in calves; mature cows commonly carry rotavirus, coronavirus, and *Cryptosporidia* and shed them into the environment, particularly around calving time. If environmental conditions are suitable for pathogen survival, environmental contamination levels can increase dramatically, elevating the risk of infection in susceptible animals. Removal of morbid calves and their dams from the healthy nursery can help minimize this environmental contamination. Once calves are identified as sick, they should be quickly isolated to prevent further environmental contamination of the healthy nursery. Once the environment is contaminated, moist cool conditions will allow survival of infectious agents for an extended period of time. *Cryptosporidia* is particularly suited to survival in this environment (Corwin 1992), and prevention of contamination in the healthy nursery is especially important.

Control for pathogenic agents of neonatal diarrhea involves segregation of sick animals from the healthy nursery to decrease environmental contamination and transmission. The practice of buying calves to "graft" onto cows or heifers that have lost their calves poses a significant risk for importing a stressed calf that potentially can shed high numbers of infective agents and contaminate the environment. Some agents of diarrhea, such as salmonella, may not be present on a ranch, and the importation of a shedding animal may be the portal of entry. If a producer must buy outside calves, the calves and their surrogate mothers must be quarantined from the rest of the herd. Specific *E. coli* and salmonella strains may be commonly introduced to a herd by the importation of a shedding animal.

Once calves are infected on the farm, they become multipliers of the infectious agent, which results in a rapid rise in environmental contamination. In addition, *E. coli* and salmonella control calls for implementing biosecurity rules to prevent the purchase and introduction of new calves or cows during the calving season.

The veterinarian should make at least one additional visit to the farm 2–3 weeks after calving has begun to assess the management and environment. Morbidity and mortality incidence levels may be established, above which the veterinarian should be called and a preliminary investigation begun. The producer should be encouraged to contact the veterinarian promptly if the established incidence of morbidity or mortality in the herd is exceeded. If computerized records on morbidity and mortality incidence are kept up to date, weekly incidence rates can be calculated to track a current measure of the effect of disease. If morbidity and mortality rates are indeed elevated, an investigation should be undertaken to identify the management factors contributing to the outbreak.

Management of the neonatal calf and calving season is a key area for controlling calf morbidity and mortality and optimizing the weaned calf crop. There are many opportunities for veterinary involvement in calving management that can be profitable both to the producer and to the veterinarian. Management decisions to control neonatal morbidity and mortality need to be made far in advance of calving season and require planning and coordination between veterinarian and producer. The veterinarian should be involved in designing and implementing management and environmental practices that will control herd morbidity and mortality.

The key areas and principles of management are discussed above. If the opportunity presents for involvement in the design of calving facilities, these principles should be applied to design an optimal calving environment. Calving corrals and nursery pastures should be large enough to prevent build-up of pathogens and increased infection pressure. Proper drainage to avoid mud build-up in the calving corral and nursery pasture is important. Drainage from the morbidity nursery to the calving corral or healthy nursery pasture, or from the calving corral to the healthy nursery pasture, should be avoided. Pens for assistance should be easily cleaned.

The economic implications of recommendations should always be examined. The veterinarian should be able to assist the producer in selecting bulls to optimize calving ease and postnatal growth based on EPDs. Ration formulation to provide proper nutrition is necessary for replacement heifer development, acceptable calving performance, and subsequent calf health and performance. Proper management of prepartum cows and heifers will decrease environmental contamination, and a well-planned vaccination program can elevate disease resistance in the herd.

PREGRAZING MANAGEMENT (BRANDING)

In many North American herds, the first opportunity to manage calves after calving occurs when the calves are aged 3–10 weeks, at the time of branding. For spring calving herds, this event occurs in late spring, before cow/calf pairs are turned out to summer pasture. Management practices at this time should be chosen to economically control morbidity and mortality or increase growth rate. Procedures to be considered include castration, growth implants, dehorning, identification, vaccination, and deworming.

CASTRATION

Castration of bull calves in early life is likely to be less stressful to calves than castration performed later, when testicular size is dramatically increased. Early castration also raises fewer concerns about humane treatment. In addition, early castration removes one stressful incident from the calf's experience at the time of weaning. In a Kansas study, early-castrated (94 days) and implanted calves gained as well during the summer grazing as did calves that were castrated at weaning (226 days). Calves castrated at weaning gained significantly less weight in the 28 days following weaning, weighing 17 lb less compared with early-castrated calves (Marston et al. 2003). Combined with the typical discount given to intact bulls at sale time, early castration and implanting resulted in the most pounds of salable product and the highest returns. In the United States, most producers that castrate calves do so by 3 months of age, but approximately 25% of producers do not castrate any calves before sale (NAHMS 1997c).

A number of methods are available for castration, including the open surgical technique, the use of rubber rings, and the Burdizzo method, where the cords are crushed. Most calves in the United States are castrated by open surgical removal of the testicles, with use of rubber bands the next most common practice (NAHMS 1997c). In one study of calves 4–11 weeks of age, surgical castrates initially exhibit more agitation compared with calves castrated with rubber rings. Salivary cortisol levels did not differ from those of controls in either group by 4 hours after castration, and both groups resumed normal behavior soon after the operation was completed (Fell et al. 1986). Intact calves are significantly discounted at market. In one Oklahoma study involving over 26,000 lots of cattle over 2 years, bulls sold for $2.24 to $3.56 per hundredweight less than steers (Smith et al. 2000).

GROWTH IMPLANTS

Castration of calves should be coupled with use of hormonal implants in steer calves to improve weight gains. Use of implants in calves at 2 months of age has been shown to increase weaning weights by 10–20 lb (Mader 1998). Economic returns are typically $5–$10 for every $1 invested. Marston et al. (2003) found that calves castrated at approximately 3 months of age gained significantly better when implanted and equaled the gain of intact bull calves.

Decisions relating to use of hormonal growth-promoting implant programs for replacement heifers must be made beginning at the spring working of calves, at 2–3 months of age. Use of growth-promoting implants in heifer calves before 30–45 days of age may have severe effects on development of normal uterine function (Bartol and Floyd 1996) and is not recommended.

Products are available for use in replacement heifer prospects from 45 days of age to weaning. Implants have been shown to increase weaning weight and frame score in heifers compared with nonimplanted control heifers. Pelvic area is increased in implanted heifers at 1 year of age, but most of their advantage over nonimplanted heifers is lost by calving, and there is no difference in calving ease.

The effect of implanting replacement heifers on their subsequent reproductive performance has been variable. Application of a single implant has resulted in an average depression in yearling pregnancy rate of 1% (range +7% to −14%). Application of two or more implants resulted in pregnancy rate depressions as high as 42% (Deutscher 1994). Some of the depression in pregnancy rate may be overcome by ensuring that implanted heifers have sufficient nutrition to support a higher rate of gain (Deutscher 1994). In general, implanting heifers at 2–3 months of age results in little long-term advantage. Implanting late-born heifers that are not replacement heifer prospects, however, may be a valuable practice.

Because a variety of implants is available for calves over 45 days of age, the decision to implant may be more important than which implant to use. Typically, an implant of lower potency containing estrogen or estrogen and progesterone is used for young suckling calves. Examples include Ralgro, Synovex-C, Calfoid, or Compudose. Despite the clear value of implanting steer calves, the NAHMS Beef '97 survey indicates that only a limited number of producers take advantage of this management technique. In the Beef '97 survey, 82% of producers reported that they did not implant any calves (NAHMS 1997b). Producers of larger herds were more likely to implant both replacement heifer candidates and other calves than were producers of small herds. Among herds of 300 or more cows, 29% of producers implanted replacement heifers one or more times, and 59% implanted other calves one or more times (NAHMS 1997b). All steer calves and heifer calves that are not candidates for breeding should be implanted.

Dehorning

Dehorning is also less stressful when performed early in life, rather than later, when the horns have increased in size. Early dehorning also removes one stressful incident from the calf's experience at the time of weaning. According to the NAHMS Beef '97 survey of U.S. producers, 28% of calves have horns, and only 61% of horned calves were dehorned before leaving the ranch. Approximately 28% of horned calves were dehorned by 3 months of age, and 52% were dehorned after 5 months of age (NAHMS 1997a) Horns in young suckling calves are generally not a problem for the cow/calf producer, who may have little incentive to devote labor to dehorning. Horns are mostly a problem for the feeding period, when horned calves require more bunk space and may cause more bruising in their pen mates than do calves without horns. Such problems are best managed by polled breeding or early dehorning. Horned calves are significantly discounted at market. In one Oklahoma study involving over 26,000 lots of cattle over 2 years, horned steers sold for $3.00 per hundredweight less than dehorned steers, and horned heifers sold for $2.00 per hundredweight less than dehorned heifers (Smith et al. 2000).

Identification

Individual identification of cows and calves is the basis for selecting based on performance. In the United States in 1996, approximately half of operations used no method of individual identification for calves, but producers with large herds were more likely to individually identify calves (78%) than were those with small herds (41%; NAHMS 1997a). Plastic ear tags were the most common method of individual identification in herds of all sizes. Approximately half of herds had no method of identification (NAHMS 1997a). Producers of large herds predominantly used hot-iron branding for herd identification. However, branding as a method of identification is coming under increasing scrutiny for product quality and animal welfare reasons.

Current country of origin labeling requirements, and the need to identify cattle for trace-back in the event of an outbreak of foreign animal disease, have raised animal identification to the top of the discussion list in the beef industry. The U.S. National Cattleman's Beef Association has supported a voluntary national cattle identification system to support value-based marketing and rapid trace-back. Ultimately, all premises in the United States will be identified with a unique premise ID, followed by identification of all livestock by a unique group, or individual ID. Individual identification of calves is useful in evaluating individual cow performance for culling and breeding decisions. Commercial products are currently on the market that allow individual electronic identification. Such initiatives may eventually replace current identification systems. Electronic capture of ID and data has the potential to decrease the labor costs and time requirements of data collection in cow/calf production systems, although initial start-up costs may be significant.

Vaccination Procedures

Young calves are not commonly vaccinated for respiratory pathogens from 21 days of age to weaning (NAHMS 1997c). Vaccines are available for viral (IBR, BVD, parainfluenza 3, bovine respiratory syncytialvirus) and bacterial (pasteurella, hemophilus) respiratory pathogens, as well as for IBK and clostridial disease. Data on the effectiveness of individual vaccinations in decreasing calf morbidity and mortality are hard to obtain. No field trial data are available that assess the effect of vaccination at branding on the morbidity, mortality, or performance of suckling calves. Pneumonia incidence is typically low during the summer grazing period, and the clinical effectiveness of a vaccination program against respiratory disease would likely be difficult to show. Residual passive immunity in young calves may limit the detectable antibody response to vaccination at an early age; however, calves vaccinated at branding time are sensitized to the antigens and respond anamnestically when given a booster vaccine at arrival in the feedlot (Parker et al. 1993). Primary sensitization to increase subsequent vaccination response preweaning may be a major benefit of such a vaccination program. Recommended vaccination programs include clostridial and viral respiratory vaccination at the time of branding. A number of "value-added calf" programs have been initiated, some of which require a vaccination program at branding time. Until recently, only vaccines with killed BVD and IBR fractions were approved for use in suckling calves. At present, one modified live virus vaccine (BoviShield 4) is available for use in suckling calves, provided the cow herd is appropriately vaccinated with a similar product from the same manufacturer.

IBK, or "pinkeye," can be a significant problem in suckling calves during the summer months, but its control is difficult. The bacteria most commonly associated with IBK has been *Moraxella bovis*. Response to vaccination with *M. bovis* bacterins has shown variable results. Typically, challenge with a homologous strain of *M. bovis* following vaccination has provided some level of control (Lepper 1988). Challenge with a heterologous strain, however, seems to result in little protection (Smith et al. 1990). Hereford cattle with pigmented eyes may show some resistance to IBK (Pugh et al. 1986), and cattle with good pigment are anecdotally thought to have lower incidence of IBK. IBK may carry a significant price tag at sale; calves with a "bad eye" may be discounted $7 to $8 per 100 lb (Brazle et al. 1988).

Control of face flies is important in interrupting the transmission of *M. bovis*. Face fly control is difficult, however, because of the limited amount of time the fly spends on cattle. Fly numbers on calves can be reduced by 50%–70% using dust bags and insecticide-impregnated ear tags. Feeding chlortetracycline in a mineral supplement may also decrease the incidence of IBK; however, consumption of mineral by calves is not reliable. IBR may also cause ocular lesions or predispose calves to IBK (George et al. 1988), and a proper vaccination program for IBR may be helpful. Table 11.1 outlines possible vaccination programs for calves.

Spring Parasite Control

The economic and production value of treating suckling calves for internal parasites has not been adequately studied. Egg burdens of young calves are typically low at spring branding but may rise significantly by midsummer (Reinemeyer 1997). There are few studies that have examined the effects of deworming calves and not cows at branding time, and the results are inconsistent. In Oklahoma, some studies have shown positive effects of deworming (Stacey et al. 1995), and others have shown no effect (Stacey et al. 1997). In a Colorado study, deworming cows with albendazole in late spring before turnout to grass resulted in increased weaning weights in calves weighing 2–8 kg. Further, dewormed cows conceived, on average, 7 days earlier than did control cows (Boyd and Kniffen 1992). Deworming of calves in addition to cows in late spring does not appear to confer significant additional benefit (Stacey et al. 1996, 1997).

External parasites of cattle are estimated to be an important cause of economic loss as well. Studies have shown a weight gain response of 10–20 lb in suckling calves when cows were tagged with insecticide-impregnated ear tags in Texas (Cocke et al. 1989) and Iowa (Quisenberry and Strohbehn 1984). In Tennessee, however, studies conducted over 7 years using a variety of fly control methods were variable in their effect on calf gains (Gerhardt and Shrode 1990).

The most common method for fly control is the use of insecticide-impregnated ear tags. Maximal fly control probably requires application of two ear tags to each cow and calf. Cows are commonly the only animal tagged, often with only one tag. With the widespread use of pyrethroid insecticides in ear tags, emerging resistance has become a problem. Organophosphate resistance has begun to emerge in some areas as well. Alternating insecticides from different chemical families has been recommended for control of emerging resistance, as resistance levels in fly populations decline rapidly when the pyrethroid exposure is removed (Krafsur et al. 1993). In addition, insecticide tags should be removed from cattle at the end of the grazing season to remove the selective effect of sublethal doses of insecticide on fly populations.

Insecticide sprays and back rubbers can also be effective in controlling flies, but cattle must be forced to use them. Their use can provide the same level of

Table 11.1. Summer calf management.

Time	Plan 1	Plan 2	Plan 3
Late spring (branding time)	Brand Castrate Dehorn Seven- or eight-way[a] clostridial Four-way respiratory (killed or MLV[b] BVD, IBR) Implant (steers and late heifers)	Brand Castrate Dehorn Seven- or eight-way[a] clostridial Implant (steers and late heifers)	Brand Castrate Dehorn Seven- or eight-way[a] clostridial Four-way respiratory (killed or MLV[b] BVD, IBR) Implant (Steers and late heifers)
Preweaning (at least 2–3 weeks before weaning)		Seven- or eight-way[a] clostridial booster Four-way respiratory (Killed or MLV[b] BVD, IBR) Pasteurella[a] Leptospira[a] Deworm	Seven- or eight-way[a] clostridial booster Four-way respiratory (killed or MLV[b] BVD, IBR) Pasteurella[a] Leptospira[a] Deworm Reimplant (if calves are retained)
Weaning	Seven- or eight-way[a] clostridial booster Four-way respiratory (MLV) Pasteurella[a] Leptospira[a] Reimplant (if calves are retained)	Four-way respiratory (MLV) Pasteurella[a] Leptospira[a] Reimplant (if calves are retained)	Remove cows from calves Leave calves in familiar surroundings with fenceline contact with cows Do not work to minimize weaning time stress

[a] Optional depending on local circumstances and disease history.
[b] Requires vaccination of the cow herd in accordance with vaccine label directions.
MLV = modified live vaccination, BVD = bovine viral diarrhea, IBR = infectious bovine rhinotracheitis.

control at a much reduced cost than that of ear tags (Sheppard 1987, Knapp and Webb 1992). Insecticide sprays can be very useful at controlling fly populations, but they require regular application for effective control. Data indicate that pyrethroid sprays select for fly resistance at a slower rate than do pyrethroid ear tags (Sheppard 1987).

GRAZING AND WEANING MANAGEMENT

The majority of calf morbidity and mortality from birth to weaning occurs before 21 days of age, but the majority of growth occurs after 21 days. Optimizing the pounds of calf weaned and cow/calf productivity requires managing calves for growth during the summer grazing period. Diarrhea becomes a less important cause of morbidity from 21 days of age to weaning, but according to producer survey data, respiratory disease and IBK become increasingly important (NAHMS 1997b) at this time. Once the cows and calves are turned out to pasture for the summer, the herd is generally not worked again until fall, just before or at weaning time. Factors to consider during the summer grazing period are nutrition, weaning management, preweaning and weaning vaccination, weaning parasite control, and early weaning.

NUTRITION

Suckling beef calves are generally not supplemented during the summer grazing period. Their diet consists primarily of milk and an increasing intake of forage. Trace mineral deficiency may be a concern for calves in some areas. Supplementation of calves is difficult, as trace mineral mix intake is sporadic at best. Proper trace mineral nutrition of the cow herd before calving is necessary to ensure that the calf has adequate trace mineral status. Cows transfer trace minerals to the fetal liver in late gestation to provide the calf with adequate stores. Poor trace mineral nutrition in the cow herd will result in a calf without adequate reserves at birth. Proper trace mineral nutrition is important for proper immune response.

An excellent review of creep feeding in beef calves is available (Herd et al. 1998). High-protein, high-energy, and limit-fed creep-feeding programs have been studied. The response of beef calves to various programs has been highly variable, as has the economic return. Economic returns are related to the price of calves, the price of feed, and the feed conversion of the calves. Feed conversions reported have been as poor as 31 and as good as 2.6 lb of feed per pound of gain. Average results are approximately 10 lb of feed per pound of gain for unlimited intake creep rations. Gains are generally better when forage quality is decreased and calf gains are low; mature low-protein forage may be effectively supplemented by a high-protein creep. Creep feeding results in decreased dry-matter intake of forage, but milk consumption remains the same (Cremin et al. 1991).

When forage quality and quantity are acceptable and cows are milking adequately, creep feed is an expensive substitute for available forage. It may have a sparing effect on forage for use by cows, but it does not reduce lactation pressure on cows. When forage quantity or quality is limited, supplemental feed is more efficiently used to feed calves than to feed cows to produce milk for calves. Early weaning may be a more valuable management practice under these circumstances, to regain cow condition and adequately support calves' nutritional needs. Use of high-protein, limited-intake creep feeds can be very efficient when forage is plentiful but of low quality. A summary of four studies conducted in Oklahoma, Florida, and Tennessee using high-protein, limited-feed creep on low-quality pasture showed an average feed conversion of 2.6:1 (Herd et al. 1998). Limited-intake, high-protein consumption should be controlled to 0.5–1.0 lb per head per day by mixing with salt.

Creep feeding of potential replacement heifers can have negative effects on long-term production. Increased preweaning gains in heifers can result in deposition of excess fat in the udder. This extra fat can decrease functional mammary parenchyma and subsequent milk production. The effect of fat deposition in the udder appears to be most significant from 3 to 8 months of age. During this time, heifers that gained 2 lb per day subsequently showed decreased milk production and decreased calf weaning weights when compared to heifers that gained 1.2 lb per day (Johnsson and Obst 1984).

Creep feeding calves for 3–4 weeks before weaning may be an effective way to reduce stress and disease at weaning. Creep-fed calves are already acclimated to feed and achieve acceptable intakes more quickly than do non-creep-fed calves. Decreased nutritional stress from diet change and higher intakes may also decrease morbidity at weaning. Small-framed, creep-fed calves may exhibit depressed feed efficiency following weaning, but large-framed calves generally do not show depressed feedlot efficiency (Herd el al. 1998). In some studies, creep feeding has allowed more calves to achieve choice quality grade by 14 months of age (Faulkner et al. 1994). The effect of calf creep feeding on cows has also been quite variable: Most studies have not shown a significant effect of creep feeding on production measures of cows the following year. It may have a sparing effect on the BCS of cows and on the amount of supplemental nutrition needed to get cows in adequate condition by calving.

Weaning Management

At weaning, the calf is removed from its mother and forced to adjust to a different diet. The population density of the calves is generally increased, allowing for increased disease exposure and transmission. The management approach to weaning should attempt to minimize the amount of stress placed on the calves. Management practices should be designed to limit stress while ensuring proper nutritional support and immunologic preparation.

Castration and dehorning are stressful events that should certainly be performed well before weaning time. By the middle of the summer grazing season, calves are consuming a significant amount of feed from forage in addition to milk. They may be consuming a limited amount of a trace mineral mix, and a quality trace mineral mix should always be available for the cows and calves. Recently, weaning programs that allow fenceline contact between calves and cows have been investigated. Fenceline contact decreased behavioral indices of distress and minimized losses in weight gain following weaning as compared to totally separated calves (Price et al. 2003). If calves are to be retained after weaning, they should receive a growth-promoting implant to improve postweaning gains. Performance of implanted calves will be poor if adequate nutrition is not provided.

Preweaning and Weaning Vaccination

Vaccination practices at preweaning and weaning are quite variable. Some calves may receive their first vaccination at the time of weaning, others their third vaccination. Vaccination of calves before weaning may be valuable to elevate immunity before the stress and exposure that follows weaning. Viral respiratory agents (IBR, BVD, parainfluenza 3, and bovine respiratory syncytialvirus) are available as vaccines and may be valuable when administered to calves 2–3 weeks before weaning. Vaccination of calves 2–3 weeks before weaning provides a period of time for calves to respond to vaccinations before weaning. Until recently, only killed vaccines for IBR and BVD were labeled for use in calves suckling pregnant cows. Recently, one modified live virus vaccine (BoviShield 4) has been cleared for use in calves suckling pregnant cows. The label requires that the cow herd be appropriately vaccinated with a similar product from the same manufacturer. Vaccination for pasteurella has also been recommended for inclusion in a preweaning or weaning vaccination program (Coffey et al. 1996).

Vaccination before weaning to allow time for calves to build specific immunity is intuitively attractive. However, field trial data on the effectiveness of preweaning or weaning vaccinations in decreasing postweaning morbidity and mortality are lacking (Perino and Hunsaker 1997). Available data do indicate that calf buyers are willing to pay higher prices for calves that have undergone a preweaning health program (King et al. 1995, 1996, 1997). Vaccinations given at weaning do not allow the calf time to respond before the stress of weaning, and their immune response may be limited by the stress associated with weaning. Immunity develops during the period of highest risk immediately after weaning and may provide only incomplete protection. Vaccination, deworming, and implant procedures taken before weaning allow calves to be weaned without handling. This serves to remove the need of working calves through the chute and decreases the stress imposed on the calves.

Weaning Parasite Control

Calves often develop significant internal parasite burdens by mid- to late summer. In one study, deworming early-weaned calves in midsummer resulted in a 10-lb increase in weaning weight (Purvis et al. 1996). The expense of deworming calves at or around weaning time must be recouped in increased gains if ownership is retained, or by sale premium if cattle are sold shortly after weaning.

Early Weaning

Beef calves can be early weaned from 30 to 170 days of age. When forage is limited, early weaning may result in more efficient use of feed resources by directly supplementing calves to maintain weight gain rather than supplementing cows to produce milk for calf growth. Early weaning at 30–60 days can improve reproductive performance of cows and heifers and is a valuable management tool in moving to an earlier calving season or in times of drought. Cows and heifers will cycle and rebreed earlier in the calving season, and pregnancy rates are higher, in a limited breeding season following early weaning at

30–60 days (Lusby et al. 1981, Guyer 1983, Arthington and Kalmbacher 2003).

Early-weaned calves, however, require substantially increased management inputs, and the earlier they are weaned, the more management is required. Early weaning at 100–150 days of age will decrease lactation stress on cows when forage resources are limited and result in improved cow condition. Reproductive performance in the breeding season will not be affected by weaning at this time. When forage resources are limited, weaning calves to decrease the nutritional requirements of the cows will allow the cows to regain condition before winter with less or no supplementation.

The nutritional needs of the weaned calves must be carefully met to ensure acceptable health and performance. When properly managed, calves weaned at 30–60 days can equal the weight gain of calves weaned at 7 months of age (Lusby et al. 1981). Early-weaned calves can perform well in the feeding period as well (Meyers et al. 1999, Barker-Neff et al. 2001). Creep feeding the weaning ration to early-weaned calves before weaning may be helpful to ensure early consumption postweaning. The postweaning ration should be carefully balanced to provide adequate nutrition to the calves. The final ration should provide 15%–16% protein, 0.80–0.85 Mcal net energy gain/lb, and 0.50 Mcal net energy for maintenance per pound, and be balanced for vitamins, macrominerals, and microminerals. Gains should be targeted at no more than 2–2.5 lb per day to avoid early fat deposition, causing the calf to finish at an unacceptable slaughter weight. Calves can be supplemented more efficiently than can cows, and depending on forage resources, feed costs, and market prices, early weaning may be an economic management consideration.

REFERENCES

Anderson K.J., J.S. Brinks, D.G. LeFever, K.G. Odde. 1993a. The factors associated with dystocia in cattle. *Veterinary Medicine* 88(8): 764–776.

Anderson K.J., J.S. Brinks, D.G. LeFever, K.G. Odde. 1993b. A strategy for minimizing calving difficulty. *Veterinary Medicine* 88(8):778–781.

Arthington, J.D., R.S. Kalmbacher. 2003. Effect of early weaning on the performance of three-year-old, first-calf beef heifers and calves reared in the subtropics. *Journal of Animal Science* 81:1136–1141.

Azzam S.M., J.E. Kinder, M.K. Nielsen. 1993. Environmental effects on neonatal mortality of beef calves. *Journal of Animal Science* 71:282–290.

Bagley C.P., J.C. Carpenter, J.I. Feazel, F.G. Hembry, D.C. Huffman, K.L Koonce. 1987. Influence of calving season on beef cow-calf productivity. *Journal of Animal Science* 64:687–694.

Barker-Neff, J.M., D.D. Buskirk, J.R. Blackt, M.E. Doumit, S.R. Rust. 2001. Biological and economic performance of early-weaned Angus steers. *Journal of Animal Science* 79:2762–2769.

Bartol, F.F. and J.G. Floyd. 1996. Critical periods, steroid exposure and reproduction. *Proceedings of the Society for Theriogenology*, August 15–17, 1996, Kansas City, MO, pp. 101–111.

Bellows R.A., M.A. Lammoglia. 2000. Effects of severity of dystocia on cold tolerance and serum concentrations of glucose and cortisol in neonatal beef calves. *Theriogenology* 53:803–813.

Bellows R.A., D.J. Patterson, P.J. Burfening, D.A. Phelps. 1987. Occurrence of neonatal and postnatal mortality in range beef cattle. II. Factors contributing to calf death. *Theriogenology* 28:573–586.

Besser T.E., O. Szenci, C.C. Gay. 1990. Decreased colostral immunoglobulin absorption in calves with postnatal respiratory acidosis. *Journal of the American Veterinary Medical Association* 196:1239–1243.

Boland, W., V. Cortese, D. Steffen. 1995. Interactions between vaccination, failure of passive transfer, and diarrhea in beef calves. *Agri-Practice* 16(4):25–88.

Boyd, G.W., D.M. Kniffen. 1992. The effects of deworming beef cows in the spring on cow/calf productivity. *Proceedings of the Western Section, American Society of Animal Science*, July 8–10 (43):247–250.

Brazle, F., J. Mintert, T. Schroeder. 1988. The effect of physical characteristics on the price of stocker and feeder cattle. *Kansas State University Cattlemen's Day Report* Vol. 539. Manhattan: Kansas State University, pp. 65–70.

Carstens G.E. 1994. Cold thermoregulation in the newborn calf. *Veterinary Clinics of North America. Food Animal Practice* 10(1):69–106.

Church, T.L., and E.D. Janzen. 1978. A system for calving heifers on a large commercial ranch. *Bovine Practice* 13:40–44.

Clement, J.C., M.E. King, M.D. Salman, et al. 1995. Use of epidemiologic principles to identify risk factors associated with the development of diarrhea in calves in five beef herds. *Journal of the American Veterinary Medical Association* 207:1334–1338.

Cocke, J., R. Knutson, D.K. Lunt. 1989. Effects of horn fly control with lambda cyhalothrin ear tags on weight gains in weaning calves in Texas. *Southwestern Entomologist* 14:357–362.

Coffey, C., J. Pumphrey, J. Brightwell, editors. 1996. *Value Added Cattle Guidelines for Cow-Calf Stocker Feeder*, pp. 1–68. Ardmore, OK: Noble Foundation.

Corah L.R., T.G. Dunn, C.C. Kaltenbach. 1975. Influence of prepartum nutrition on the reproductive performance of beef females and the performance of their progeny. *Journal of Animal Science* 41:819–824.

Corah L.R., C.L. Wright, J.D. Arthington. 1998. Applied aspects of vitamin E and trace-mineral supplementation. *Compendium on Continuing Education for the Practicing Veterinarian* 20:866–874.

Cornaglia, E.M., F.M. Fernandez, M. Gottschalk, et al. 1992. Reduction in morbidity due to diarrhea in nursing beef calves by use of an inactivated oil-adjuvenated rotavirus-*Escherichia coli* vaccine in the dam. *Veterinary Microbiology* 30(2–3):191–202.

Corwin, R.M. 1992. Cryptosporidiosis: A coccidiosis of calves. *Compendium on Continuing Education for the Practicing Veterinarian* 14:1005–1007.

Cremin, J.D. Jr., D.B. Faulkner, N.R. Merchen, et al. 1991. Digestion criteria in nursing beef calves supplemented with limited levels of protein and energy. *Journal of Animal Science* 69:1322–1331.

Dawes M.E., J.W. Tyler, D. Hostetler, J. Lakritz, R. Tessman. 2002. Evaluation of a commercially available immunoassay for assessing adequacy of passive transfer in calves. *Journal of the American Veterinary Medical Association,* 220:791–793.

Deutscher G.H. 1985. Using pelvic measurements to reduce dystocia in heifers. *Modern Veterinary Practice* 66:751–755.

Deutscher, G.H. 1994. Growth promoting implants on heifer reproduction: A research review. *Proceedings of the Society for Theriogenology*, August 25–27, 1994, Kansas City, MO, pp. 76–85.

Doornbos D.E., R.A. Bellows, P.J. Burfening, et al. 1984. Effects of dam age, prepartum nutrition and duration of labor on productivity and postpartum reproduction in beef females. *Journal of Animal Science* 59:1–10.

Duenas, M.I., M.J. Paape, R.P. Wettmann, L.W. Douglass. 2001. Incidence of mastitis in beef cows after intramuscular administration of oxytetracycline. *Journal of Animal Science* 79:1996–2005.

Dufty J.H. 1981. The influence of various degrees of confinement and supervision on the incidence of dystocia and stillbirths in Hereford heifers. *New Zealand Veterinary Journal* 29:44–48.

Faulkner, D.B., D.F. Hummel, D.D. Buskirk, et al. 1994. Performance and nutrient metabolism by nursing calves supplemented with limited or unlimited corn or soyhulls. *Journal of Animal Science* 72:470–477.

Fell, L.R., R. Wells, D.A. Shutt. 1986. Stress in calves castrated surgically or by the application of rubber rings. *Australian Veterinary Journal* 63:16–18.

Ganaba, R., M. Bigras-Poulin, D. Belanger, Y. Couture. 1995. Description of cow-calf productivity in Northwestern Quebec and path models for calf mortality and growth. *Preventive Veterinary Medicine* 24:31–42.

George, L.W., A. Ardans, J. Mihalyi, et al. 1988. Enhancement of infectious bovine keratoconjunctivitis by modified-live infectious bovine rhinotracheitis virus vaccine. *American Journal of Veterinary Research* 49:1800–1806.

Gerhardt, R.R., R.R. Shrode. 1990. Influence of face and horn fly control on weight gain in pastured cow-calf groups. *Journal of Agricultural Entomology* 7(1):11–15.

Guyer, P.Q. 1983. Management of early weaned calves. *University of Nebraska Cooperative Extension Guide* G655.

Herd, D.B., S.E. Wikse, G.E. Carstens. 1998. The role of creep feeding in beef cattle production. *Compendium on Continuing Education for the Practicing Veterinarian* 20:748–759.

Hjerpe C.A. 1990. Neonatal enteric disease vaccines. *Veterinary Clinics of North America. Food Animal Practice* 6(1):234–246.

Hough, R.L., F.D. McCarthy, H.D. Kent, D.E. Eversole, M.L. Wahlberg. 1990. Influence of nutritional restriction during late gestation on production measures and passive immunity in beef cattle. *Journal of Animal Science* 68:2622–2627.

Johnson, S.K., G.H. Deutscher, A. Parkhurst. 1988. Relationships of pelvic structure, body measurements, pelvic area, and calving difficulty. *Journal of Animal Science* 66:1081–1088.

Johnsson, I.D., J.M. Obst. 1984. The effects of level of nutrition before and after 8 months of age on subsequent milk and calf production of beef heifers over three lactations. *Animal Production* 38(1):57–68.

King, M.E., T.E. Wittum, K.G. Odde. 1996. The effect of value added health programs on the price of beef calves sold through seven superior livestock video auctions in 1995. *Colorado State University Beef Program Report* 14:167–173.

King, M.E., T.E. Wittum, K.G. Odde. 1997. The effect of value added health programs on the price of beef calves sold through nine superior livestock video auctions in 1996. *Colorado State University Beef Program Report* 15:159–165.

King, M.E., T.E. Wittum, M.D. Salman, K.G. Odde. 1995. The effect of value added health programs on the price of beef calves sold through five video auctions in 1994. *Colorado State University Beef Program Report* 13:7–13.

Knapp, F.W., J.D. Webb. 1992. Use of dichlorvos in a portable automatic sprayer-mineral station to control face flies (*Diptera: Muscidae*) and pyrethroid-resistant horn flies (*Diptera: Muscidae*) on beef cattle. *Journal of Agricultural Entomology* 9:273–281.

Krafsur, E.S., A.L. Rosales, Jr., J.F. Robison-Cox, et al. 1993. Bionomics of pyrethroid-resistant and susceptible horn fly populations (*Diptera: Muscidae*) in Iowa. *Journal of Economic Entomology* 86(2):246–257.

Kroker, G.A., L.J. Cummins. 1979. The effect of nutritional restriction on Hereford heifers in late pregnancy. *Australian Veterinary Journal* 55:467–474.

Larson R.L., W.O. Herring. 1998. Cattle breeding. Part I. Expected progeny differences. *Compendium on Continuing Education for the Practicing Veterinarian* 20:S76–S80.

Larson R.L., D.W. Moser. 1998. Replacement heifer development: Selection. *Compendium on Continuing Education for the Practicing Veterinarian* 20:652–659.

Lepper, A.W.D. 1988. Vaccination against infectious bovine keratoconjunctivitis: Protective efficacy and antibody response induced by pili of homologous and heterologous strains of *Moraxella bovis*. *Australian Veterinary Journal* 65(10):310–316.

Lusby, K.S., R.P. Wetteman, E.J. Turman. 1981. Effects of early weaning calves from first-calf heifers on calf and heifer performance. *Journal of Animal Science* 53:1193–1197.

Lowman B.G., M.S. Hankey, N.A. Scott, et al. 1981. Influence of time of feeding on time of parturition in beef cows. *Veterinary Record* 109:557–559.

Mader, T.L. 1998. Implants. In *Feedlot Medicine and Management, Veterinary Clinics of North America Food Animal Practice*, Vol. 14, edited by G.L. Stokka. Philadelphia: W.B. Saunders, pp. 279–290.

Makarechian M., R.T. Berg. 1983. A study of some of the factors influencing ease of calving in range beef heifers. *Canadian Journal of Animal Science* 63:255–262.

Marston, T.T., D.A. Llewellyn, L.C. Hollis, J.W. Homm. 2003. Effects of castration age and a growth implant during suckling on weaning and preconditioned weights. *Kansas State University Cattleman's Day Report of Progress* 908:69–71.

McDermott J.J., D.M. Alves, N.G. Anderson, S.W. Martin. 1991. Measures of herd health and productivity in Ontario cow-calf herds. *Canadian Veterinary Journal* 32:413–420.

Meijering A. 1984. Dystocia and stillbirth in cattle: A review of causes, relations and implications. *Livestock Production Science* 11(2):143–177.

Meyers, S.E., D.B. Faulkner, F.A. Ireland, D.F. Parret. 1999. Comparison of three weaning ages on cow-calf performance and steer carcass traits. *Journal of Animal Science* 77:323–329.

Mihura, H.E., M. King, K. Odde. 1997. Effect of colostral supplements on disease protection in neonatal calves born to two-year-old beef heifers. *Revista Argentina de Produccion Animal* 17(4):421–429.

Naazie A., M. Makarechian, R. Berg. 1989. Factors influencing calving difficulty in beef heifers. *Journal of Animal Science* 67:3243–3249.

National Animal Health Monitoring System. 1997a. *Beef '97 Part I: Reference of 1997 Beef Cow-Calf Management Practices*. N233.697. USDA:APHIS:VS, CEAH, pp. 1–55. Fort Collins, CO: National Animal Health Monitoring System.

National Animal Health Monitoring System. 1997b. *Beef '97 Part II: Reference of 1997 Beef Cow-Calf Health and Health Management Practices*. N238–797. USDA:APHIS:VS, CEAH, pp. 1–38. Fort Collins, CO: National Animal Health Monitoring System.

National Animal Health Monitoring System. 1997c. *Beef '97 Part III: Reference of 1997 Beef Cow-Calf Production Management and Disease Control*. N247.198. USDA:APHIS:VS, CEAH, pp. 1–42. Fort Collins, CO: National Animal Health Monitoring System.

Newman, M.A., L.L. Wilson, E.H. Cash, R.J. Eberhardt, T.R. Drake. 1991. Mastitis in beef cows and its effects on calf weight gain. *Journal of Animal Science* 69:4259–4272.

Odde, K.G. 1988. Survival of the neonatal calf. *Veterinary Clinics of North America. Food Animal Practice* 4(3):501–508.

Paape, M.J., M.I. Duenas, R.P. Wettemann, L.W. Douglass. 2000. Effects of intramammary infection and parity on calf weaning weight and milk quality in beef cows. *Journal of Animal Science* 78:2508–2514.

Parker, W.R., M.L. Galyean, J.A. Winder, et al. 1993. Effects of vaccination at branding on serum antibody titers to viral agents of bovine respiratory disease (BRD) in newly weaned New Mexico

calves. *Proceedings, Western Section, American Society of Animal Science* 44:132–134.
Patterson D.J., R.A. Bellows, P.J. Burfening, and J.B. Carr. 1987. Occurrence of neonatal and postnatal mortality in range beef cattle. 1. Calf loss incidence from birth to weaning, backward and breech presentations and effects of calf loss on subsequent pregnancy rate of dams. *Theriogenology* 28:557–571.
Perino, L.J. 1997. A guide to colostrum management in beef cows and calves. *Veterinary Medicine* 92:75–82.
Perino, L.J., B.D. Hunsaker. 1997. A review of bovine respiratory disease vaccine field efficacy. *Bovine Practice* 31(1):59–66.
Perino, L.J., R.L. Sutherland, N.E. Woollen. 1993. Serum γ-glutamyltransferase activity and protein concentration at birth and after suckling in calves with adequate and inadequate passive transfer of immunoglobulin G. *American Journal of Veterinary Research* 54:56–59.
Perino, L.J., T.E. Wittum, G.S. Ross. 1995. Effects of various risk factors on plasma protein and serum immunoglobulin concentrations of calves at postpartum hours 10 and 24. *American Journal of Veterinary Research* 56:1144–1148.
Price, E.O., J.E. Harris, R.E. Borgward, M.L. Sween, J.M. Connor. 2003. Fenceline contact of beef calves with their dams at weaning reduces the negative effects of separation on behaviour and growth rate. *Journal of Animal Science* 81:116–121.
Prior, R.L., D.B. Laster. 1979. Development of the bovine fetus. *Journal of Animal Science* 48:1546–1553.
Prior, R.L., R.A. Scott, D.B. Laster, D.R. Campion. 1979. Maternal energy status and development of liver and muscle in the bovine fetus. *Journal of Animal Science* 48:1538–1545.
Pugh, G.W., T.J. McDonald, K.E. Kopecky, et al. 1986. Infectious bovine keratoconjunctivitis: Evidence for genetic modulation of resistance in purebred Hereford cattle. *American Journal of Veterinary Research* 47:885–889.
Purvis, II, H.T., C.R. Floyd, K.S. Lusby. 1996. Performance of early weaned stocker calves or suckling calves treated with a mid-summer application of Ivomec-Pour on. *Animal Science Research Report Agricultural Experiment Station, Oklahoma State University* P-951: 245–248.
Quisenberry, S.S., D.R. Strohbehn. 1984. Horn fly (Diptera: Muscidae) control on beef cows with permethrin-impregnated ear tags and effect on subsequent calf weight gains. *Journal of Economic Entomology* 77:422–424.
Radostits, O.M., K.E. Leslie, J. Fetrow. 1994. *Herd Health Food Animal Production Medicine*, 2nd ed. Philadelphia, PA: W.B. Saunders.
Reinemeyer, C.R. 1997. The economics of parasite control for beef cattle. *Proceedings of the American Association of Bovine Practitioners* 30:117–123.
Rice, L.E. 1994. Dystocia related risk factors. *Veterinary Clinics of North America. Food Animal Practice* 10:53–68.
Sanderson, M.W., D.A. Dargatz. 2000. Risk factors for high herd level calf morbidity risk from birth to weaning in beef herds in the USA. *Preventive Veterinary Medicine* 44:99–108.
Schuijt, G. 1990. Iatrogenic fractures of ribs and vertebrae during delivery of perinatally dying calves: 235 cases (1978–1988). *Journal of the American Veterinary Medical Association* 197:1196–1202.
Schumann, F.J., H.G.G. Townsend, J.M. Naylor. 1990. Risk factors for mortality from diarrhea in beef calves in Alberta. *Canadian Journal of Veterinary Research* 54:266–372.
Sheppard, D.C. 1987. Differential pyrethroid resistance selection in horn fly populations treated with pyrethroid cattle ear tags and pyrethroid sprays. *Journal of Agricultural Entomology* 4(2):167–178.
Simpson, R.B., D.P. Wesen, K.L. Anderson, et al. 1995. Subclinical mastitis and milk production in primiparous Simmental cows. *Journal of Animal Science* 73:1552–1558.
Smith, P.G., T. Blankenship, T.R. Hoover, et al. 1990. Effectiveness of two commercial infectious bovine keratoconjunctivitis vaccines. *American Journal of Veterinary Research* 51:1147–1150.
Smith, S.C., D.R. Gill, T.R. Evicks, J. Prawl. 2000. Effect of selected characteristics on the sale price of feeder cattle in Eastern Oklahoma: 1997 and 1999. *Animal Science Research Report Agricultural Experiment Station, Oklahoma State University* P-980: 14–19.
Spire M.F. 1997. Managing replacement heifers from weaning to breeding. *Veterinary Medicine* 92:182–192.
Stacey, B.R., K.C. Barnes, K.S. Lusby. 1995. The effect of Ivomec R on weight gains of spring-born calves nursing untreated cows in Eastern Oklahoma. *Animal Science Research Report Agricultural Experiment Station, Oklahoma State University* P-943: 85–87.
Stacey, B.R., K.C. Barnes, G.E. Selk. 1996. Evaluation of deworming strategies in cows and calves in Eastern Oklahoma. *Animal Science Research Report Agricultural Experiment Station, Oklahoma State University* P-951: 95–97.

Stacey, B.R., K.C. Barnes, G.E. Selk. 1997. Evaluation of deworming strategies in cows and calves in Eastern Oklahoma. *Animal Science Research Report Agricultural Experiment Station, Oklahoma State University* P-958: 60–62.

Stott, G.H., E.J. Reinhard. 1978. Adrenal function and passive immunity in the dystocial calf. *Journal of Dairy Science* 61:1457–1461.

Tyler, J.W., T.E. Besser, L. Wilson, D.D. Hancock, S. Sanders, D.E. Rea. 1996. Evaluation of 3 assays for failure of passive transfer in calves. *Journal of Veterinary Internal Medicine* 10:304–307.

Van Donkersgoed, J., C.S. Ribble, H.G.G. Townsend, E.D. Janzen. 1990. The usefulness of pelvic area measurements as an on-farm test for predicting calving difficulty in beef heifers. *Canadian Veterinary Journal* 31:190–193.

VanVleck, L.D., L.V. Cundiff. 1998. Across-breed EPD tables for 1998 adjusted to a 1996 base. *Proceedings Beef Improvement Federation 30thAnnual Research Symposium and Annual Meeting*, June 30–July 3, 1998, Calgary, Alberta, Canada, pp. 196–212.

Ventorp, M., P. Michanek. 1992. The importance of udder and teat conformation for teat seeking by the newborn calf. *Journal of Dairy Science* 75:262–268.

Waldhalm, D.G., R.R. Hall, W.J. DeLong, D.P. Olson, D.O. Everson. 1979. Restricted dietary protein in pregnant beef cows. 1. The effect on length of gestation and calfhood mortality. *Theriogenology* 12:61–68.

Wittum, T.E., L.J. Perino. 1995. Passive immune status at postpartum hour 24 and long-term health and performance of calves. *American Journal of Veterinary Research* 56:1149–1154.

Wittum, T.E., M.D. Salman, M.E. King, R.G. Mortimer, K.G. Odde, D.L. Morris. 1994. Individual animal and maternal risk factors for morbidity and mortality of neonatal beef calves in Colorado, USA. *Preventive Veterinary Medicine* 19:1–13.

12
Beef Cattle Economics and Finance

Grant Dewell and Thomas Kasari

INTRODUCTION

In the past, the disciplines of accounting, economics, and finance may not have been used routinely by veterinarians in their endeavors to manage the health and well-being of their client's beef herds. However, in an era in which production agriculture is under increasing cost pressures and shrinking revenues, veterinarians who know how to marry these disciplines with veterinary medicine will be able to differentiate themselves from their colleagues by providing value-added services to their clients who are struggling to maintain profitability in the beef industry. In this chapter, we will introduce basic accounting, economic, and financial concepts that are useful for veterinarians involved with beef cattle production. In addition, tools that can be used to determine the profitability of the operation or evaluation of a decision will be presented.

ECONOMICS

Basic Economics

The study of economics is commonly divided into two entities: macroeconomics and microeconomics. Macroeconomics describes the study of economics of the entire economy. Macroeconomists are usually concerned with national and global matters such as inflation, interest rates, unemployment, taxes, exchange rates, trade, and so forth.

In contrast, microeconomics is the study of the individual within the whole economy. These individuals may be industries, firms, managers, consumers, or markets. The aggregate of individuals determines the level of production in the entire economy. Agricultural economics is a specialization of microeconomics that meshes biological models into economic theories.

Economics can be broadly defined as the study of how limited resources can best be used to fulfill unlimited human wants. This is simply defined as the allocation of scarce resources. The economic value of an item is established by its scarcity or availability to the consumer. If an item is unlimited and readily available, then its value is low and has little or no meaning.

Resources serve as inputs in the agricultural production process and can be broadly categorized as natural or societal. Natural resources such as oil, land, and water are easy to envision as inputs with value. These natural resources are vitally important to members of the agricultural community. Natural resources are tangible, and producers are used to recognizing them as valuable assets to be protected and cared for.

Societal resources are often more conceptual in nature than are natural resources. Members of this category include the available labor to complete a task and money invested in capital. Other, more abstract items that may concern producers include skill obtained through job experience as well as knowledge and formal education.

Successful beef cattle producers are effective managers and stewards of the societal and natural resources that are available to them. Good managers must effectively and wisely apportion these resources in a balanced manner. The wise and efficient allocation of available resources allows an operation the greatest opportunity to be profitable. A rational manager will allocate these resources to the most appropriate activities.

An example of allocation of natural resources would be the amount of hay ranchers have available for winter feeding of cattle. The producer usually has a finite quantity of hay available and must allocate this resource among the herd. Often, unless there is tremendous excess, the producer will distribute the hay differently among his herd of cattle. If the hay is allocated wisely, the mature cows and bulls may receive the average and lesser-quality hay because their productivity would not likely suffer. The prudent rancher would reserve the higher-quality hay for growing calves and heifers and to supplement cows around calving. Whether the producer realized it or not, economic principals were fundamental in the decision process. The scarcity of the resource—cheap, high-quality hay—forces the producer to rationally appropriate it to produce the best possible outcome.

This simple, commonsense example demonstrates a fundamental of economics: Economics seeks to explain, not dictate, the behavior of consumers and individuals. An analogy would be Newton's Law of Gravity. Newton did not invent gravity—he explained it.

Opportunity Costs

One tool that economists (and the disciplines of finance and managerial accounting) use to describe behavior is the principle of opportunity costs. This is an attempt to apply value to an otherwise immeasurable process. Opportunity costs are derived from the next best use of a specified resource. This concept seeks to define the difference in value of a resource used in different ways. Opportunity costs are forward looking. They are the anticipated (estimated) foregone benefits from actions that could, but will not be undertaken (i.e., the benefit foregone as a result of choosing one course of action rather than another). Therefore, the notion of opportunity cost is crucial in decision making.

There are numerous examples in agriculture in which opportunity costs are significant. The opportunity cost for a cow/calf manager would be the salary difference available in another occupation or place of employment requiring the same skill level.

One could also apply opportunity costs to other resources such as land. The opportunity cost of running cows on pasture would be the revenue gained from the land if it could be leased to someone else or used for another purpose. This opportunity cost may explain why some ranchers decide to sell parcels of land for subdivisions or other businesses.

An opportunity cost may also be applied to the assets of the operation. Assessment of applicable opportunity costs related to capital investments, for example, can define the difference in value gained or lost if the capital invested in a ranch was devoted to some other enterprise. It could be used to answer the question, "What would be a reasonable return if the capital was invested in some activity other than ranching?" In many agricultural businesses, opportunity costs negatively affect the analysis because greater return could be gained elsewhere. However, most operations are willing to "pay" the opportunity cost to maintain a "way of life." As stated before, economics and economic analysis seek to explain, not dictate, behavior.

Production Function

The relationship between inputs and output is traditionally described graphically with the neoclassical production function. The production function allows you to visualize several important features (Figure 12.1). Most notable is the law of diminishing marginal returns. Diminishing returns implies that

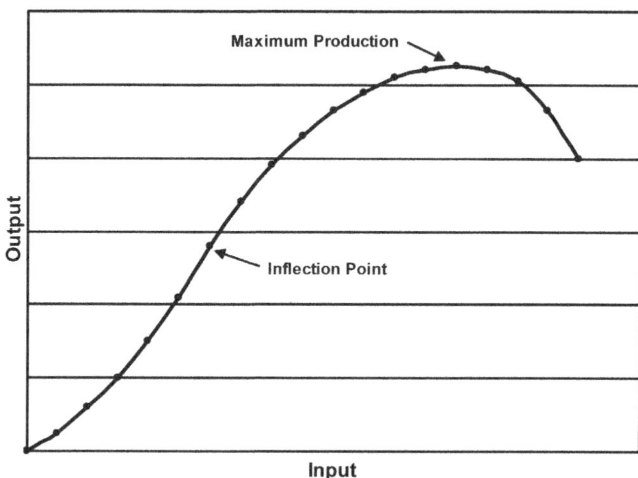

Figure 12.1. Production function.

incremental increases in one input will result in decreasing increments of output.

For example, let's consider a pen of feeder steers. Refer to Figure 12.1, where the y-axis is average daily gain (ADG) (output) and the x-axis is pounds of corn (input). Initially a pound of corn causes an increase in ADG at an increasing rate, meaning that each additional unit of corn results in a greater increase in ADG. Eventually, an inflection point is reached, signaling the beginning of diminishing marginal returns. At this point any additional corn still results in an increase in ADG, but at a smaller magnitude than did the previous increment (margin). As input is increased, the production function reaches a maximum. This maximum point defines the highest obtainable ADG. At this point, the marginal return equals zero. Any additional units of input cause marginal returns to become negative, and production begins to decline. In our example, this is the point at which the ration becomes too "hot" and cattle begin to go off feed and experience episodes of acidosis, liver abscesses, and so forth.

An example more pertinent to cow/calf producers would be bull selection for calf birth weight. If the average birth weight of calves born to heifers was reduced to 60 lb, dystocia would be minimal. However, resulting productivity of these calves would also be reduced. If a bull was selected for more moderate birth weights, a rapid increase in resultant calf performance could be expected. Similar to the previous example, as the birth weights increased, weaning weights would also increase, although with less magnitude than before (diminishing returns). At some point, death loss from dystocia would negate the positive effects of selecting for greater growth gains.

The goal of economics is to then establish the area in a production function where it is most profitable to operate. Initially, one might surmise that operating at the maximum level would be ideal. However, to identify the optimal level, cost of production must be added to the equation (Figure 12.2). Because the unit cost of an input remains constant in the short run, the "total cost curve" is described by a straight line. This total cost curve can then be overlaid on the production function, which has been converted from pounds to dollars.

In our feedlot example, one can recognize that the production function best applies to the "average" steer. However, there will be some steers representing points further along the curve that are developing acidosis and damaging their performance. So our optimum production level would be further back on the curve, where the likelihood for a steer to develop acidosis is significantly decreased. This optimal operating level is located where the distance between the total cost curve and the total production curve is greatest. Mathematically, this location is defined as the point where the marginal cost (unit cost of production) equals the value of the marginal product (value of incremental output).

Initially, it may not appear optimal to operate where input costs equal production. Intuitively, we recognize that one would not want to go beyond the point where additional units of input cost more than what is returned. Why would a producer strive to operate at zero profit? To capture the profit to the last

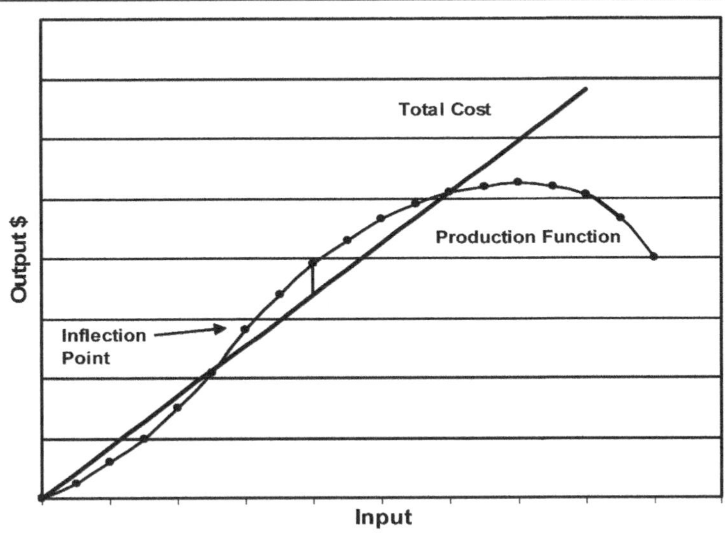

Figure 12.2. Production and cost function.

penny, a producer must be operating where there is zero profit left. At any time before this point, the return can be expected to be greater than the per unit cost of production.

Costs

When completing an economic analysis, it is important to account for various parameters and inputs. Accurate evaluation and measurement of costs is vital to a truthful analysis. Costs are usually described as either variable or fixed costs.

Variable costs are associated with inputs that can be modified during the current time period. The time period, often referred to as a "run," can range from an ultrashort run to the long run. In the ultrashort run, all inputs are fixed. This time period is rarely evaluated because all inputs are already locked in. In most instances, producers are working in the short run, where some inputs are variable but most are fixed. In agricultural circles the short run is usually about 1 year, or the length of the production cycle. During this time, producers can vary the inputs into the nutrition, herd health, and breeding programs. They can also hire more or less labor as well as regulate supplies. These are all considered variable costs.

Fixed costs describe inputs that cannot be easily modified in a current time period. Examples of fixed costs include land, buildings, cattle herds, and the ranch's equipment base. None of these inputs could be easily modified or varied for an ultrashort or short time period. In long-range planning, which involves more than one production cycle, all inputs can be categorized as variable. In the long run, all costs (variable and fixed) can be modified. Examples of this include expansion or contraction of the operation as well as diversification into other enterprises.

Economies of Size

"Economies of size" refers to the concept that increasing the size of the operation results in an increase in revenue. Often, this concept is viewed as a method to spread fixed costs out over more production units. Potentially, variable costs can be reduced with increased purchasing power. For example, the ability to purchase items in bulk may decrease a portion of the handling and packaging charges.

We may also observe diseconomies of size. In this scenario, increased size actually decreases the operation's profitability. Usually, diseconomies of size can be attributed to increased or unmet management requirements. The complexity of increasing herd size or land mass will increase the magnitude and scope of the management's responsibilities. If the operation's management skills are not sufficient, profitability suffers. This decrease in profitability may originate from increased death loss at calving, lowered nutritional plane, and so forth. The goal when considering expansion possibilities is to determine whether the operation can benefit from expansion or whether management is already at capacity.

Demand and Supply

Whether selling calves at auction, hamburger in a grocery store, or production consultation to clients, understanding the intricacies of the marketplace is essential. Activities within a market are reflected by the demand of buyers and the supply of sellers.

The law of demand explains the behavior of buyers. This law states that the lower the price of a good (service), the larger the quantity consumers wish to purchase. The demand curve is graphically expressed as a negatively sloped line (Figure 12.3) with price of the good or service on the y-axis and quantity on the x-axis. This model assumes that all other factors governing purchasing decisions are held constant, so that only the price of a product will move consumption up or down the demand curve.

This law explains why items are featured on sale. When T-bone steaks are on special, grocers can expect to sell more. In this example, demand does not change (the scenario is still operating with the same demand curve), but the quantity demanded changes to reflect the lower adjustment of price. A shift in demand occurs when some other factor influences buying decisions. In this situation, if demand has increased or shifted out at the same price, more product would be demanded (Figure 12.4). Demand shifts are usually attributed to changes in the income of consumers, prices of related goods, tastes and preferences, or health concerns. The recent increase in demand for beef is primarily a result of a thriving economy, giving consumers more discretionary income.

Similarly, the law of supply also facilitates the understanding of buyers' behavior. Intuitively we recognize that the higher the price of a good, the larger the quantity of the good that firms or producers will strive to turn out. When depicted graphically, the supply curve is a positive, sloped line (Figure 12.5), which, by and large, has a positive y-intercept. This intercept serves to reflect fixed costs. If the price falls below this point, it would be better to halt production and not incur additional variable costs. The main supply shifters include technology and changes in the supply of inputs. Advances in technology often

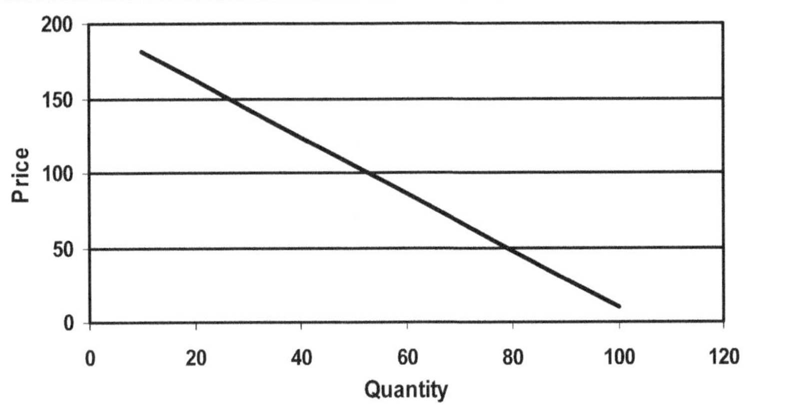

Figure 12.3. Demand.

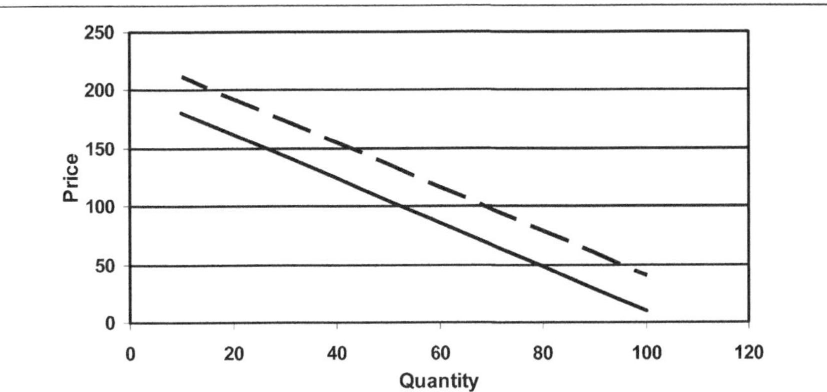

Figure 12.4. Demand shift.

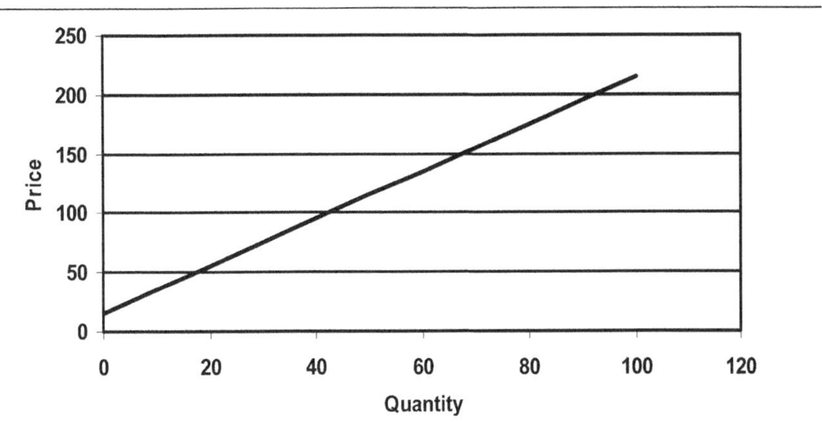

Figure 12.5. Supply.

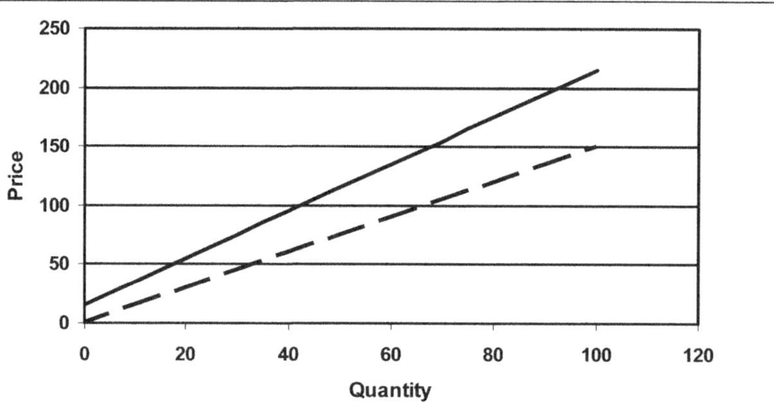

Figure 12.6. Supply shift.

result in decreased production costs. This decrease in costs causes the supply curve to be shifted out or down (Figure 12.6). For example, usage of a growth implant can produce more beef for less cost. A producer would then be able to produce more pounds of beef for the same price.

For production management veterinarians, understanding both sides of the supply and demand market is important. Beef cattle producers' economic behavior is governed by the supply side of the market. The recurring cattle cycle is an example of how supply affects producers.

The prices received by beef producers for their products are largely determined by consumer demand for beef, value added by the food-marketing sector, and farm supply of calves. The first two factors are beyond the control of producers and tend to have a "trickle down" effect on the prices they receive. The annual calf supply may be affected by uncontrollable variations in weather, disease, or imports. The biggest factors determining the annual supply of calves are the production decisions made by individual producers.

In general, these decisions have been heavily influenced by current calf prices. These current calf prices, and the resultant decisions, lay the foundation for the familiar cyclical pattern of the cattle inventories. The beef cattle cycle is characterized by a 10-year span in which cattle numbers rise and then fall as the cattle inventory tracks or responds to changes in calf prices (Figure 12.7).

The length of the cattle cycle is dependent on biological and psychological lag times. The biological lag time is the amount of time between the point at which producers decide to expand or contract production and the point at which supplies of cattle actually change. The psychological lag is the length of time necessary for producers to change production levels in response to higher or lower prices.

The cycle commences when nationwide cow numbers have reached their lowest point in the current cycle. Most commonly, this occurs at or about the beginning of a decade. The low supply of cattle relative to demand results in increased prices. During this period of profitability, producers will often expand their herd size, and new producers will enter the industry to take advantage of the favorable marketing situation. This expansionary process withholds females from the market, causing prices to rise and the cowherd to expand even more. By the middle of the decade, cattle numbers usually approach a peak. Because of the biological time lag required to develop, breed, and calve a heifer, 2–3 years elapse before additional calves reach the cash market. By this time, an oversupply of calves relative to demand for beef has been created and the cash prices for beef begin to decline. Producers who have overextended themselves financially during the high market may have to liquidate herds, and many may be forced to exit the industry. These herd liquidations speed the decline in cattle numbers until shortages of calves are created and prices rebound. Inventories continue to drop until producers overcome their wariness and begin to retain heifers, signaling the end of the current cattle cycle and the beginning of a new one.

Veterinarians provide valuable applications of their production management knowledge when they help their clients position themselves to either take advantage of or minimize the negative effects of the cattle cycle. However, the cattle cycle does not limit its effects to beef producers. Beef cattle veterinarians can also be influenced by the cycle. According to the NAHMS Beef '97 survey, approximately 10% of

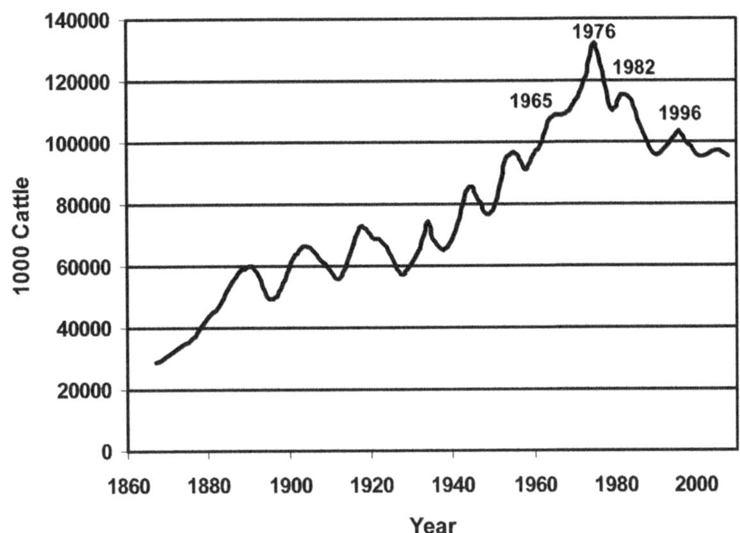

Figure 12.7. Cattle cycle.

producers throughout the United States reduced their veterinary expenses in an effort to cope with the down cattle market. The NAHMS study revealed that 7.4% of producers decreased their vaccination expenses, 7.7% decreased herd medications (de-worming, lice and grub treatment, and so forth), and 5.0% decreased individual medications for their sick animals. Of particular interest to veterinarians, 14.1% of producers decreased the level of veterinary services (e.g., pregnancy testing, bull testing, obstetrical work) provided for the operation.

Although the NAHMS study did not determine the magnitude of the producer's reduction of veterinary-related expenses, a trend in decreasing the operation's herd health program was identified. Projections for the cattle cycle should be considered before planning long-term inventory purchases or capital investments. For example, purchasing an ultrasound machine to improve cash flow with income from staging pregnancies in beef cattle would not be advisable right before the cattle market is expected to decline. This is true even though the time period just before the beginning of declining cattle prices is often when demand for veterinary services may be the greatest because calf prices will be high and most producers will be very profitable.

ACCOUNTING AND FINANCE

Although it is important to have some understanding of economic concepts, most consulting veterinarians will actually be using more accounting and finance-related analyses. Accounting is defined as the art of recording, classifying, and summarizing in terms of money the transactions or events of a business and reporting the results thereof. In contrast, the discipline of finance is defined as the practice of manipulating and managing money.

Accounting can be partitioned further into the disciplines of financial accounting and managerial accounting. The former is concerned with reporting historical financial information (e.g., financial transactions that have occurred in the previous year) in a verifiable and objective manner to an external party such as the Internal Revenue Service or, in the case of some corporate agricultural entities, to shareholders. The appropriate classifying, summarizing, and recording of this information is guided by generally accepted accounting principles promulgated by the Financial Accounting Standards Board. Financial accounting is not normally part of the routine business of providing decision-making services for future management of an entities business.

In contrast, management accounting uses this same accounting information in a forward-looking sense for decision-making purposes. Management accounting is strictly for use internally. Although the accounting information used by a management accountant for decision making is extracted from the exact same pool of information used by a certified public accountant employed by the ranch to prepare financial statements for submission to the IRS, the management accountant is not obligated to adhere to Financial Accounting Standards Board rules to collate the accounting information into a form useful for decision making. As a consequence, although it is vital to know what information has been used and how it has been organized for financial statement submission to the IRS, the role of the veterinarian should be that of a managerial accountant. The veterinarian should ask himself or herself whether the present accounting information can be used to help the client better manage his or her enterprise in the future. Perhaps the information should be rearranged or additional information (even nonaccounting information) be gathered to better guide decision making.

ACCOUNTING EQUATION

A specific framework of formats and procedures has been developed to facilitate the reporting of accounting information in a consistent manner. This accounting framework rests on the premise that it is possible to identify an accounting entity (the person or ranch for which a chart of accounts is maintained) and that the resources (assets) available to this accounting entity will be exactly equal to the resources (equities) provided by both nonowners (creditors) and owners. Thus, Assets = Equities. Equities are commonly partitioned further into liabilities and owners' equity: Assets = Liabilities + Owners' Equity. This equation is referred to as the "accounting equation."

The partition of equities into liabilities and owners' equity recognizes that there is a fundamental difference between the obligations that the ranch has to outsiders (e.g., the bank) and the obligations that it has to the owners who have invested their capital. As a consequence, the accounting processes of observing, measuring, and reporting are always carried out with the objective of maintaining this fundamental equality of assets and equities.

Assets typically comprise all things of value that the ranch has a right to use. These consist of financial resources (e.g., cash and cash balances in banks, amounts owed by customers, marketable securities), physical resources (e.g., land, buildings, equipment, inventories of unsold cattle), and miscellaneous other resources having value to the entity.

Equities represent claims on these assets. Some claimants (nonowner creditors) provide monetary and other resources and expect to be repaid, whereas others (owners) contribute monetary and other resources and thereby become participants. Equities of nonowners include amounts owed to suppliers, amounts owed on short-term loans, and loans represented by other miscellaneous debts.

As mentioned previously, Assets = Liabilities + Owners' Equity. As a consequence, if the resources owned by the ranch increase, then the accounting equation says that the new resources originated from a source that has a claim against them. Likewise, if the resources of value owned by the ranch decrease, obligations to nonowners were paid or the equity of owners was reduced.

ELEMENTS OF FINANCIAL STATEMENTS

Common financial statements are an invaluable tool to use to assess the financial health of a beef enterprise and to delineate where action is needed to treat specific financial ills. As a consequence, an understanding and working knowledge of basic financial sheets is required to advise clients competently on these matters. The nature and format of financial statements are derived directly from the basic accounting equation.

Balance Sheet

The "balance sheet" (or "statement of financial position") presents a ranch's financial position as of a specific date, based on measurements made in accordance with generally accepted accounting principle rules. The balance sheet contains separate sections for listing all assets, liabilities, and owners' equity. According to the accounting equation, the measured amount of total assets is always equal to the measured sum of liabilities and owners' equity. Except for monetary amounts such as cash, accounts receivable, and accounts payable, the measurements of each classification will rarely be equal to the actual current fair market value or cash value shown. This is because most measurements in accounting are made at the time of a transaction, and these historical costs are retained in the accounts, even though the values of assets and some obligations may increase or decrease with the occurrence of events or passage of time.

The asset portion of the balance sheet is divided into current assets and noncurrent (long-term) assets. Current assets are items that will be realized, sold, or consumed during the normal operating cycle of the ranch or within 1 year if the operating cycle is shorter than 1 year. Examples of current assets include cash on hand and in a bank, investments such as a money market account, feed on hand, livestock for sale, and prepaid expenses. (The last are expenditures that have been made to acquire future benefits or services, such as prepaid winter protein supplement, but for which benefits have not yet been obtained as of the date of the balance sheet.) Noncurrent assets typically comprise tangible assets such as land, machinery, and breeding livestock used in the ranch's operations. Land is almost always included at its original cost, whereas other assets are stated at their original cost less the proportion of original cost that has been included in the expenses of prior periods' operations as depreciation (net value, referred to as "book value"). Depreciation is relevant only to purchased breeding livestock; raised breeding livestock is nondepreciable. Assets that are not part of the operation ("nonoperating" assets) are also included here as "investments."

On the other side of the accounting equation are "liabilities" and "owners' equity." Liabilities are obligations that the ranch must satisfy by transferring assets to, or performing services for, another party in the future. Current liabilities are those obligations that the ranch expects to satisfy with current assets in the course of normal operations. Examples of current liabilities are accounts payable (e.g., chemical or fertilizer bill), annual operating notes, and the interest and principal due in the current year on longer-term loans. Noncurrent liabilities are items that are not due during this time period; noncurrent liabilities would be the interest and principal due on loans beyond the current year.

Another important differentiation on the balance sheet is "cost basis" vs. "market value" when reporting assets and liabilities. The cost basis tracks the real cost of an individual item. For breeding females, this would be the costs associated with raising or, alternatively, buying (use book value) a replacement heifer. Market value reflects what current value for an asset would be if sold. Cost basis requires more extensive record keeping but can be advantageous for the producer in terms of knowing what the true cost of production is for each product.

"Owners' equity" (or "shareholders' equity" or "stockholders' equity") reflects the residual claims to ranch owners. Owners' equity is always equal to the monetary amount that remains after deducting the total liabilities of the ranch from its total assets. Owners' equity is also referred to as the "book value" of the enterprise. For a single proprietorship and partnership, retained earnings are the only item to be classified here. A corporation will list capital stock that is issued and outstanding in addition to retained earnings. Initially, retained earnings represent the original capital invested by the owner or owners to start the business. Subsequently, net income (loss) is added to retained earnings each year. Thus, retained earnings reflect the accumulated net income of the ranch from its origin to the present after deducting family living withdrawals or, if a corporation, deducting dividends to shareholders.

Income Statement

The "income statement" ("statement of income" or "profit-and-loss statement") represents the difference between revenues and expenses of the operation for a given time period, usually the calendar year. There are two different formats used in producing income statements (i.e., cash-basis or accrual-basis accounting). The most commonly used accounting method is a cash-basis system. This system records income when it is actually received and expenses when they are actually paid. This system is typically used for IRS tax purposes. Unfortunately, it does not assess the true profitability of the operation because unsold current assets (inventory items such as replacement heifers that are potential sources of cash) and their associated costs are not recognized in a cash-based income statement. Accrual accounting, however, records immediately all sources of revenue and expenditures (matched to this revenue), irrespective of whether a sale has occurred or not. Thus, this method tracks changes in revenue sources and associated costs that would be missed with cash accounting.

Revenues result from selling products or services to customers. Immediately below the revenue line in the income statement lies cost of goods sold; that is, the costs that can be traced directly (referred to as "direct costs" or "variable costs") to the production of goods (such as calves) being sold. "Gross profit" simply represents the difference between revenues and cost of goods sold. Below gross profit are listed the various operating expenses such as selling, administrative, and general expenses (sometimes referred to by the acronym SG&A). These expenses represent general overhead expenses that the beef enterprise incurred during the designated periods of time. Deducting these general overhead expenses from the gross profit yields the ranch's operating income, which is thus a measure of the income derived from the principal operations of the ranch.

Below operating income, other income and expense items must be added and subtracted before the "net income" total is determined. An example of an income source that should be listed here is crop or livestock support or disaster payments. Admittedly, most ranches probably will not show large amounts of nonoperating income. However, it should be shown separately in the income statement to enable a reader of the financial report to observe where income originated.

Cash Flow Statement

The "statement of cash flows" ("cash-flow statement" or CFS) details the reasons why the amount of cash changed in response to the fundamental operations of the ranch during the accounting period. The CFS summarizes the three categories of activities: increase or decrease in cash flow because of operating (production of primary product) activities, investing (acquisition or sale of assets) activities, or financing (changes in debt or capital stock) activities. The operating section is the most important of the three sections because it describes how cash is being generated or used by the core business activity of the ranch.

Regarding the operating section of the CFS, cash inflows and outflows can be reconstructed using a direct or indirect format. The ending number is the same, but the starting point between the formats differs. Comments will be restricted to those of the indirect format. The indirect format does not show these items directly but assumes that most of these cash inflows and outflows are already summarized in net income. Thus, starting with net income, adjustments are made for everything that is not a true representation of actual cash inflow and outflow in net income. For example, adjustments to reconcile net income to net cash provided by operating activities involve adding back the noncash items of previously deducted depreciation and any amortization expense. Typical changes in other accounts affecting operations include (increase)/decrease in accounts receivable, (increase)/decrease in inventories, (increase)/decrease in prepaid expenses, increase/(decrease) in accounts payable, and increase/(decrease) in taxes payable (parentheses in this example, per accounting methodology, mean to subtract the amount).

The investing section of the statement of cash flows summarizes activities associated with purchases and sales of noncurrent assets (e.g., land or equipment). The financing section of the statement of cash flows lists all transactions pertaining to long-term liabilities as well as owners' equity activities (e.g., family living withdrawal or stock dividend if a corporation).

Analysis of Financial Statements

For a decision maker, accounting information is often best used on a comparative basis (e.g., measurements in a prior period or periods, against budgeted amounts, or against industry norms). However, deciding which comparison(s) will best answer a question or series of questions about a particular business practice or practices on a ranch can be challenging. In some instances, the decision maker may find it necessary to summarize (or develop) nonfinancial data in association with financial data to answer some questions.

Common-Size Format of Balance Sheet and Income Statement

Creation of "common-size financial statements" can be valuable to better detect changes in the financial structure or nature of operations of a ranch over time. Common-size financial statements involve translating each line in the statement of financial position and the statement of income into a percentage format. In a common-size balance sheet, each asset, liability, and owners' equity amount is expressed as a percentage of total assets (total assets = 100%). In a common-size statement of income, sales are set at 100%, and each item is expressed as a percentage of sales. The ability to view balance sheet and income statement items as a proportion of assets and sales, respectively, allows earlier detection of subtle deviations from normal and, hopefully, earlier corrective action, if necessary. Common-size financial statements are also a good method to use to compare ranches of differing size.

CFS Analysis

The CFS is one of the most useful financial statements available to a veterinarian to enable him or her to get a true picture of the financial health of a ranch. At least 2, and preferably 3, years' worth of cash flows should be prepared for comparison.

A convenient place to start an examination of a CFS is to review net income across years. Initial attention should focus on whether income or losses over the last few years have occurred and whether income (or losses) are growing or shrinking. The operating section of the CFS is where most cash inflows to the ranch should come. Although an occasional year of negative cash flow from operating activities is not necessarily bad, the trend should be toward positive cash flows. It is a bad sign to see that most cash flow is coming from either investing (e.g., selling off assets) or financing (e.g., operating loans). The former creates concern that the ranching entity is "shrinking" and may not remain viable, whereas the latter may indicate that the ranch is forced to take on more debt or use more owner equity to operate.

Assuming cash flow is positive from operating activities, an assessment should be made as to whether it is adequate to fund important expenditures to the ranch, particularly replacement of fixed assets needed to produce the primary product for the ranch, as well as for living withdrawals. The amount of annual depreciation expense can be used as a crude proxy for the amount of cash needed to fund the replacement of fixed assets. To take the subject of fixed assets and depreciation expense one step further, to ensure that the ranch is kept "whole" and is not shrinking, the portion of investing activities related to the purchase of fixed assets should also exceed that of depreciation.

In the investing section of the CFS, one can review whether the ranch is generating or burning cash in its investing activities. A healthy ranching enterprise is expected to purchase more capital (fixed) assets than it sells. As a consequence, negative cash flows are generally expected to occur from investing activities.

As mentioned previously, positive cash flow from investing activity from one year to the next or the one after is a bad sign, indicating that the ranch is shrinking and losing viability.

Cash flows from financing activities can be positive or negative during the course of normal ranching operations. Neither trend is necessarily good or bad on face value but must be analyzed in concert with the other previously mentioned operating and investing activities to get a true impression of what is happening. For example, because of leverage and cost of capital considerations, a ranch may make the conscious decision to finance operations through debt rather than through cash flows from operations. However, a bad trend to see is a ranch that is forced to generate funds from other sources because low or negative cash flows are coming from operating activities.

Financial Performance Ratios

Provided that accounting data exist, ratios can be constructed to evaluate virtually any business aspect of a ranch. However, only a few will be discussed here, examining liquidity, solvency, profitability, and leverage. "Liquidity" is a term used to describe the ease with which assets available to the ranch can be turned into cash to satisfy liability obligations. Obviously, current assets are more liquid than are noncurrent (long-term) assets. Therefore, liquidity ratios measure the ranch's short-term ability to pay its maturing obligations to creditors. "Solvency" refers to the ability of a ranch to pay its debts as they mature; lower solvency is associated with a ranch that carries a higher level of long-term debt relative to assets. Ranches with higher debt are relatively more "risky" because more of their assets will be required to meet these fixed obligations such as interest and principal payments. "Profitability" ratios measure the extent to which a ranch has succeeded (failed) monetarily over a given period of time. Leverage is a term used to describe the degree to which the activities of the ranch are supported by liabilities and long-term debt as opposed to owner contributions (equity capital). A ranch that has a high proportion of debt to owners' equity contributions is considered to be highly leveraged. A highly leveraged enterprise runs a higher risk of being forced into insolvency if profits and cash flows are insufficient to service the cost of interest (and principal) on debt.

Several words of caution about using ratios are worthy of mention. No one ratio is more useful than another. Ratios should not take precedence over other financial information (e.g., CFS) in determining a course of action but should supplement that information. Also, rather than zeroing in on each specific ratio for that single point in time, veterinarians and their clients should concentrate on evaluating the changes observed in ratios from year to year. Finally, any desire to recommend or take immediate action prompted by hitting a particular threshold ratio number should be resisted until investigating the reasons why the change has occurred. Remember, the ratio is only as good as the data on which it is based.

Liquidity: Current Ratio

The current ratio is calculated as current assets (e.g., cash, accounts receivable, inventories) divided by current liabilities (e.g., accounts payable, operating loan payments). The current ratio is a basic indicator of the ranch's ability to satisfy short-term debt obligations if the enterprise were forced to liquidate current assets. The goal is to have a positive ratio greater than 1.0, but the size of the current ratio reflects the trade-off between the rapidity with which cash inflows must occur and the demands for cash payments to creditors.

Solvency: Equity-to-Asset Ratio

The equity-to-asset ratio is calculated as total equity divided by total assets. The ratio measures what proportion of total ranch assets is owed. Explained another way, the ratio identifies the proportion of assets owned that can be used to settle debt obligations. A reasonable goal is to maintain an equity-to-asset ratio greater then 70%.

Profitability

Return on Assets The return on assets ratio is calculated as after-tax net income (with any interest expense added back) divided by total assets. Return on assets assesses profitability per dollar of assets invested; that is, net income is a function of the investment in all financial resources (debt and equity) at the command of the owner without consideration of how the financial resources were obtained and financed. The goal is to have a 5%–8% annual return on assets.

Return on Equity The return on equity ratio is calculated as after-tax net income divided by the cumulative amount of capital (owners' equity) invested by the owners. Return on equity is an index measurement of the efficiency with which an owner's original invested capital and earnings retained in the ranch have been used. The percentage return on equity for each ranching enterprise will likely vary depending on owner desires. However, the rate of return on equity should be larger than the rate of return on assets for borrowing to be advantageous to the ranch.

Operating Profit Margin The operating profit margin ratio is calculated as after-tax net income (with interest expense added back and less all family living withdrawals) divided by gross revenues. Operating profit margin indicates the proportion of revenues that are available to service debt and grow the ranch (add to equity). This financial performance ratio probably has limited usefulness to beef producers. Normally, beef is sold as a commodity. As a consequence, producers are not in a position to dictate selling price (and expected revenue) but, instead, are usually price takers. Therefore, absent the ability to exercise some form of risk management contractually, control over the amount of revenue received is limited. Even if a ranch is closely managing its cost of production and decides to use percentage operating profit margin for management purposes, expect volatility in this ratio because of potential wide swings in the price received for the product.

Du Pont Model for Return on Equity The Du Pont model for return on equity determination is actually formulated using several ratios:

Return on equity = Net income/Sales × Sales/Assets × Assets/Owners' equity.

The objective of calculating this ratio is to evaluate how certain individual business decisions made by the ranch affected the overall goal of increasing returns to owners. The first ratio describes profit margin. It can be used to focus attention on the relationship between the price and cost of the product sold. As described above, profit margin targets are easier to reach by controlling the cost of production. The second ratio describes asset turnover, a measure of how efficiently ranch assets generate revenues. The last ratio, assets over equity, evaluates how efficiently the ranch's assets are used to provide maximum return to owners.

Leverage: Debt Ratio

The debt ratio is calculated as total debt divided by total assets. As the ratio increases incrementally higher, so does the degree of leverage. As mentioned previously, a highly leveraged enterprise runs a higher risk of being forced into insolvency if profits and cash flows are insufficient to service the cost of interest (and principal) on debt.

FINANCE

As already discussed, the discipline of finance is concerned with the practice of manipulating and managing money. "Opportunity cost" and the "time value of money" are two critical factors to take into account when making any financial decision. The concept of opportunity cost has already been discussed in the section on economics. The concept of the time value of money is that a quantity of money received sometime in the future is worth less than the same amount of money received today. Alternatively, more money must be received in the future to equal the same amount of money in hand today: But how much more? Consider the following scenario: Is it better to have $1000 today or $1200 in 5 years? The answer lies in what a person could do with that money in the intervening time. If one could generate more than $200 in today's dollars before 5 years elapse, then one would be better off having the money today. If, however, one would generate less than $200, then it would be wise to wait 5 years for the $1200. Therefore, the discount (interest) rate chosen by the holder of this money ultimately determines the rate of return that must be achieved on any money invested to convince him or her not to invest elsewhere.

A standardized formula (referred to as net present value, or NPV) has been developed to help a decision maker take into consideration the time value of money when determining whether an investment or capital budgeting opportunity (e.g., purchase of a long-term asset) is a good one. Data necessary for an NPV calculation include the relevant initial outflow of cash (purchase price and associated costs), yearly net cash inflows over the period of time under consideration, and a discount rate. The discount rate reflects a reasonable return to management of the asset.

Historically, 8% has been used in production agriculture for a long-term investment. The formula is:

$$\text{NPV} = (-\text{initial cash outflow}) + \sum \frac{\text{net cash inflow}_n}{(1+R)^n}$$

where R = discount rate and n corresponds to the year number associated with each net cash inflow. The discounted cash flows for each year are summed and compared to the initial outflow of cash. The investment or capital budgeting opportunity is acceptable if NPV is greater than 0, whereas it should be rejected if NPV is less than 0. Indifference to the decision occurs if NPV is equal to 0. If several different scenarios are under consideration, the one with the highest NPV is considered the best opportunity to pursue.

PROFITABILITY

In general, there are three factors that determine the profitability of an operation. These include production (pounds of beef produced), cost of production, and marketing, or the price received when the beef is sold. All of these areas can be managed to stabilize the financial situation of beef cow producers. Veterinarians can play a vital role in helping producers make sound decisions to maximize profitability.

Production

Historically, cow/calf producers have measured their herd's productivity by the pounds of calf they were able to wean. Efforts by producers to improve have been focused on increasing the productivity of their herd. Toward this goal, continental breeds have been introduced to increase size and growth, growth implants have been developed for calves, and nutrition programs have been enhanced. All of these practices have increased farm productivity, requiring fewer cows to produce more beef. This increased productivity, in conjunction with a decreased consumer demand, led to the long-term decline in cattle numbers, as depicted in Figure 12.7, where cattle numbers peaked at 132 million versus 103 million in 1996. Overall, however, profitability has not followed the enhanced productivity.

Cost of Production

Optimum production for a cow/calf operation would be profit maximization, not production maximization. To achieve higher productivity, a cost must be paid. Many producers are able to wean 600-lb calves using high-growth genetics, creep feeding, and growth implants. However, as discussed previously about the production function, the 500-lb calf may be more profitable when compared to its 600-lb contemporary. For this reason, many people have focused on cost reduction as a way to increase profitability. Being a low-cost producer does not necessarily guarantee profitability, however. If production is adversely affected by decreases in spending, then the advantages of a cost-savings scheme are negated.

A rational approach to cost reduction would be to trim any fat or excess from the operation. This will probably not result in huge savings for the producer. Many people focus on feed costs as a means to reduce expenditures. Although, in the United States, feed costs are the largest portion of expenses (50%–60%), the necessity of proper nutrition requires a delicate balance to prevent problems. Recommendations for reductions or savings in nutrition should only be considered after the feed has been analyzed and a ration balanced to meet or exceed the animals' requirements. If possible, cows should be divided into contemporary groups based on age, body condition, and stage of pregnancy to fine-tune the feeding program and prevent wastage.

Management decisions designed to decrease intangible costs result in improved long-run profitability. Recent research has shown that a preweaning case of respiratory disease results in an average 20-lb performance loss when measured at weaning. Combining the loss of dollars resulting from decreased performance with additional treatment costs and subsequent death loss results in each case of respiratory disease costing the producer $50.

Depending on the prevalence of disease, a change in the health program may be warranted. To properly assess a herd's health program, accurate baseline information regarding prevalence, fatality rate, treatment costs, and production losses must be obtained. These measures can then be used in a partial budget analysis to determine the cost-effectiveness of various interventions.

Another intangible, but very significant, cost is the depreciation of replacement stock. This depreciation cost is the difference between the purchase price of the stock and the salvage value. If a herd's replacement rate is 20%, then the average female weans only three calves. Because it generally takes 5 years for a heifer to pay for herself, half of the replacement stock do not break even in their short tenure. Management strategies designed to increase the longevity of replacement heifers have great potential to improve profitability.

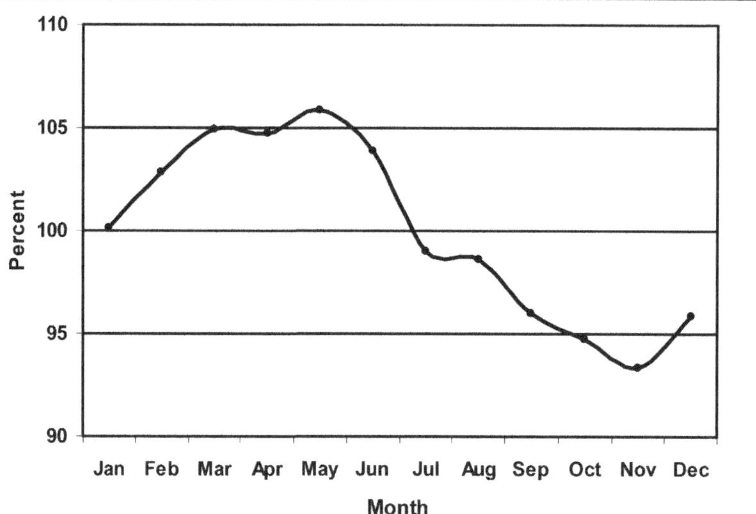

Figure 12.8. Seasonal feeder calf price index.

Marketing

The last area that can be modified to increase profitability is developing a marketing scheme to enhance the price received for products. Futures and options can be used to minimize some market risk. The introduction of stocker cattle futures in 1998 added this commodity class to the financial management tools available for cow/calf producers. Futures contracts can also be used for inputs such as corn, or soybean meal. Many cow/calf producers are unfamiliar with the futures market and are uncomfortable using this tool. Veterinarians with a working knowledge of risk management and marketing can be instrumental in aiding clients with marketing decisions and in working with their broker.

In addition to the fluctuations in cattle prices from the recurrent cattle cycle mentioned earlier, there are also seasonal variations throughout the year. Like the cattle cycle, these price fluctuations are a result of changes in product supply and demand. In the fall of the year, when most calves and cull cows are marketed, prices will be lower as compared to the rest of the year (Figure 12.8). This difference may have potential for some producers to enhance their profitability. Veterinarians should be an integral part of this decision process. The margin of profitability with this type of venture is usually tight enough that there is little room for morbidity and mortality losses. Attention to cull cow marketing can also be beneficial. In general, cull cows account for one-fourth to one-third of the operation's revenue. Helping your clients identify which cows should be marketed early, marketed later, or retained into the herd can be a valuable asset.

A recent trend in the beef industry has been a migration toward integrated marketing schemes. These schemes range from alliances based on targeted end performance to more integrated contracts that control genetic and management decisions of the operation. Many of these programs are incorporating veterinarians within the information chain. For some veterinarians, this is an opportunity to become more involved with information management. For others, lack of knowledge or the fear of compromising their client's trust has prevented them from advocating specific programs. Other veterinarians have started marketing programs for their clients. The long-term potential for alliances and marketing programs is unknown. Historically, increased emphasis has been placed on marketing schemes during downturns in the market. To take advantage of these programs, the operation should be prepared for this before the market turns.

ANALYSIS OF HERD PERFORMANCE

Once the veterinarian has attained an understanding of economics and finance, this can then be used to analyze a client's operation or individual management decisions. Before an analysis begins, the goals for the operation must be firmly in place. This "goal setting" allows the best evaluation and most appropriate financial tools to be used, resulting in more accurate recommendations.

Standard Performance Analysis

Standard performance analysis (SPA) was developed from an initiative of the National Cattlemen's Beef Association in 1992. As the name implies, SPA is an enterprise analysis tool and not an accounting or record-keeping system. However, proper financial statements and production records are necessary to complete an accurate analysis. Because SPA has a standardized format, ranches can be more readily compared. The standard format also forces producers to accurately input data. In addition to allowing the individual producer to analyze his or her operation, SPA is designed to collate the data nationally into a database. The data can then be analyzed to identify trends in the industry. It also enables the producer to compare his or her operation to others throughout the United States.

Typically, the SPA analysis is divided into a production (SPA-P) and financial (SPA-F) portion. The SPA-P encompasses the entire production cycle. This production cycle spans a time frame of approximately 18 months from the beginning of the breeding season through calving, ending when calves are weaned. Any events that occur after weaning are not considered to be a part of the cow/calf enterprise analysis. Most of the performance evaluations use the number of cows exposed for the denominator. For example, calf crop would be:

$$\text{Percentage Calf Crop} = \frac{\text{Calves weaned}}{\text{Exposed females}} \times 100$$

Or in other words, the percentage of calves that were weaned by the cows that had a chance to become pregnant, calve, and wean a calf.

The adage "garbage in, garbage out" is certainly true when using an SPA analysis to measure producer performance. To correctly complete a useful and informative SPA-P analysis, producers must be willing and able to provide accurate and reliable information. The power of a SPA-P is derived from the final ratio, which is pounds of calf weaned per exposed female. This relatively simple but important equation summarizes the reproductive performance, calving ability, and weaning performance of the herd.

The SPA-F portion is designed to evaluate the operation's financial performance. Like most financial reports, the SPA-F encompasses a 12-month fiscal year. Important information to complete a SPA-F includes beginning and ending balance sheets, an income statement, depreciation schedule, and pertinent IRS tax schedules. Because most producers use a cash accounting method, an accrual adjusted income statement using inventory adjustments will usually be necessary.

The SPA-F reports the financial position and performance of the operation. The financial position includes items such as assets per breeding cow, debt per breeding cow, and percentage equity in the operation. The financial performance is reported as dollars per breeding cow or per hundredweight of calf. Items of interest would include feed cost, operating cost, revenue, income, and economic return. The two most widely used measures of financial position reported by the SPA-F are the unit costs of production. When reported on a per cow basis, this equates to the annual cow cost. The unit cost of production per hundredweight of calf is equivalent to an operation's break even point.

To be most effective, a SPA analysis should be completed for 3–5 years in a row. The first year usually identifies what data are missing or incomplete. By the third year of analysis, results will be more dependable and, when combined with multiple years of analysis, serve as an effective and reliable source of information for decision making.

Unfortunately, most producers have not readily accepted SPA. At present, less than 5% of producers have completed a SPA. For some producers and veterinarians, the SPA analysis can be intimidating and frustrating. Another reason for lack of widespread use of SPA analysis is the dearth of trained SPA facilitators. Originally, extension agents were tapped to administer the analysis. Unfortunately, budget cuts and other priorities have limited the ability of extension agents to administer a SPA. There are many veterinarians who have received certification to be SPA facilitators. This requires 20 hours of education covering the program and case studies. For more information regarding SPA certification, contact Dr. Jim McGrann at Texas A&M University, Department of Agricultural Economics, 2124 TAMU, College Station, TX 77843-2124.

Partial Budgeting

"Partial budgeting" is a valuable tool that can be used to estimate the effect that a management decision may have on the operation's profitability. As the title implies, this tool only examines a small part of the business. When developing the partial budget, only the revenue and expense items affected by the potential change are considered. The resulting calculations then determine whether the proposed change would be more or less profitable than the status quo. The partial budget is divided into two sections: added returns and added costs. The added returns section consists of the increased revenue and decreased costs of the management change. The added costs section is made up of decreased revenue and increased costs.

Let us consider an example in which a producer is trying to determine whether to split the cow herd into two feeding groups before calving. It is estimated that, by focusing on nutritional requirements, the productivity of the cow herd will be improved. The pregnant cows will be split 2 months before calving, with the replacement heifers, 3-year-old cows, and cows with a body condition score less than 5 placed into one group and the remaining cows in the other group. It is predicted that the performance of the herd will be increased by preventing two calves from dying at birth because of poor calf vigor, and the pregnancy rate will be increased by three percentage points.

The partial budget would potentially look like this:

- Added Returns
  - Increased Revenue
    - Two additional calves for sale $______
    - Three additional heifer calves for sale (decreased replacement rate) $______
  - Decreased Costs
    - Less feed used $______

  Subtotal $______
- Added Costs
  - Decreased Revenue
    - Three fewer open culls $______
  - Increased Costs
    - Labor for feeding $______
    - Cross fence (yearly depreciation) $______

  Subtotal $______
- Added Returns − Added Costs $_______

If the difference is a positive number, then split feeding is considered to be a profitable venture. More detail can be added to this simplistic design. For example, because replacement rate is decreased, there would be decreased costs associated with heifer development. Other possible additional costs or revenues should also be considered and might include increased fuel and maintenance requirements of equipment, purchase of additional feeders, and increased weaning weights from calves being born earlier in the calving season.

The goal of partial budgeting is to develop a management change, identify potential effects, and compare scenarios. Problems arise when too much detail is added. Often, more detail adds complexity to a model with unknown information. In the example above, increased weaning weight is a potential added revenue, but quantifying the true value may be difficult.

NPV

For the cow/calf producer, the principle of NPV is most often used with replacement heifers. In this scenario, the producer has a large upfront cost and then a stream of revenue over the life of the cow. Data necessary for NPV include purchase price, depreciation, projected revenue, yearly cow costs, and a discount rate. The discount rate reflects a reasonable return to management of the asset. Historically, 8% has been used for a long-term investment. The formula is:

$$\mathrm{NPV} = \frac{\mathrm{ANCF}_1}{(1+R)^1} + \frac{\mathrm{ANCF}_2}{(1+R)^2} + \frac{\mathrm{ANCF}_3}{(1+R)^3} + \cdots \frac{\mathrm{ANCF}_n}{(1+R)^n} + \frac{\mathrm{SV}}{(1+R)^n}$$

Table 12.1. What a beef cow is worth—economic analysis 2003–2009.

Calves	Year	Calf Price	Net Income	Discount Factor	Present Value
1	2003	$ 94.84	$ 171.62	0.9259259	$ 158.91
2	2004	$ 98.49	$ 191.70	0.8573288	$ 164.35
3	2005	$ 100.18	$ 200.99	0.7938322	$ 159.55
4	2006	$ 93.04	$ 161.72	0.7350299	$ 118.87
5	2007	$ 86.42	$ 125.31	0.6805832	$ 85.28
6	2008	$ 81.74	$ 99.57	0.6301696	$ 62.75
7	2009	$ 78.26	$ 80.43	0.5834904	$ 46.96
		Cull cow value	$ 350.00	0.6500000	$ 227.50
		Total net income	$ 998.64	Net present value	$ 1024.13

where ANCF is the annual net cash flow (ANCF = Revenue − Annual cost − Depreciation), SV is the salvage value, and R is the discount rate.

The NPV is then compared to the purchase price to determine which has the most value today. Using this system requires excellent knowledge of operational costs and forecasting of future prices. This methodology can also be effectively used in decision making for retained heifers or for retaining open cows. In that case, the first year would be limited to costs associated with heifer development without any revenue generated. The NPV would then be compared to the fair market value of the heifer at weaning.

Interestingly, the compounding effect of the discount rate reduces the effect of the profit in the long run. This makes the first 3 years critical for the success of the venture (Table 12.1). This table illustrates the effect that the cattle cycle as well as the discount factor have on future profitability. One must remember that even if NPV shows that the investment may be profitable, the operation's cash flow may not be sufficient to service debt or other costs associated with the venture. There are spreadsheets available that model the NPV system. Many of these add complexity such as considering taxes, steer versus heifer calves, and principal and interest payments, and they also account for the weaning percentage of the operation.

Veterinarians interested in providing service to beef producers in today's market must have a basic understanding of accounting, economics, and finance. Having some appreciation of these principles increases a veterinarian's ability to relate to his or her clients. Principles and tools outlined in this chapter will serve as a basis for the beef cattle practitioner to service his or her clients outside the realm of traditional veterinary practice. Beef cattle practitioners willing to expand into this type of consulting provide an added value to their clients.

REFERENCES

Browning, E.K., M.A. Zupan. 1996. *Microeconomic Theory and Applications*. New York: HarperCollins.

Bruns, W.J., J.H. Hertenstein, S.M. McKinnon. 1992. *Reading Financial Reports*. Boston: Harvard Business School Publishing.

Food and Agricultural Policy Research Institute. 2001. *FAPRI 2001 U.S. Baseline Briefing Book*, pp. 1–58. Columbia: FAPRI, University of Missouri-Columbia.

Kohls, R.L., J.N. Uhl. 1972. *Marketing of Agricultural Products*. New York: MacMillan.

McGrann, J.M. 1996. *Partial Budget for Beef Cattle Management*. L-2212. College Station: Texas Agricultural Extension Service.

McGrann, J.M. 2000. *Cow-Calf Standardized Performance Analysis Handbook and Software User Manual*, pp. 1–175. College Station: Texas Agricultural Extension Service.

Morris, D.L. 1995. Standardized performance analysis of beef cattle operations. *The Veterinary Clinics of North America: Food Animal Practice* 11(2):1–391.

Myers, S.C. R.A. Brealey. 2000. *Principles of Corporate Finance*. Boston: Irwin McGraw-Hill.

National Animal Health Monitoring System. 1998. *Reference of 1997 Beef Cow-Calf Production Management and Disease Control*. Part III. USDA:APHIS:VS. Fort Collins, CO: National Animal Health Monitoring System.

Zimmerman, J.L. 2000. *Accounting for Decision Making and Control*. Boston: Irwin McGraw-Hill.

13
Beef Quality Assurance

D. Dee Griffin

INTRODUCTION

The purpose of the Beef Quality Assurance (BQA) program, begun voluntarily in 1982 by the beef industry, is to identify and avoid activities in beef production operations that can cause a quality or safety defect. The program encourages beef operations to seek all sources of information needed to accomplish the BQA goals and objectives. Although the BQA program started in the finish cattle feeding side of the industry, BQA today is a cooperative effort among beef producers in all segments, veterinarians, nutritionists, extension staff, suppliers, and other professionals. The program asks everyone involved with beef production to follow the government guidelines for product use and to use common sense, reasonable management skills, and accepted scientific knowledge to avoid product defects at the consumer level. The goal of the BQA program is to assure the consumer that all cattle shipped are healthy, wholesome, and safe, and that their management has met all government and industry standards. The BQA objectives for the cow/calf operation should include

- Setting production standards that can be met or exceeded
- Establishing systems for data retention and record keeping—record keeping systems, which meet government and industry guidelines, allow validation of management activities and fulfill the program goal
- Providing hands-on training and education for participants in order to meet or exceed the guidelines of the BQA program and to realize the benefits of such program
- Providing technical assistance through cattlemen's associations, veterinarians, and university staff; the veterinarian should serve as the facilitator of the BQA program and trainer in proper production management techniques that meet BQA standards.

QUALITY AND SAFETY CHALLENGES

The importance of BQA is obvious when analyzing the top nine quality challenges within the beef industry. These quality challenges include injection site blemishes, rib brands, excessive external fat, excessive seam fat, dark cutters, inconsistent size of meat cuts, inconsistent cuts, nonuniform cattle, and *Escherichia coli* O157:H7.

According to the National Beef Quality Audit, injection site blemishes cost the beef industry $188 million annually and cost producers approximately $7.05 per animal. Brands and other hide defects such as parasite damage cost the beef industry more than $648 million annually. Typically, this loss is passed along to all cattle that are sold in the industry by a reduction in live cattle price, or "built-in discounts." This is equivalent to $24.30 per head.

Of the U.S. Department of Agriculture (USDA)–Food Safety Inspection Service (FSIS) three targeted safety areas, the USDA-FSIS reports beef to be below their chemical (residue) and physical (broken needles) defect targets. For the biological safety target, *E. coli* O157:H7 presents a very difficult problem, and at present, effective on-farm control techniques have not been found. As soon as even a partially effective technique has been developed to control potential bacterial hazards, the technique will likely be incorporated into the BQA guidelines.

The BQA challenge to improve beef quality includes eliminating side and multiple brands, removing horns, improving parasite control, improving red meat yield, improving handling/transport techniques, eliminating intramuscular (IM) injections, measuring traits that affect value, and eliminating genetic and management systems that negatively affect tenderness, juiciness, and flavor.

History of BQA in the United States

Consumers have always wanted safe food. Because of concerns about additional governmental regulation and potential loss of modern production tools, in 1980 cattle producers began investigating ways to ensure that their production practices were safe and would satisfy consumer concerns.

In 1982, the USDA-FSIS began working with the beef industry in the United States to develop the Preharvest Beef Safety Production Program. Not wanting any additional governmental regulatory programs, the beef industry adopted the term "Beef Quality Assurance." BQA provided cattle producers with an important key for avoiding additional governmental regulation. At present, 47 states and more than 90% of the U.S. beef production industry are involved with the voluntary program. Industry self-regulation has proven very successful and will continue to allow the industry the flexibility needed to produce safe and wholesome food in an economical manner. Success of the effort is clear; violative chemical residues have almost disappeared in fed beef cattle, and injection site lesions have been reduced by almost 90%.

The Pillsbury Company developed and implemented a quality control program for the space industry. Their program, Hazard Analysis Critical Control Point (HACCP), gained universal acceptance and is presently the dominant outline for quality assurance programs in processed foods. Per USDA regulation, all U.S. packing plants have developed or will be developing HACCP programs.

The concepts of HACCP are the same as those of BQA. It is a process of determining what could go wrong, planning to avoid it, and documenting what has been done, with the additional step of validation. USDA's HACCP program for the packing industry includes foodborne bacterial pathogen control, a problem that currently is not controllable at the farm, ranch, or feedlot level. When preharvest bacterial pathogen control measures become available or government regulations mandate control measures, the current valuable experience working with the present BQA programs will help producers adapt to any new requirements.

IMPORTANT POINTS TO REMEMBER FOR THE SUCCESS OF BQA

First, the veterinarian, producer, and employees cannot foresee all potential problems. Participants in the BQA process must identify one area to focus on initially, and then develop and implement a plan for ensuring quality and safety in that area of production. The experience gained will make it easier to develop quality assurance in other areas of the operation.

Second, violative residues and product-related defects can be avoided if animal health products are administered according to government and industry standards and if BQA record-keeping standards are maintained.

Third, there are a number of safeguards built into the beef feeding system that help avoid quality defects. These include handling of animals on an individual basis, length of time required to produce a finished product, and quality and safety built into modern health-related technologies used in beef production.

Fourth, every producer and employee must be trained to know, understand, and identify areas in which possible contamination with violative residues may occur or where quality or hazardous physical defects may occur. Anyone who supplies services, commodities, or products to a producer must understand the producer's quality and safety assurance objectives.

Fifth, the producer must be able to document all the steps of production. Good production records allow for documentation, analysis, and improved financial decisions.

Sixth, there are points in production that must be monitored to ensure that no residue violations or carcass defects occur. The critical points include, but are not limited to, cattle treated with any product, incoming products and commodities, cattle handling, and evaluation of outgoing cattle, particularly cull cows and bulls.

Seventh, there are production areas that have higher residue and carcass defect risks than do others. High-risk production areas include but are not limited to nonperforming cattle, cull cows and bulls, unusual single-source feed ingredients, and suppliers of nonstandard supplies.

BQA GUIDELINES AND AGREEMENTS

FEEDSTUFFS

To comply with BQA guidelines for feedstuffs, producers should

- Maintain records of any pesticide/herbicide use on pasture or crops that could potentially lead to violative residues in grazing cattle or feedlot cattle
- Ensure that adequate quality control programs are in place for incoming feedstuffs—programs should be designed to eliminate contamination from molds, mycotoxins, or chemicals in incoming feed ingredients, and supplier assurance of feed ingredient quality is recommended
- Analyze suspect feedstuffs before use
- Never feed prohibited protein sources, per Food and Drug Administration (FDA) regulations
- Support the feeding of by-products ingredients with sound science.

FEED ADDITIVES AND MEDICATIONS

Only FDA-approved medicated feed additives will be used in rations. To comply with BQA guidelines for feed additives and medications, producers should

- Use medicated feed additives in accordance with the FDA Good Manufacturing Practices (GMP) regulations
- Follow judicious antibiotic use guidelines
- Remember that extra-label use of feed additives is illegal and strictly prohibited, and never engage in such use
- Adhere strictly to withdrawal times to avoid violative residues
- Where applicable, keep complete records when formulating or feeding medicated feed rations
- Keep records a minimum of 2 years
- Ensure that all additives are withdrawn at the proper time to avoid violative residues.

PROCESSING/TREATMENT AND RECORDS

All FDA/USDA/Environmental Protection Agency guidelines for products must be followed. To comply with BQA guidelines for treatment and records, producers should

- Use all products per label directions
- Keep extra-label drug use (ELDU) to a minimum, and use only when prescribed by a veterinarian working under a valid veterinary client/patient relationship
- Adhere strictly to extended withdrawal periods (as determined by the veterinarian within the context of a valid veterinary client/patient relationship)
- Maintain treatment records with the following recorded:
  1. Individual animal or group identification
  2. Date treated
  3. Product administrated and manufacture's lot/serial number
  4. Dosage used
  5. Route and location of administration
  6. Earliest date animal will have cleared withdrawal period
- When cattle are processed as a group, identify all cattle within the group as such, and record the following information:
  1. Group or lot identification
  2. Date treated
  3. Product administered and manufacturer's lot/serial number
  4. Dosage used
  5. Route and location of administration
  6. Earliest date animal will have cleared withdrawal period
- Check all cattle shipped by appropriate personnel to ensure that animals that have been treated meet or exceed label or prescription withdrawal times for all animal health products administered
- Transfer all processing and treatment records with the cattle to the next production level. Prospective buyers must be informed of any cattle that have not met withdrawal times.

INJECTABLE ANIMAL HEALTH PRODUCTS

To comply with BQA guidelines on injectable animal health products, producers and veterinarians should:

- Administer products labeled for subcutaneous (SQ) injection in the neck region
- Give all products labeled for IM use in the neck region only (no exceptions, regardless of age)
- Avoid all IM administration, if possible, because all products cause tissue damage when injected IM
- Choose products cleared for SQ, intravenous, or oral administration whenever possible

- Give products with low dosage rates (as recommended), and follow guidelines for proper spacing
- Administer no more than 10 cc of product per IM injection site.

Care and Husbandry Practices

To comply with BQA guidelines for care and husbandry, producers or veterinarians should follow the National Cattlemen's Beef Association "Quality Assurance Herd Health Plan," which conforms to good veterinary and husbandry practices. In addition, they should:

- Handle and transport all cattle in such a fashion so as to minimize stress, injury, or bruising
- Inspect facilities (fences, corrals, load-outs, and so forth) regularly to ensure proper care and ease of handling
- Strive to keep feed and water handling equipment clean
- Provide appropriate nutritional and feedstuffs management
- Strive to maintain an environment appropriate to the production setting
- Evaluate biosecurity measures
- Keep records for a minimum of 2 years (3 for restricted use pesticides).

BQA GOOD MANAGEMENT PRACTICES

Good management practices include feed-handling facilities designed to reduce the risk of feed contamination from chemicals, foreign materials, and disease-causing infectious agents. All chemicals (pesticides, lubricants, solvents, medications, and so forth) should be stored away from feed supplies. Feed-handling equipment should be routinely checked for fluid leaks. Producers should avoid storing feedstuffs around electrical transformers. Dual-purpose equipment, such as loaders (including shovels), which may handle feed and other materials (such as manure or dead animals), should be thoroughly cleaned before handling feed. No vehicles other than feed-handling equipment should be allowed into feed storage areas such as silage pits. The most common source of infectious agent contamination is animal or human feces. It is important to protect feedstuffs, feed troughs, and water supplies from contamination with chemicals, foreign material, and feces. Control of rodents, birds, and other wildlife is important to avoid fecal contamination from many common infectious agents.

FEEDSTUFFS AND COMMODITY SOURCES GMP

It is essential to monitor feed sources. Producers purchasing outside feeds should maintain a sampling program to test for quality specifications of feedstuffs. This could include moisture, protein, foreign material, and so forth. Suppliers should be informed that sampling of delivered products will occur. A good business practice is to require all products to be accompanied by an invoice that includes the date, quantity, and signatures of both the person who delivered the product and the person who received the product. Grain suppliers should understand that protectants could have withdrawal times. Most good suppliers have a quality control testing program of their own. Bonded suppliers often test for polychlorinated biphenyls, chlorinated hydrocarbons, organophosphates, pesticides and herbicides, heavy metals, and microbes (*Salmonella*). As part of the BQA program, the producer should ask for the test results.

A quality control program for feedstuffs aids in preventing chemical residues and ensures high quality feeds. Visual inspection of feeds can be effective in avoiding some problems. Create a checklist that includes such items as color (typical, bright, and uniform), odor (clean and characteristic), moisture (free flowing, no wet spots, moisture tested), and temperature (no evidence of heating). The checklist should also note whether there is evidence of foreign material or bird, rodent, or insect contamination.

It is neither efficient nor economically feasible to test every load of grain or forage for contaminants. However, a logical alternative is to obtain and store a representative sample of each batch of newly purchased feed. Commonly, a thorough investigation of suspected feed-related problems is not possible because no representative sample is available for testing. If feed sampling and storage is done on a routine basis and a suspected feed-related problem occurs, samples for appropriate laboratory testing will be available. A recommended sampling method for purchased grains, supplements, or complete feeds is to randomly sample each batch of feed in from five to ten locations and pool the individual samples into a larger sample of 2–5 lb. The pooled sample should be placed in a paper bag or small cardboard box, labeled, and frozen. (Dry samples can be labeled and kept in a dry area. Higher-moisture samples should be frozen.) A feed tag should be attached to the sample for future reference if needed.

Forage samples should also be collected and stored. If multiple bales of hay are purchased, representative samples should be obtained from several bales and mixed together before storage. Core samples are preferred over "grab" samples, particularly from large bales of hay. Most hay samples can be placed in a labeled paper bag and kept in a clean, dry area.

High-Risk Feeds

High-risk feeds are single loads or batches that will be fed to cattle over a prolonged period of time, thereby exposing large numbers of animals. Examples of high-risk feeds include fats, rendered byproducts, plant byproducts, supplements, and additives. Typically, these feedstuffs are only a small percentage of the total diet and are very expensive to test. Suppliers should understand the producer's BQA concerns and should provide quality specifications with the product. It is best to do business with a bonded supplier, so producers should find dependable suppliers and stay with them.

Ruminant Byproducts

Because of bovine spongiform encephalopathy concerns, certain ruminant-derived protein sources such as meat and bone meal cannot be fed per FDA regulations. Some other ruminant-derived products, such as tallow, are acceptable under the BQA program. However, federal and state regulations concerning the feeding of ruminant-derived feed to cattle are subject to change. Producers must stay abreast of current regulations to remain in compliance. Pure porcine and equine meat and bone meals can be fed to cattle.

Potential Feed Toxins

It is important that producers and employees have some knowledge about the relative toxicities of chemicals to livestock so that highly toxic chemicals, such as soil insecticides, can be handled and stored properly. All chemicals should be treated as potential hazards and should be stored away from feed storage and mixing areas. If a feed-related poisoning is suspected, it is critical for the veterinarian to contact a diagnostic laboratory for assistance in confirming the suspicion. Some poisoning incidents may be reportable to the appropriate federal or state/provincial government agencies.

Naturally occurring mycotoxins also pose a threat to quality beef production. Mycotoxins can be found in grains and forages and, if present in sufficient concentrations, can cause reduced feed consumption, poor production, and adverse health effects. Mycotoxins can be produced in feedstuffs before harvesting or during storage. More commonly found mycotoxins include aflatoxin, vomitoxin, zearalenone, and fumonisins in grain, primarily corn and slaframine in red clover. Ergot alkaloids can be found in either grain or grass hays.

Fats

Steps should be taken to ensure that purchased fats and oils do not contain a residue. Discuss the quality of products with suppliers and request information concerning the sources, quality, stability, efficacy, and consistency of the product. Producers may be approached by brokers who offer a cheaper source of feed-grade fats; however, the potential for contamination increases with cheaper sources of fats.

A reputable dealer should be testing fats for such contaminants as polychlorinated biphenyls, chlorinated hydrocarbons, pesticides, heavy metals, salmonella, and tall oil (hydrocarbon). Verification of testing should accompany the product.

ANIMAL TREATMENTS AND HEALTH MAINTENANCE GMP

Management

As discussed earlier, producers and/or veterinarians should follow the National Cattlemen's Beef Association Quality Assurance Herd Health Plan, which conforms to good veterinary and husbandry practices. Below are guidelines for various segments of beef production.

For All Cattle and Production Segments

Quality Assurance Herd Health Plan minimum guidelines require that producers and/or veterinarians should

- Provide appropriate nutritional feedstuffs
- Handle cattle to minimize stress and bruising
- Administer all injections in front of the shoulder
- Individually identify any animals treated to ensure proper withdrawal time
- Make records available to the next production sector
- Always read and follow label directions

- Keep records of all products administered, including product used, serial number, amount administered, route of administration, and withdrawal time.

For All Heifers and Purchased Breeding Stock Entering the Herd

Quality Assurance Herd Health Plan minimum guidelines require that producers and/or veterinarians should

- Vaccinate in front of the shoulder for viral and clostridial diseases, giving two vaccinations, 2–3 weeks apart
- Control external and internal parasites

For the Entire Cow Herd

Quality Assurance Herd Health Plan minimum guidelines require that producers should

- Control external and internal parasites
- Administer annual booster vaccinations in front of the shoulder
- Consult with the veterinarian for additional health procedures appropriate to the producer's area

For Cattle At Preweaning, Weaning, or Backgrounding

Quality Assurance Herd Health Plan minimum guidelines require that producers and/or veterinarians should

- If implanting, administer implants properly in a sanitary manner
- Vaccinate in front of the shoulder for viral and clostridial diseases, administering two vaccinations, 2–3 weeks apart
- Perform all surgeries such as dehorning and castration in a humane manner
- Control external and internal parasites
- Consult with the veterinarian for additional health procedures appropriate to the producer's area
- Keep records of all products administered including product used, serial number, amount administered, route of administration, and withdrawal time
- Wean cattle (45 days recommended) to ensure cattle health and producer return on health management investment.

The beef industry, largely through the efforts of feedlots, has done an excellent job of controlling violative drug residues. This has been accomplished by placing emphasis on the identification of treated cattle, along with good record-keeping practices, which includes identifying each animal treated, accurately recording the products used, noting treatment date and treatment dosage, and following prescribed withdrawal times. The cow/calf industry is responsible for providing residue-free animals to the feeding industry. Of particular note are drugs with long withdrawal times such as gentamicin. Gentamicin is under a voluntary ban in use instituted by the Academy of Veterinary Consultants and the American Association of Bovine Practitioners.

Treatment Protocol Book

As part of overall animal treatment and herd health management, the veterinarian should provide a "treatment protocol book" specific to the producer. The treatment protocol book should be reviewed regularly and updated at least every 90 days, or more often if appropriate. One copy of the treatment protocol book should be kept at the operation's headquarters, and a second should be readily available at the working facilities. A written treatment protocol, along with current prescriptions, is an important document that the operation must have to outline drug usage procedures and residue avoidance plans. The treatment protocol book and prescriptions should meet the Animal Medicinal Drug Use Clarification Act specifications. Of greater significance, the treatment book provides written guidelines for animal health programs, thus minimizing chances of mistakes or misunderstandings.

Injections

A critical part of a BQA program is the proper administration of animal health products. All injections should be given in front of the shoulders—never in the rump or back leg. When possible, IM injections should be avoided altogether. However, some animal health products are labeled for IM use only. If IM medications must be used, always administer them in the neck and never exceed 10 cc per IM injection site. As a rule, personnel should not inject more than 10 mL subcutaneously in each injection site; however, some products are labeled for use up to 20 mL per injection site and do not violate BQA guidelines.

Bent and Broken Needles

The veterinarian should train all producers and employees on good cattle handling, cattle restraint, and

Table 13.1. Guidelines for needle selection for cattle.

	Route of Administration and Cattle Weight								
	Subcutaneously 1/2–3/4-in needle			Intravenously 1 1/2-in needle			Intramuscularly 1–1 1/2-in needle		
Injectable Viscosity	<300	300–700	>700	<300	300–700	>700	<300	300–700	>700
Thin, example: virus, vaccine	18 gauge	18–16 gauge	16 gauge	18–16 gauge	16 gauge	16–14 gauge	20–18 gauge	18–16 gauge	18–16 gauge
Thick, example: oxytetra-cycline	18–16 gauge	18–16 gauge	16 gauge	16 gauge	16–14 gauge	16–14 gauge	18 gauge	16 gauge	16 gauge

Select the needle to fit the cattle size (the smallest practical size without bending).

proper injection technique. Improper animal restraint is the cause of most bent needle problems. If a needle bends, it should never be straightened and reused. Although it happens only very rarely, a needle can break off in the muscle. A broken needle is an emergency, and time will be of the essence. Broken needles migrate in tissue and, if not immediately handled, will be impossible to find, requiring the animal to be destroyed. Under no circumstances can animals with broken needles be sent to market. The veterinarian should outline procedures for handling such cases in the treatment protocol book. Purchasing high-quality needles, changing and discarding damaged needles, and providing proper restraint are all preventative measures. Guidelines for needle selection are found in Table 13.1. Figures 13.1 and 13.2 show proper injection site techniques. Always remember that cattle are never too young or old for producers or veterinarians to cause a quality defect.

SQ Injections in Dewlap Region

As beef producers adopt BQA practices, one of the most obvious results is to administer all injections to the neck area in front of the shoulder. IM injections must be given in the triangle region of the animal's neck only. SQ injections must remain ahead of the point of the shoulder.

It is common at branding time for calves to receive several preventative injections. However, as the calves are treated and worked, the restrainer's leg is usually covering much of the approved injection area on a calf. Rather than trying to reposition the holder's leg, or risking human injection, producers can give an SQ injection in the dewlap (see Figure 13.3).

BQA Dewlap Techniques

The dewlap is the flap of skin found along the jaw and neck of the calf that follows the neck down to the brisket region (fold of skin between front legs). An SQ injection in the dewlap region is an approved Nebraska Cattleman–BQA practice, so long as the injection site remains ahead of the point of the shoulder (see Figure 13.3).

The producer or veterinarian should

- Restrain the calf on its side while pulling the front leg back and locate the dewlap, which is ahead of the point of the shoulder
- Grab the skin, using the tenting technique, and conduct the SQ injection
- Also remember to use the correct needle size for the job. An 18 × 5/8-in needle works excellently to administer most products, but a 16 × 5/8-in needle may also suffice.

RESIDUE AVOIDANCE

Avoiding violative residues is dependent on using FDA-approved medications; following label or prescribed directions; establishing ELDU parameters for withdrawal times appropriate for the dose, medication, and route of administration; not exceeding dose per injection site recommendations; and screening cattle that may not have cleared the antibiotics normally.

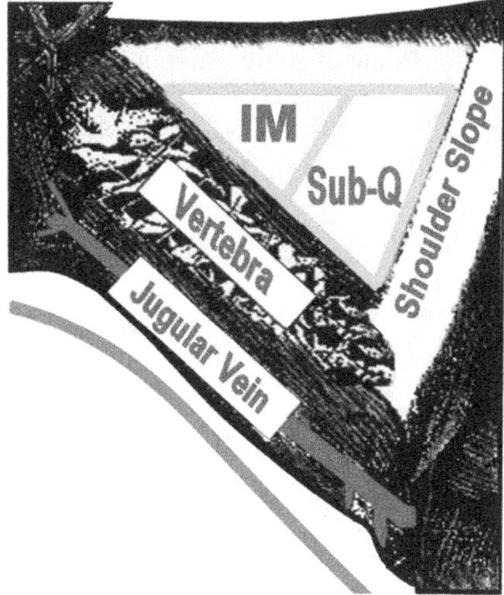

Figure 13.1. Injection site diagram.

- Needles should have metal hubs
- Change needles frequently (10–15 head)
- Change needle if contaminated or damaged. Never straighten a needle—the second time it bends there is a chance it will break
- 16-gauge, $^1/_2$–$^3/_4$-in needles work well for subcutaneous injection
- 16-gauge, 1–1$^1/_2$-in needles work well for intramuscular.

Never inject anything behind the slope of the shoulder. Never exceed 10 cc per intramuscular injection site.

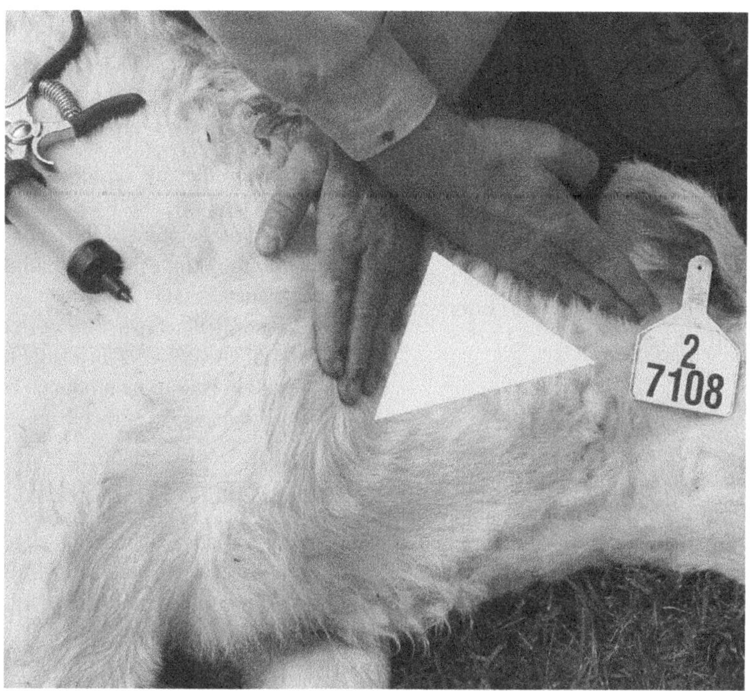

Figure 13.2. Neck injection site for IM and SQ injections.

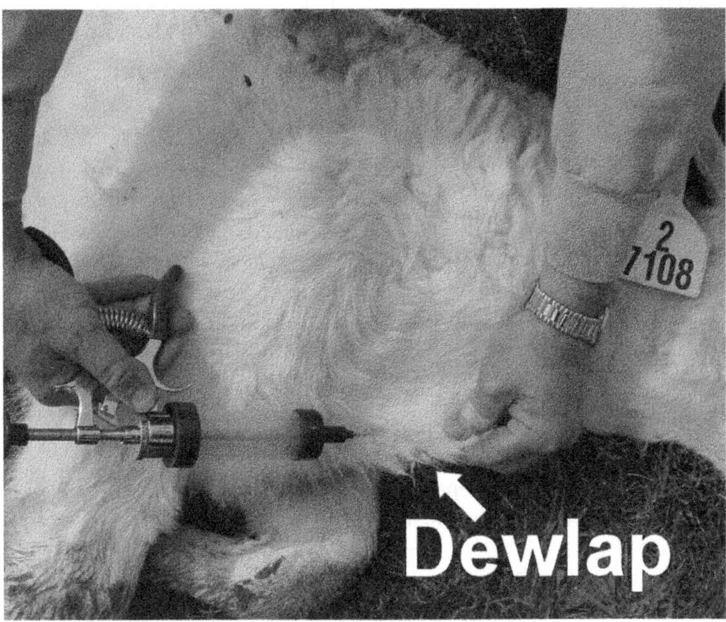

Figure 13.3. Dewlap injection technique.

The Need to Screen Selected Cattle Before Marketing

Nonperforming animals, and cull cows and bulls that have been treated with an antibiotic, should be considered to be at "high risk for antibiotic residue violation" because of the potential for their poor performance associated with organ (liver or kidney) dysfunction. Liver and kidney function is vital to clearing antibiotics. Also, ELDU requires a veterinarian by law to adjust the withdrawal time from the label indications to a time more appropriate to the dose used. Such occurrences of ELDU include using SQ injection when labels indicate IV or IM, and increasing the dose above the label dose. In addition, doses above 10 cc per site may depot and not be eliminated as rapidly as required to meet assigned withdrawal time. Perivascular injections are a frequent cause of suspected residue violations.

Antibiotic Residue Avoidance Strategy

To avoid antibiotic residue violations, producers and/or veterinarians should:

- Identify all animals treated
- Record all treatments: date, identification, dose, route, who administered treatment, and withdrawal time
- Strictly follow antibiotic use guidelines
- Use newer technology antibiotics when possible, which will reduce unwanted depot effects; select low-volume products when available; "don't step over a dollar trying to pick up a dime" (i.e., don't try to save by using old products—they may cost you more in the long run as a result of injection-site lesions)
- Select antibiotics with short withdrawal when antibiotic choice is equivalent
- Avoid ELDU of antibiotics; use label dose and route of administration
- Avoid using multiple antibiotics at the same time
- Avoid mixing antibiotics in the same syringe, especially if given IM or SQ
- Check ALL medication/treatment records before marketing:
    1. Cattle should not be marketed that have fewer than 60 withdrawal days without examining the treatment history
    2. Cattle should not be marketed that have relapsed unless the producer and veterinarian have examined the treatment history
    3. Cattle should not be marketed that have suspected liver or kidney damage unless the producer and veterinarian have examined the treatment history

4. Cattle should not be marketed with antibiotic injection site knots unless the producer and veterinarian have examined the treatment history

- Screen the urine for antibiotics of all cattle identified in the steps immediately above; at present, the PHAST (Preharvest Antibiotic Screening Test) is the only screening test that mirrors the antibiotic residue test used by the USDA-FSIS.

Introduction to the PHAST

The PHAST is a microbial inhibition test (an indirect assay) that measures substances in the urine that inhibit the growth of the test microbe, *Bacillus megaterium*. *Bacillus megaterium* is classified by the USDA among bacteria that are generally accepted as safe. The organism will not cause disease in humans or domestic animals.

Because *B. megaterium* is frequently more sensitive to target antibiotics than the FDA-established tolerance for the target antibiotic, and because there are a number of microbial inhibition substances that are not antibiotics, false positives are the most common problem with these types of tests. False negatives are thought to be rare but are dependent on the sensitivity of the test organism to the antibiotic relative to the FDA tolerance to the target antibiotic. *Bacillus megaterium* is very sensitive to penicillin-type antibiotics and intermediately sensitive to aminoglycosides and sulfa drugs.

The reliability of the PHAST for detecting violative residues in cattle has not been investigated. The most common false positives should be associated with antibiotics such as oxytetracycline and Naxcel that have an established FDA tolerance level above the level detectable in the kidney or cleared in the urine. Specific *B. megaterium* sensitivities to commonly used antimicrobials and their relationships to FDA tolerances are listed in the USDA-FSIS Bioassay Residue Screening Test Evaluation table located at the end of this chapter. Please see the Resources section for information on ordering PHAST supplies.

The PHAST is not a reliable test to evaluate the residue status of an animal that has not met the withdrawal time specified on an antibiotic label. Never use the PHAST to evaluate the residue status of such an animal.

The PHAST is useful in evaluating cattle that have undergone prolonged treatment, been treated with multiple antibiotics, failed to perform normally following antibiotic therapy, or have suspected organ (kidney or liver) damage that might interfere with excretion and elimination of an antibiotic.

PHAST Procedure

Collecting Urine

Collecting urine from male cattle requires a screw-top tube (plastic tubes are safest). While an animal is restrained in a chute, a person should grasp the hairs on one side of the prepuce and slide the opened screw top test tube into the prepuce. The rough rings of the test tube sliding into the prepuce will stimulate most male cattle to urinate. Sliding the test tube back and forth inside the prepuce may stimulate urination in cattle that did not immediately urinate when the tube was first slid in the prepuce. If the animal does not void at least a few milliliters of urine, at least one-half of the tube can be left in the prepuce by pulling the hair from the tip of the sheath around the test tube and securing it with a rubber band. The animal can then be turned out to urinate. It should not take long before he can be brought back through the chute and the tube filled with urine collected.

Streaking the Plate

A cotton swab, soaked in PHAST (*B. megaterium*) spores is used to completely streak a Muller-Hinton agar plate. The veterinarian or other tester should:

1. Soak a cotton swab in PHAST spores (*B. megaterium*). Although PHAST spores will not cause disease in humans or animals, handle the spores and resultant test cultures with care.
2. Remove the lid from Muller-Hinton agar. Beginning at one side, completely streak the entire surface of the agar plate. Rotate the plate 45 degrees and repeat. The objective is to cover the plate with a carpet of *B. megaterium*. Use light, gentle strokes. Excessive pressure can break the agar, which interferes with bacterial growth and test results.
3. Discard the swab in a biohazard container. Do not reuse a swab.

Positioning the N5 Control Disc

The N5 disc contains 5 μg of neomycin, an antibiotic that stops *B. megaterium* spores from germinating and growing. The disc serves as the control to show that the test is working normally. After the plate is streaked and incubated, the bacteria should grow on the gel but not around the N5 disc. The clear zone around the control disc should be at least 5/8 in (16 mm) for the zones around the PHAST screening swabs to be considered valid. The dispenser containing the N5 discs should be stored below 5°C (42°F). The veterinarian or other tester should:

1. Open the Muller-Hinton agar plate that was previously streaked and dispense one N5 control disc on to the inside surface of the agar plate lid. **Do not touch** the disc with the fingers because the antibiotic could contaminate the urine-soaked swab(s) that will be subsequently handled and cause the test to give false-positive results. If the disc is accidentally touched, the tester should wash and rinse his or her hands well before continuing the PHAST.
2. Pick up the N5 disc from the inside surface of the agar plate lid with clean tweezers or forceps and gently place it on the surface of the Muller-Hinton agar that was previously streaked with *B. megaterium* spores. Gently press the disc in place with the forceps. Do not break the surface of the agar. Do not reposition the disc: Growth will be inhibited where the disc touches the agar. If this occurs, begin again with a new Muller-Hinton agar plate.
3. Replace the lid on the agar plate and return the N5 dispenser to the refrigerator.

Testing the Urine

Now that the test plate is prepared, you are ready to test the urine. The veterinarian or other tester should:

1. Remove a fresh sterile cotton swab from its wrapper and dip the swab into the tube of urine.
2. Gently shake off excess urine.
3. Break the shaft of the urine-soaked swab as close to the cotton tip as possible without touching the cotton tip, and discard the shaft.
4. Remove the lid of the PHAST plate. Hold the remaining shaft of the urine-soaked cotton swab with your fingertips, and carefully place it on the surface of the Muller-Hinton agar plate that has had a carpet of *B. megaterium* spores streaked on its surface not closer than 1 in (25 mm) from the N5 disc.
5. Gently press the cotton swab shaft with the fingertip to firmly seat the swab tip on the agar. **Caution**: Discard the plate and prepare another one if the swab should roll across the gel or if the gel cracks. Both events could invalidate the test, giving incorrect results.
6. Replace the cover on the plate.
7. To test a second urine sample, take out another swab and repeat steps 1 through 5, placing the second swab tip on the other side of the N5 disc in a rabbit-ear configuration.
8. Replace the cover, and place the PHAST plate, agar-side down, into an incubator that has been preheated to 37°C (100°F). Note that unlike typical bacterial cultures, the PHAST plate must be incubated agar-side down to prevent the urine-soaked swabs from falling off the agar surface.
9. Incubate the PHAST plate for 12–24 hours at 37°C (100°F).

Interpreting the Results

The warmth of the incubator allows the *B. megaterium* spores to germinate and grow. The bacterial growth will make the gel appear opaque and cream colored or grayish instead of clear, but the presence of an antibiotic in the urine-soaked swab interferes with bacterial growth, creating a transparent area around the swab. Using Table 13.2 and Figure 13.4 the veterinarian or other tester should:

1. Remove the PHAST plate from the incubator, remove the agar plate lid, and examine the agar. Where bacteria have grown, the agar will have turned creamy or grayish and opaque.
2. Check the N5 disc. Examine the areas around the cotton tips and N5 disc, using Figure 13.4 as a guide. The area around the N5 disc should be transparent. This clear area is called a "zone of inhibition" because bacterial growth has been stopped or inhibited by the antibiotic neomycin in the N5 control disc.
3. Determine whether the test is positive or negative. A clear zone of inhibition around the cotton swab tip indicates a positive test (Figure 13.4). This means a microbial inhibitor such as an antibiotic is present in the animal's urine. Animals should not be marketed until repeated PHAST procedures produce negative results. Bacterial growth completely surrounding and touching the cotton swab tip indicates a negative test. This indicates that no antibiotic residues are probably present in the animal's urine (Figure 13.4).
4. Measure the zone around the N5 control disc. A clear zone less than 5/8 in (16 mm) in diameter indicates a problem with the test. The results are unreliable, and PHAST must be repeated. The animal must not be marketed until a negative test has been obtained, with a zone of inhibition around the N5 disc that is at least 5/8 in (16mm) in diameter.

Table 13.2. Interpreting test results.

	Antibiotic Negative	Antibiotic Positive	Test Inconclusive	Antibiotic Positive	Antibiotic Negative	Antibiotic Positive
If you have . . .	Opaque bacterial growth right up to the swab tips	Clear zones around swab tips	Opaque bacterial growth right up to swab tips	Clear zones around swab tips	Opaque bacterial growth right up to the swab tips	Clear zones around swab tips
And you have . . .	A clear zone around the N5 disc 5/8–15/16 in (16–24 mm)	Clear zone around N5 disc between 5/8 and 15/16 in (16–24 mm)	Clear zone around N5 disc that is less than 5/8 in (16 mm)	Clear zone around N5 disc is less than 5/8 in (16 mm)	A clear zone around the N5 disc greater than 15/16 in (24 mm)	Clear zone around N5 disc greater than 15/16 in (24 mm)
Then take this action	Animal is ready for market	Animal has antibiotic residues; retest	Rerun test	Animal has antibiotic residues; retest	Animal is ready to market	Animal has antibiotic residues; retest

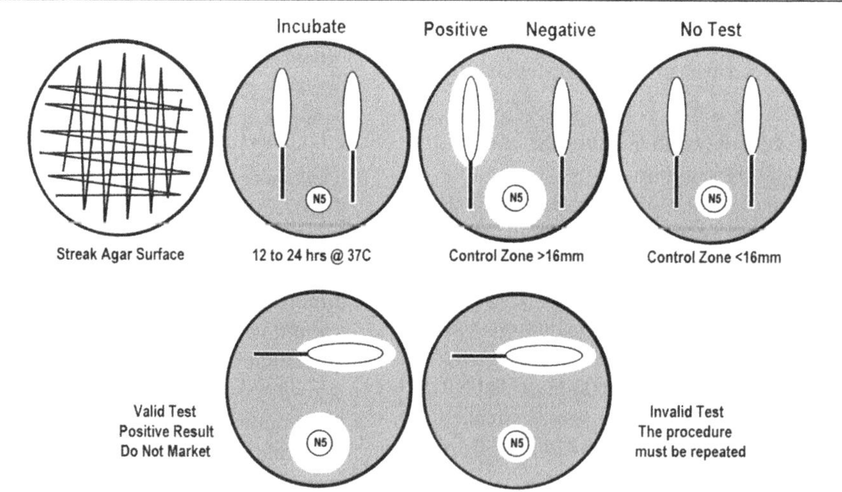

Figure 13.4. Preharvest Antibiotic Screening Test agar preparation and interpretation.

PUTTING BQA TO WORK THROUGH THE HACCP SYSTEM

"Build on what you know" is the operative phrase in any BQA program. The BQA program road map is the HACCP system. Veterinarians, nutritionists, producers, employees, and other specialists must look for what could go wrong, and then determine ways to avoid the problem. Operations should institute practices that allow checking, verifying, and documenting what has been accomplished. Everyday working techniques should be designed to avoid problems, especially those problems that can cause a safety or quality defect. This includes evaluating safety problems that can affect employees.

HACCP: Five Preliminary Steps

Assembling the HACCP Team

Bring together your HACCP resources. Everyone involved should help identify areas in production in which quality problems and defects could occur. This process should include the veterinarian, the nutritionist, extension and university specialists, suppliers, and employees who can help develop the HACCP plan. All participants must be willing to make a commitment to brainstorm about aspects of the operation where problems might occur and how to avoid them. They must also be willing to review the final HACCP plan to make sure all the pieces fit and nothing has been overlooked. There should be no "most valuable players"–everyone's ideas must be heard.

Describing the Production System and How the Cattle Are Managed

Each operation will have some unique aspects of its HACCP system. There is a built-in margin of safety for withdrawal times in the beef industry. As most withdrawal times for animal health products are shorter than the feeding periods, producers must be aware of high-residue risk situations such as marketing cull or nonperforming cattle. Nonperforming cattle (i.e., chronics) might have organ damage, which would prevent the normal clearance of a product. They could have passed the withdrawal time and still have violative drug residues present.

Identifying Who Buys the Cattle and Communicating with Them

Many cattle producers send their cattle (including nonperforming realizers) directly to packers. There is no room for error in preslaughter withdrawal times or physical injury that causes bruising. Nonperforming or cull cattle, although not a large percentage of the cattle sold, present serious safety and quality concerns for the industry. A summary of the treatment (including the names of medications used) and vaccination should be transferred with all cattle sold.

Developing and Verifying a Process Flow Diagram

Outline all of the steps of production. The members of the HACCP team should evaluate each step for its potential to cause quality defects. The defects include bacterial contamination that can cause infectious disease in cattle or employees, chemical usage/contamination that can lead to violative residues, and physical damage such as injection site damage, bruising, or broken needles in animal tissue.

Meeting the Requirements for the BQA GMP and Standard Operating Procedures

Be sure to include sanitation standard operating procedures (finding ways to prevent or minimize fecal/oral contamination).

Seven Specific HACCP Steps

Identifying Potential Problems or Hazards: Bacterial, Chemical, and Physical

Conduct a production analysis to identify potential problems that could occur in the production process. The use of a production flow diagram/outline/list provides the best guide to ensure no area has been overlooked.

Identifying CCPs

CCPs must be identified in the production process where potential problems could occur and be prevented or controlled. The use of the production flow diagram/outline/list provides the best guide for identifying where potential problems might occur and points where training/management activities may reduce or eliminate the hazard.

Establishing Critical Limits for CCPs

Limits must be established for preventative measures associated with each critical control management point. Determine how to identify when a production activity is not being conducted properly. Some limits are easy to establish, some are not. For example, giving all injections in the neck is a CCP. Any injection not given in the neck is outside the critical limits (CLs). Providing clean water for cattle to drink might be a CCP, but within minutes of cleaning a water trough, recontamination of the water (outside CLs) can occur. The best that can be done is to follow a reasonable cleaning schedule. Proper handling of cattle (another CCP) is important, but establishing a well-defined CL is difficult.

Establishing CCP Monitoring Procedures

All CCPs must be monitored to ensure they stay within the established limits set by management. Supervision on a timely basis is the key. This step outlines assigning a person or persons in the cattle operation to regularly check to ensure the activity is being carried out in a manner that meets the cattle operation's (management's) objectives (CCP).

Simply stated, is anyone checking to make sure that activities are being carried out as they were intended to be, or at least as well as they can be under the circumstances? A scheduled checklist is useful. The CCP monitoring list will help keep employees from overlooking important items. A CCP monitoring list may have items scheduled for checks on a daily, weekly, monthly, or yearly basis. Examples of BQA and HACCP checklists are shown in Figures 13.5 through 13.8.

Establishing Corrective Action

Corrective action must be taken when monitoring determines that a critical control point is not within established CLs. Corrective action should include what will be done in the future to prevent the problem

Example: BQA Cow-Calf Feed Checklist

Beef Operation ____________ Date _________ Evaluator ____________

Pasture Maintenance and Raised Feeds
- Water source protected and checked yearly for contamination.
- Pastures protected from contamination.
- Training for handling pesticides.
- Pesticides stored in protected area away from feed or health products.
- FDA/USDA/EPA guidelines followed for all product use.
- All pesticide handling equipment checked before each use for delivery accuracy and contamination.
- Cattle or harvest withdrawal time established (if needed) before allowing cattle to graze.
- Proper disposal of used containers.

Purchased Feeds
- Evaluation, sampling, and sample storage protocol developed/used.
- Receiving/Inventory Log/Record: source (verified), date, description (name, invoice number).
- Training for evaluating received/purchased feeds.
- Feed storage inspected for contamination before receiving new loads of ingredients.
- Feed storage area only used to store feed ingredients (no pesticides, solvents, etc.).
- Procedures in place to protect feed-handling equipment contamination.
- All feed-handling equipment checked before each use for contamination.

Feed Additives
- Receiving Log Record: source (verified), date, description (including serial/lot number).
- Stored separately from other feedstuffs.
- Use Log Record: date, dose per ton, ID of animals.
- Physical Inventory Log (can be column in use log).
- Training for using feed additives.

Feed Formulas
- Record of all feed formulas.
- Medicated feed formulas checked by nutritionist or veterinarian for accurate dosing.
- Directions for use, including withdrawal time.
- Training for mixing and quality control sampling/testing for feed mixing.

Batch/Load/Feed Delivery
- Batch/Delivery Log/Load, (delivery matches feeding plan if needed).
- Minimum/Maximum and exception table or chart for ingredients and mixing.
- Training (see above).

Cattle Release
- Withdrawal checked on All feed records.

Figure 13.5. Beef Quality Assurance cow/calf feed checklist.

Example: BQA Cow-Calf Product Use Checklist

Beef Operation____________ Date__________ Evaluator____________

Cattle Handling Facilities

- Inspected for proper function for cattle and human safety before each use.
- Handling facilities and equipment properly designed, maintained, and used.

New Cattle Entering the Operation

- Receiving Log Record: source (verified), date, description.
- Appropriate health/import/transfer/movement records.
- Cattle handling training.
- Basic Quality Control:
  --Holding pens and handling alleys properly designed and maintained.
  --Clean feed and water as needed available to cattle on arrival.
  --Visual inspection of cattle on arrival.

Health Management, Mass Medication, and Pesticide Products (Receiving, Storage and Use)

- Receiving/Inventory Log/Record: source (verified), date, description (name, serial/lot number).
- Stored in protected area: refrigerated as needed, sunlight controlled, locked if required.
- Use (Health Management/Treatment) Records for all cattle: date, animal IDs, diagnosis/reason, product, dose, withdrawal, and release date.
- Cattle Product Use Maps: (includes product and serial/lot number).
- Minimum/maximum and exception table or chart for product use.
- Product handling and use training (including MSDS/product inserts/etc.).
- All injections should be given in the neck region, injectables given SQ if possible.
- Supplier agreements and veterinary drug orders (as appropriate).
- Signed use protocols (health maintenance, treatment, premise pesticides).
- FDA/USDA/EPA guidelines followed for all product use.
- Equipment for delivery properly designed, maintained, and used.
  --Cattle: chutes, snakes, holding pens, syringes, needles.
  --Feed and pesticides: scales, mixers, delivery system.
- Proper disposal of used containers.
- Withdrawal time established and estimated date for release, (for injectables, see above).
- Residue screening of non-performers (exceptions: reproduction and lameness if no Rx).
- Training for processing, health management, mass medication, and pesticide products.

Feed Management

- Withdrawal time established, release date estimated.
- Feed management, mixing, and delivery training.
- FDA/USDA/EPA guidelines followed for all product use.
- Training for feed management.

Cattle Release

- Withdrawal checked on all products used (health management and treatment) records.
- All withdrawal times met and PHAST of all non-performers (except no Rx: repro and lame).

Release/transfer form signed.

Figure 13.6. Beef Quality Assurance cow/calf product use checklist.

Brief Example of a HACCP (CCP) Master Plan Outline for Feed

The following table was reviewed with ______________ (owner/manager)

To identify those areas appropriate to ____________ (beef operation)

Min: Potential site of *minor* problem (s=safety, p=production, q=quality).
Maj: Potential site of *major* problem (s=safety, p=production, q=quality).
CCP: Problem will exist if not controlled at this point (s=safety, p=production, q=quality).

BQA Hazard	***CCP***	***Preventive Measures***	***Monitoring Procedures***	***Corrective Action***	***Records***	***Verification***
Pasture maintenance pest/herb use	Min	Follow label directions	Observe Record Inspect	Keep cattle off until WD met	Pest/Herb use Records	Evaluate Records
Raised feed	Min	Employee Training Approved products Withdrawal if require, from grazing treated pastures Disposal of product containers	Sample and test Visual inspection	Quarantine Store until cleaned EPA approved disposal	Production log and test sheets	Evaluate Records
Purchased feed	Min	Employee Training	Sample and test Storage, Visual inspection	Reject load Quarantine Store	Receiving log, test sheets and Invoices	Check records Invoices
Receiving Feedstuffs	CCP-1	Source verified Invoice date, description Employee Training Only approved products	Sample every load, test and/or store, Visual inspection Product and truck	Reject load Quarantine, Store until cleaned EPA approved disposal	Receiving log, test sheet, and Invoices	Check records Invoices(receiving log-test log-in) by feeding mgmt and nutritionist evaluation
Feed Facilities	Min	Regular examination	Visual Inspection	Clean and Inspect	Inspection Record	Records Mill and Main office
Feed Additives	CCP-2	Invoice date, description and numbers Employee Training approved products	Additives inventory Against Inventory balance	Notify - nutritionist Quarantine Withdrawal adjusted for group if need	Receiving log Invoices Use log	Check records and Invoices (receiving log-use log, withdrawal report before releasing
Feed Formulas	CCP-3	All formulas managed by Nutritionist	Checked by Nutritionist Estimated DOF against withdrawal, Batch checked	Withdrawal errors Max level chart	Formulation sheets, batch sheets Feeders log	Check records (for-batch-log) as used and before releasing
Batch/Load	Maj	Establish rout and Sequence, Balance min Establish Min/Max and exception chart, Employee Training	Batch check list Accumulation and total Batch/load sheets, Regular audits	Withdrawal errors Max level chart	Batch logs Truck log Feeders log	Balance logs
Feed delivery	Min	Employee training Establish rout Loads match call	Balance load total against feeders log	Assign delivery balance load against delivery	Load records: Group feed log	Check records (delivery-call) Re-checked before releasing
Cattle release	CCP-4	All withdrawal times met	Records of transferred cattle reviewed and balanced	PHAST - RELEASE	Release form signed	All forms examined before release

Figure 13.7. Brief example of a Hazard Analysis Critical Control Point master plan outline for feed.

Brief Example of a HACCP (CCP) Master Plan Outline for Product Use

The following table was reviewed with _______________ (owner/manager)

To identify those areas appropriate to _____________ (beef operation)

Min: Potential site of *minor* problem (s=safety, p=production, q=quality).
Maj: Potential site of *major* problem (s=safety, p=production, =quality).
CCP: Problem will exist if not controlled at this point (s=safety, p=production, q=quality).

BQA Hazard	*CCP*	*Preventive Measures*	*Monitoring Procedures*	*Corrective Action*	*Records*	*Verification*
Cow/Bull Health	CCP-5	Health/ Nutrition appropriate to operation	Vet Diagnosis of problem Palpation/BCS/ Others	Adjust as Vet and examine inject-SQ neck	Production Records	Records checked by operator and vet
Calf Health (Birth)	Min	Cow in optimum condition at calving, Environment appropriate to optimum calf Assist as needed Colostrum mgmt. Employee Training approved product	Calving management	Adjust as Vet and examine injections SQ neck	Calf health records	Check records To see all treatments and procedures recorded.
Calf Health Early Management	Maj	Individual ID Health appropriate to operation Withdrawal established Employee Training approved products	Vet Diagnosis of problem Date, Product, ID, withdrawal	Set withdrawal, SQ neck	Processing records	Check protocol against invoices of products and processing records
Pre-wean Health	Maj	Individual ID Health appropriate to operation Withdrawal established Employee Training approved products	Vet Diagnosis of problem Date, Product, ID, withdrawal	Set withdrawal SQ neck	Production group Receiving/ Pen/Yard sheet	Check protocol against invoices of products and processing records
Cattle Receiving	CCP-6	Assume Contaminated	Observe/record variation	Sort and examine	Receiving Record	Records foreman Office
Processing products	Min	Group ID Name/serial number's Withdrawal time Employee Training approved products	Set clear dates Inventory	Establish min sale date for group	Production group Receiving/ Pen/Yard sheet	Check records (receiving-feed-hospital-main) before releasing
Pesticides management	Min	Employee Training Pesticide use plan approved products	Inventory	Lock and separate	Use records: 1) Individual 2) Premise	Check records before releasing
Mass Med	Maj	Group ID Withdrawal time approved products	Projected sell date against withdrawal, Inventory	PHAST test non-performers	Production group Receiving/ Pen/Yard sheet	Check records (receiving-feed-hospital) before releasing
Sickness	CCP-7	ID, Date, Product Protocol and established withdrawal Employee Training approved products	Vet Protocol, Date, Product, ID, withdrawal, est. sell date	Monitor and set withdrawal SQ neck, LAST non-performers	Health records	Check protocol against health records
Transfer, Culling or shipment	Min CCP-8	Check Withdrawal	Check withdrawal Records available	Withdrawal check Last test non-performers	Release form signed	Check records before releasing

Figure 13.8. Brief Example of a Hazard Analysis Critical Control Point master plan outline for product use.

from happening again. Producers and other participants in the process need to decide "what do we do if" something goes wrong. There are two reasons it is important to establish corrective action before a problem occurs: the problem will be corrected more quickly when it occurs, and discussing the corrective action needed may help emphasize the seriousness of the problem to employees and other participants. Understanding the potential consequences of not following CCPs may change how prevention of the problem is viewed by employees. The HACCP team should help develop possible and appropriate corrective actions.

Establishing Verification Procedures

Testing and other measurements must be used to verify that the program is working properly. For example, fecal examination is necessary to assess the adequacy of the parasite control program.

Establishing Record-Keeping Procedures

Good records document that the management system is being monitored and is working correctly. Any record system will work as long as appropriate to meet the needs of the HACCP system developed for the operation.

The concept of BQA is as simple as considering what can go wrong in production that could cause a quality defect, and figuring out how to prevent it from going wrong. You cannot monitor what you do not measure, so it is very important to document and verify the steps you take to avoid having problems occur.

RESOURCES

PHAST Supplies

Bacillus megaterium spores (ATCC 9885) may be acquired from Med-Tox Diagnostics Lab, Burlington, NC 27215 (phone: 800-334-1116). They come diluted to 1×10^6 spores per milliliter. Also, the UNL-GPVEC will make *B. megaterium* spores available to veterinarians. PHAST (*B. megaterium*) spores may be ordered from UNL-GPVEC by calling 402-762-4500 or sending an e-mail request to Dee Griffin (dgriffin@gpvec.unl.edu).

The N5 (5 mcg neomycin) antibiotic sensitivity disc used for the control (item number 617189) and Muller-Hinton agar plates (item number 1008) can be obtained from Physicians Lab Supply, P.O. Box 853, Rochester, MI 48308 (phone: 800-445-6507), http://www.pls-medical.com. Please note that wrapping agar plates in plastic wrap (Handi, Glad, etc.) will dramatically extend the shelf life of the plate.

Screw-top plastic tubes, such as the Becton-Dickenson FALCOM 2027 tube, work well for collecting urine and are available from Fisher Scientific (phone: 800-766-7000), http://www.fishersci.com. Any sterile tube smaller than 10 mL with a rough top edge will work for stimulating male cattle to urinate.

Any sterile cotton tipped applicator will work as long as the stick behind the cotton tip breaks easily.

Other Web Sites of Interest

REFERENCES

Beef Quality Assurance: http://www.bqa.org

Food and Drug Administration, Center for Veterinary Medicine: http://www.fda.gov/cvm/default.html

Food Safety Inspection Service: http://www.fsis.usda.gov/index.htm

Temple Grandin Livestock Handling: http://www.grandin.com/

National Cattleman's Beef Association: http://www.beef.org/

14
Cow/Calf Welfare Considerations

Peter J. Chenoweth

INTRODUCTION

Animal welfare is a complex issue, and the many definitions of the term reflect a wide spectrum of philosophies and viewpoints. A pragmatic approach suggests that, for livestock at least, it embraces "proper housing, management, nutrition, disease prevention and treatment, responsible care, humane handling, and euthanasia" (American Veterinary Medical Association [AMVA] 1991). An "acceptable" state of welfare for cattle would probably include freedom from disease, hunger, or thirst; adequate protection from extreme weather conditions; and freedom from pain or injury (Morrow-Tesch 2001); this view is derived from the "new" five freedoms described by the Farm Animal Welfare Council in the U.K. (http://www.fawc.co.uk/freedoms.htm). This, in turn, implies that we need to understand both normal and abnormal behavior of cattle. Although all of these considerations are relevant for the cow/calf industry, it has generally been subjected to less criticism in relation to animal welfare concerns than have other segments of the livestock industries. This is possibly because it most closely approximates the social ethic of animal husbandry: Most cattle producers do not like the idea of inflicting unnecessary pain on their animals; most see themselves as stewards of land and animals, and as people who perpetuate a lifestyle that upholds commendable values (Rollin 1995).

Despite this, welfare concerns have been raised about a number of practices in the cow/calf sector of the cattle industry. Those problems identified are generally of long standing and reflect solutions to management issues that have been sanctified by time, custom, and culture (Rollin 1995), as well as by economic considerations. However, some are of more recent origin and are associated with newer technologies, such as cloning. Some criticism may reflect inaccurate perceptions and may be addressed via appropriate communication, whereas other criticism may represent more valid concerns. In either case, the raising of such concerns provides us with the opportunity to objectively scrutinize controversial practices and procedures and to ask the questions below.

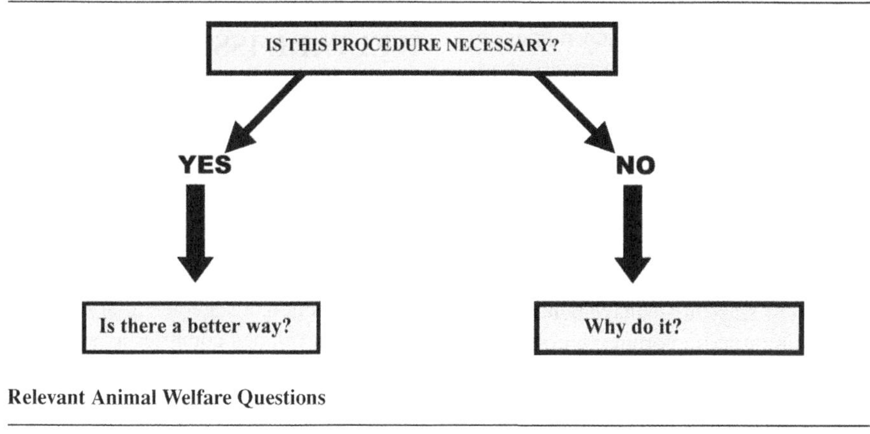

Relevant Animal Welfare Questions

STOCKMANSHIP

Human–animal interactions are a key feature of modern livestock production. There is compelling evidence that such interactions can markedly affect livestock productivity and welfare (Hemsworth and Coleman 1998). The identification of key characteristics that influence the behavior and interactions of human and animal partners allows manipulation of animal production systems for the benefit of both. For example, fear is a strong stressor, with differences in handling response being observed between cattle raised extensively and those raised more intensively (Grandin 1998).

The term "stockperson" has been used to describe a professional manager of animals, integral to both animal performance and welfare, with "stockmanship" being the skills, knowledge, and attitudes implied in such professionalism (Hemsworth and Coleman 1998). Although good stockmanship may well have an innate component, there is little doubt that screening of personnel before their potential employment with livestock, as well as providing appropriate education and training, can improve human–animal relationships to maximize mutual benefits. Here, the Farm Animal Welfare Council states (http://www.fawc.co.uk) that "stockmanship, plus the training and supervision necessary to achieve required standards, are key factors in the handling and care of livestock . . . without competent, diligent stockmanship, the welfare of animals cannot be adequately safeguarded."

In the cow/calf industry, there is a wealth of anecdotal evidence for the positive effects that accrue from good stockmanship (and, conversely, for the adverse effects of poor stockmanship), although few studies have quantified such effects. This research deficit is partly a result of the extensive nature of many cow/calf operations and of the relative lack of human–animal contact compared with that of more intensive industries. However, some generalities apply across species and systems. For example, short-term stressors imposed during either handling or transport can depress both immune and reproductive functions (Grandin 1998). In cattle, both transport and restraint have been shown to lower immune function, whereas use of electric prods has been associated with depressed reproduction in females (Grandin 1998). In addition, the appropriate training of individuals who perform stressful management procedures (such as castration and dehorning) is critical, so these procedures are conducted with the best equipment and care and at the most appropriate time in the animal's development (Morrow-Tesch 2001).

GENETICS

The temperament of an animal is a major factor in determining its response to handling. Temperament, in turn, is highly associated with fearfulness (Grandin 1998). A heritability estimate of 0.46 has been obtained for temperament in cattle (Grandin and Deesing, accessed December 18, 2004, at www.grandin.com), which supports anecdotal observations that there is a strong genetic influence on this trait. The observation that cattle of *Bos indicus* background are generally more excitable than are *Bos taurus* cattle is one example of probable genetic effects on temperament. (Temperament is also influenced by previous experiences and handling, as discussed in chapter 7.) Improvements can be made in animal temperament by selection, a fact long acknowledged by cattlemen. Selection for temperament can be an important tool, in concert with personnel training, animal handling, management, and facility design, in improving welfare on the ranch or farm.

HEALTH

In much of the world, livestock today are protected from many destructive diseases that hitherto caused severe animal pain and distress. This protection is achieved via movement restriction, vaccination, and biosecurity measures, as well as by informed management procedures. Here, the veterinary profession has made major contributions to animal health (and thus animal welfare) by reducing or eliminating the effects of diseases that otherwise would cause untold animal suffering. Although animal health, and its veterinary underpinnings, is an important aspect of welfare, it is an often neglected topic in animal welfare discussions.

SPECIFIC ANIMAL WELFARE ISSUES

Despite being relatively untargeted in terms of animal welfare issues to date, the cow/calf industry cannot assume that this situation will persist indefinitely. A number of animal advocate organizations have been showing an increasing interest in farm animals (Rollin 1995). This author also makes the point that U.S. society is extremely naive about the nature of animal production, a statement that is presumably true for other developed countries as well. The corollary of this statement is that many citizens of such countries are susceptible to suggestion, as well as misinformation, concerning aspects of animal agriculture.

Some specific welfare issues have been raised in relation to the cattle-breeding industry. These include branding, castration, spaying, weaning, dehorning, cattle handling and transport, electroejaculation (EE) and serving-capacity testing of bulls, adverse environmental effects, treatment of cull cows, and surgical preparation of teaser bulls. A number of these issues are addressed elsewhere in this book. Biotechnical procedures such as cloning and transgenics may also raise welfare concerns. Veterinarians and farm advisors should be aware of such concerns and be prepared to provide informed, objective responses when they hear them. There are valid reasons for many procedures in the cattle industry, and those that might appear to be inhumane to those outside the industry can often be justified on grounds that include sound welfare considerations. However, procedures that cause unnecessary pain or suffering should be scrutinized closely. Those that are unjustifiable should be discarded or should be conducted using more appropriate alternatives. In this chapter, we focus on several animal welfare issues of recent concern in the cow/calf industry.

Pain

Before pain can be managed, it must first be recognized and assessed (Underwood 2002, George 2003). The cerebral cortex, the thalamus, and the limbic system are involved with pain processing, so reactions to specific painful stimuli vary according to breed, species, temperament, and rearing (George 2003). Increasing emphasis is being placed on pain management in livestock, even when procedures have previously been conducted without such consideration (George 2003). Whether cattle experience pain similarly to humans is beyond the scope of this discussion. However, signs of pain in cattle may be difficult to recognize (Underwood 2002). As a "prey" species, cattle have evolved to minimize signalling that may attract predation, such as overt signs of serious pain or distress. Internal signs (e.g., heart rate, blood pressure, temperature) often show more marked changes than do external signs (e.g., vocalization, grunting, teeth grinding, reluctance to move, and inappetence). Any abnormal behavior, such as excessive vocalization, changes in posture and physical activity, reduced eating, or reduced drinking may reflect impaired welfare status in food animals (Gonyou 1994).

A number of standard, routine agricultural practices and procedures cause pain and distress to animals (as discussed below). When such practices and procedures are performed, veterinarians and producers should be cognizant that pain minimization requires that they are done for the right reason, by the best method, using correct equipment, at the right time, and to the right class of animal (Bath 1998). In addition, personnel should be appropriately trained and should monitor the aftermath of the procedures (Underwood 2002).

Handling

Principles underlying good handling in cattle, including movement of animals at pasture, chute and tub design, use of back-gates, restraint (including squeeze chutes), floor surface, and loading ramp guidelines are discussed in chapter 7. Poor handling causes both physiological and psychological stress in animals, as well as increased likelihood of injury. Bad facility design and use will often result in increased usage of coercive stimulation (e.g., electric prods, dogs) that coercive additional pain or distress. In concert with facility design and proper use, informed, empathic stockmanship is critical in reducing handling stress.

Animal Identification

Interest in individual animal identification systems has intensified in North America since the detection of several bovine spongiform encephalopathy–positive animals. At present, a large number of methods are employed to identify cattle, reflecting a wide spectrum of management systems and needs in the cow/calf sector. Selection of a particular identification system is related to the questions below:

- Is individual animal, or group, identification necessary? (In countries requiring source verification, this question is probably irrelevant.)
- Should identification be permanent or temporary?
- Should it be legible at a distance?
- Is the method employed cost-effective?
- Does the method employed cause unnecessary stress or pain to the animal?

Identification methods that raise questions welfare concerns include hot-iron branding, severe forms of "ear splitting," and the practice of cutting wattles and jug handles (J. Swanson, personal communication). The previous U.S. regulatory practice of hot-iron branding brucellosis-suspect females on the cheek would also fall into this category (Rollin 1995).

Hot-iron branding produces permanent, unalterable identification that can be recognized from a distance. It is "low-tech" and does not require an inordinate amount of skill to perform. In fact, the end result can be legible even when imperfect. In a number of areas, this is the only legally recognized form of cattle identification. Despite such considerations, hot-iron branding does inflict pain and stress, as would be expected with any third-degree burn. In addition, it reduces the value of each disfigured hide significantly (Rollin 1995). There might be greater incentive for ranchers to seek alternative means of identification if they received direct financial benefit from the marketing of cattle with nondisfigured hides.

Conventional alternatives to hot-iron branding for cattle identification purposes include eartags, neck chains, tattooing of ears, horn brands, and freeze branding. All have advantages and disadvantages. Freeze branding, for example, requires access to liquid nitrogen, demands a higher degree of skill in application than does hot-iron branding, and is impractical for use on animals with light-colored skins. Eartags are often lost, as are neck chains, which, along with horn brands and tattoos, are difficult to read from a distance.

This discussion occurs at a time when permanent animal identification has become an issue in the quest to track individual animals from ranch and country of origin to the dinner table. Impetus has been given to this quest by the recent identification of bovine spongiform encephalopathy in Canada and the United States. High-tech alternatives such as electronic (microchip) identification systems, genetic fingerprinting, and nose prints are undergoing serious scrutiny, as are systems of enforcement. In the United States, a number of leading agricultural data service companies joined forces to create the Beef Information Exchange. These companies are involved with the U.S. Animal Identification Plan, in which rapid and secure sharing of data will be necessary for national identification and trace-back requirements. Future development will undoubtedly lead to even greater exchange of production-related data.

Dehorning and Debudding

The presence of horns on commercial cattle poses certain problems, including increased likelihood for injury, damaged hides, and bruising. It may be argued that such problems are more likely to occur under intensive management conditions, rather than the relatively extensive conditions of many cow/calf operations. However, any time horned cattle are congregated (e.g., for movement, handling, transport), there is potential for problems to occur. Thus, for example, dehorning represents a routine procedure for cattle entering confinement operations.

There has been a trend to reduce the numbers of horned cattle in general, with several options available to achieve this objective. The best approach to ensuring horn-free cattle is genetic, by breeding with homozygous polled animals only. This may conflict with other genetic goals, however, as a number of horned breeds are perceived to have desirable genetic traits. In general, there is also more genetic production data available on horned beef breeds than there is on polled breeds. A future possibility is that genetic engineering will allow the insertion of a particular favorable gene (e.g., the poll gene) without compromising other favorable traits.

Dehorning itself can be performed in different ways, some of which are more applicable to cattle at certain stages of development. Considerable controversy exists over the ideal method of dehorning (George 2003). Dehorning of commercial cattle is mostly done by cowboys or stockmen and without the use of anesthetics. Chemical debudding, whereby a caustic chemical is applied to the horn bud, is best done with very young calves. Difficulties here can occur if the chemical affects other areas of the body, a problem that might arise with poor technique or when leakage occurs from the horn bud site. Thermal debudding is also best done with young (<5 months of age) animals. Here, accurate placement and duration of application of the device are important in ensuring efficacy without causing damage to other structures.

A number of dehorning devices work on the basis of cutting and gouging the horn base and its surrounds. Again, this is best done with younger animals, and it requires both a sharp implement and good technique to ensure efficacy. For older animals, in which horns are well established, the horns may be removed by surgical means, or they may be trimmed with the use of a saw, embryotomy wire, or guillotine. It is recommended that anesthesia, analgesia, or both be provided for adult animals being dehorned (Underwood 2002, George 2003).

All of these procedures, when conducted without anesthetics, cause some degree of pain and stress with adverse effects (such as reduced growth) that can persist for some time. Ways to alleviate at least the immediate pain include the use of local anesthetics (via a ring or corneal nerve block) or systemic agents, such as Xylazine.

Castration

Young calves have traditionally been castrated for several reasons that generally favor economics over welfare, although there are areas of overlap. Such reasons include the avoidance of indiscriminate breeding, improvement of meat quality, removal of selected animals from the breeding pool to improve genetics, and the reduction of unwanted behavior problems. There is a belief that intact, sexually mature males are prone to inflict or incur injury. Animals castrated at a young age do not grow as rapidly or develop the same muscle mass as do those with intact testicles. Thus, it is perhaps paradoxical that hormonal growth promotants are often employed to restore these attributes.

Welfare considerations arise when considering the pain and discomfort associated with different castration procedures. Most common is surgery, which is often conducted without analgesia or anesthetic. Other techniques include elastration (banding, rubber ring) which causes scrotal necrosis and eventual sloughing of the scrotum and testes, and emasculatome neutering, in which crushing of the spermatic cord causes irreversible damage to the blood supply. Producers generally prefer performing castration early in the animal's life, as it is more easily and safely performed when calves are very young, and many believe that it is relatively painless at this stage. However, some calves in the Unites States are not castrated until they are at, or close to, the feedlot stage. In Great Britain, castration of male calves is not permitted without anesthesia after 8 weeks of age.

All forms of castration cause pain and stress, with this being reflected in both behavior (e.g., kicking, rolling, restlessness) as well as in elevated cortisol levels. Research is being conducted on forms of local anesthesia that might be swift in action and economical to use. However, even with local anesthesia, postsurgical pain and discomfort persist. In relatively unhandled cattle, more stress appears to be derived from the restraint process itself than from the actual process of castration. Potential alternatives include the use of antifertility vaccines, such as an luteinizing hormone releasing hormone (LHRH) fusion protein vaccine (Aissat et al. 2002), as well as sclerosing agents injected into the testicle.

Some advocate the abandonment of male castration in favor of raising bulls, a system widely used in countries such as Mexico. Although such a system would favor lean beef production (a possible marketing advantage), widespread adoption in the United States would require fundamental structural changes at the level of breeders, feeders, packers, and retailers.

Spaying

The primary reason for spaying female cattle is to prevent pregnancy. Thus, in the United States, the procedure is mostly employed with heifers intended for the feedlot. The average daily gain of spayed heifers is less than that of intact heifers, with this deficit being restored by appropriate implantation. In Australia, spaying is recognized as a procedure that can benefit welfare and productivity in situations where females cannot be segregated from males. In this context, it can allow cull females to survive and achieve marketable condition by preventing the added stressors of mismanaged pregnancy, calving, and lactation. The Australian Veterinary Association (www.ava.com.au, Dec. 18, 2004) recommends the Willis spay technique as the technique of choice with the provision that it be performed by a trained competent individual. This technique, similar to the Kimberling-Rupp spay method, is performed via vaginal penetration with an instrument that is used to separate the ovary from its attachments. Both techniques require considerable training and skill to master.

A number of recommendations have been made to reduce the stress and other adverse effects of spaying in cattle. These guidelines include (National Consultative Committee on Animal Welfare 1999) (http://www.affa.gov.au).

- Avoidance of unnecessary stress before, during and after spaying
- Avoidance of concurrent stressful surgical procedures (such as dehorning)
- Avoidance of spaying stressed or debilitated animals
- Use of effective restraint
- Avoidance of extreme weather conditions
- Use of hygienic procedures
- Monitoring for the first few days after spaying, and avoidance of long distance travel.

In the United States, anesthetics are recommended for spay techniques that involve a flank incision, with per-vaginal techniques (e.g., Willis, Kimberling-Rupp) being preferred (Grandin 1998).

Research is strongly encouraged to develop noninvasive and less stressful methods to ensure complete cessation of reproductive functions in both males and females. In feedlots, one alternative has been to feed melengestrol acetate to heifers to suppress estrus. This technique relies on prolonged consistent individual intake of melengestrol acetate and is not practicable for range females. Further, it does not provide the degree of effectiveness required of feedlot sterilization regimes. However, promising results in negating gonadal activity in both males and females have been obtained using immunogenic approaches, such as an anti-GnRH vaccine.

Weaning

Beef calves generally remain with their mothers until after lactation has declined significantly, permitting complete or partial weaning to occur naturally. Thus "weaning in beef cattle is generally not a welfare issue" (Morrow-Tesch 2001). Certainly the same welfare questions do not usually arise as they do in the dairy industry, where calves are weaned within the first 24 hours of birth. Despite this, weaning does cause stress in beef cattle, and ways to reduce this are described in chapters 7 and 11. It is well established that allowing young calves to become accustomed to quiet human contact and handling helps to ensure they are more easily handled subsequently. Weaning calves into an area where they can become quietly accustomed to handling facilities can achieve similar benefits (Grandin 1998).

Bull EE

Welfare concerns relating to bull breeding soundness evaluations (BSEs) have mainly focused on the use of EE to obtain a semen sample. Historically, the development of effective EE permitted routine semen collection from unhandled range beef bulls as well as from wild and endangered species. Samples from such animals cannot be collected safely by traditional means, such as with an artificial vagina (AV). In bulls, a representative semen sample obtained by EE is comparable in most important respects with that obtained from an AV (Singleton 1970). Not only does EE permit range-type bulls to be evaluated, but it allows semen to be collected from bulls in which their physical or other impairment rules out AV collection (Clark et al. 1973).

Despite this, the EE process does cause some discomfort to bulls, if not distress, although constant improvement in probe design and machine circuitry have led to a marked reduction of some of evident signs of distress. However, bulls unused to restraint, as well as those that react adversely to novel stimuli, could be expected to exhibit adverse reactions regardless of the degree of pain or discomfort caused by EE. Temperament, which is highly heritable in cattle (cited by Grandin 1998) is also a factor. One study ascertained that bull heart rates were lower when an epidural was given before EE, although no comparison was made with other activities, either painful or pleasurable (Mosure et al. 1998). In rams, based on relative aversion, EE was considered no more aversive than partial shearing (Stafford 1995). Similarly, ram blood cortisol levels were shown to be similar for shearing, EE, and lateral restraint (Stafford 1995). Other work has indicated that, although blood cortisol levels were elevated with bull EE, these were not as high as those caused by restraint alone, or by transrectal palpation (Welsh and Johnson 1981).

In addressing this issue, relevant questions include the following:

- Does EE qualify as a necessary managerial procedure?
- Does it cause more pain and distress than do other necessary procedures?
- Does it cause unacceptable pain or harm?
- Is there a better alternative for reliably collecting semen from bulls?
- Do we need to examine semen from bulls anyway?

Here, the terms of debate may differ between those countries or regions that have predominantly range- or pasture-based natural breeding systems (e.g., the Americas and Australasia), and those in which artificial insemination predominates, as in Europe (where several countries have banned EE in nonanesthetized animals). Although male fertility is, of course, essential in both systems, unhandled bulls are usually assessed in the former, as compared to the latter, in which bulls that are used to handling and AV collection are assessed. For unhandled bulls, the choices are to use a method of semen collection that poses minimal danger to both man and beast (e.g., EE), to use an alternative method for semen collection (e.g., per-rectal massage of the internal genitalia), or to avoid semen assessment altogether.

Although per-rectal massage can be effective in obtaining a semen sample from bulls, it suffers from two disadvantages. First, as semen is obtained only from more distal regions of the reproductive tract, the sample obtained may misrepresent the bull's capabilities in terms of both sperm numbers and semen quality. Second, it is difficult to repetitively perform this procedure, as it can be physically demanding. This can be a constraint in collecting semen from large numbers of bulls, particularly in poor facilities.

Complete avoidance of the collection and assessment of bull semen also presents problems. Failure to examine semen as part of the BSE would lead to unacceptable errors in bull evaluation (see chapter 9). Failure to conduct any form of bull assessment leads to even greater risk of bull-related problems. In turn, this would lead to infertility in breeding herds, adding extra economic burdens to producers and added stress to natural resources.

If it is conceded that EE represents a useful, routine managerial tool, then it is still valid to ask whether or not the pain outweighs the gain. Although objective studies are lacking, observations indicate that this procedure is no more distressful than other accepted procedures such as restraint, vaccination, or palpation, although the degree of distress is probably influenced by probe design, restraint method, and experience of the operator (Stafford 1995). Thus, bull EE may be considered an essential, routine, managerial task that causes no more distress than do other tasks in this category. However, it should be carried out as humanely as possible.

Bull Libido/Serving Capacity Testing

The use of restrained, nonestrous females as stimuli for bull sex drive testing, a feature of a number of published procedures, has led to understandable concerns for the welfare of the females so used. Public concern, both in terms of welfare and esthetics, has been expressed, especially in instances where testing has been conducted without due care or sensitivity. Such concerns are reflected in ongoing research to find more acceptable methods of assessing bull sex-drive and mating ability.

Guidelines for conduct of libido/serving capacity testing of bulls have been developed by the Australian Association of Cattle Veterinarians, as below. Here, particular emphasis is placed on avoiding undue stress and suffering in restrained females. Use of placid females, prior tranquilization, and appropriate lubrication of the female's vulva and vagina before use are all strongly recommended. Further recommendations below are adapted from *The Veterinary Examination of Bulls* (Australian Association of Cattle Veterinarians 1995):

- Service crates should be properly designed to restrain previously unhandled heifers or cows
- Females should be slightly smaller than the bulls tested
- Obstetrical lubricant should be applied to the female vulva and posterior vagina
- Mild sedation of females should be considered
- Heifers may be used for two consecutive tests (no more than 15 services)
- Precautions should be taken to prevent transfer of reproductive diseases.

Consideration should also be shown for the welfare of the bulls being subjected to such tests, with appropriate precautions as follows (adapted from Australian Association of Cattle Veterinarians 1995):

- Test bulls together in their general age groups
- Do not congregate groups of bulls in yards or pens any longer than necessary, as such congregation increases the chances of bull injury; this is particularly true for older bulls, but it also applies to mixed-age bull groups
- Separate any bulls that are showing aggressive behaviors that threaten other animals
- Avoid using service crates that have projecting edges or other design features that may injure animals
- Ensure that test area has sound footing
- Place service crates at least 7 m apart in pens/yards that allow animal movement without crowding or jostling
- Avoid distracting external stimuli during test.

Cull Cow Management and Handling

Welfare aspects of cull cow management and marketing have received attention for several reasons. First, cull cows often represent a vulnerable group of animals. Survey data in the United States indicate that a large number of U.S. cull animals have a number of defects in health, quality, or both (National Cattlemen's Beef Association 1999). Indications are that many of these animals are sent to market with neither concern for the health and well-being of the animal nor public concern about food health or quality (Spire and Holly 1995). In addition, value losses from a variety of problems associated with the management, monitoring, and marketing of cull animals were estimated to represent approximately $69 per head in the United States in 1999 (National Cattlemen's Beef Association 1999). Such problems included animals in which debilitating disease or infirmity was well progressed (obviously unhealthy animals are severely discounted), extensive bruising, injection-site lesions, and poor body condition. Recent changes to U.S. regulations in which nonambulatory animals are banned from the human food chain will undoubtedly affect the profile of cull cows being marketed.

Veterinarians can play an important role in ensuring that the culling process meets welfare as well as production needs. First, they can educate clients of the need to identify potential problems early. This allows timely shipment or intervention. The decisions to intervene medically or surgically as well as which products to choose and how to administer them also influence the marketability of animals that are sold subsequently. Good records of all such interventions and products should be maintained. In a number of cases, the best option may be to maintain the cull animal on an appropriate nutrition program until it is in suitable condition for market. This may also allow time for markets to improve.

Cull cows are often old, sick, injured, or diseased. Welfare codes for injured, sick, and disabled cattle have been developed by different organizations, with the following examples from guidelines produced by Agriculture Canada (Hurnick 1991):

- Only cattle fit to travel should be considered for transportation, unless special precautions are taken to ensure the animal does not suffer
- An animal that becomes injured, sick, or disabled during transit must be taken to an appropriate place for treatment; special precautions must be taken to minimize its suffering
- An animal that becomes injured, sick, or disabled in transit should be kept separate from other animals

A prominent welfare issue associated with transportation and marketing is that of bruising. Research indicates that presence of horns (tipped or not) as well as class of cattle (bull, steer, cow, calf) are important determinants (Wythes et al. 1985), with temperament also playing a role (Fordyce et al. 1985). Cows were most likely to suffer bruising, as were dehydrated animals or those in poor condition. Movement through sale yards, distance traveled, rough handling, and multiple loadings and unloadings also contributed to increased bruising. Many of these determinants have particular significance for cull cows, which should be handled with more care than are other classes of livestock and should be subjected to less handling and less strenuous transport schedules (Wythes et al. 1985).

Disabled and Nonambulatory Animals

The handling of nonambulatory animals (e.g., "downer cows") has received unfavorable media attention, particularly in relation to transport and slaughter. Recommendations in this respect have been made as follows (accessed December 18, 2004, at http://www.avma.org/policies/animalwelfare.asp).

Ambulatory Animals

If an otherwise healthy animal has been recently injured, and the animal is ambulatory, it should be treated, shipped directly to a state or federally inspected slaughter plant, humanely slaughtered on the farm (where state laws permit), or euthanized. Injured ambulatory animals should not be commingled with other animals during transport. Care should be taken during loading, unloading, and handling of these animals to prevent further injury or stress.

Nonambulatory Animals

The U.S. Department of Agriculture (USDA)–Food Safety Inspection Service (FSIS) has defined nonambulatory cattle as those that cannot rise from a recumbent position or that cannot walk. This includes, but is not limited to, animals with broken appendages, severed tendons or ligaments, nerve paralysis, fractured vertebral column, or metabolic conditions.

If an animal is down on a farm, and if the animal is not in extreme distress and continues to eat and drink, the producer should contact a veterinarian for assistance and provide food, water, shelter, and appropriate nursing care to keep the animal comfortable. If the animal is in extreme distress and the condition is obviously irreversible, the animal should be euthanized immediately or humanely slaughtered on the farm (where state laws permit).

If an animal is down at a nonterminal market (e.g., sale yard or auction), and if the animal is not in extreme distress, but is disabled, treatment measures should be initiated. If the animal is in extreme distress or the condition is obviously irreversible, the animal should be euthanized immediately.

If an animal is down at a terminal market (e.g., slaughterhouse or packing plant), the animal should be euthanized immediately.

Recently, the USDA has amended its rules concerning the slaughter of nonambulatory (downer) cattle as follows:

In the wake of the positive testing for BSE of an immobile Holstein cow in Washington State (December 2003), the USDA/FSIS issued three interim rules and a notice on January 12, 2004, that included banning the sale and slaughter for human food of nonambulatory cattle. These and succeeding regulations may be viewed on the USDA/FSIS Web site (http://www.fsis.usda.gov/OPPDE/rdad/FRPubs/04-002N.htm).

Euthanasia

Euthanasia represents a welfare issue, as it is an option for releasing an animal from severe pain and distress, where other options either do not exist or are inappropriate. By definition, it represents "an easy death or means of inducing one" (Merriam-Webster 1981). The euthanasia method employed is also a welfare consideration. Here, techniques used with cattle fall into one of the following categories (American Association of Bovine Practitioners [AABP] 1999):

Table 14.1. Bovine euthanasia methods (adapted from American Association of Bovine Practitioners 1999).

Method	Human Safety Risk	Skill Required	Cost	Esthetic Concerns	Suitability for Mass Euthanasia
Gunshot	High	Moderate[b]	Low	Moderate: some blood and motion	High
Captive bolt	Moderate	Moderate[b]	Low	Moderate: as above	Moderate
Barbiturate overdose[a]	Low	Moderate[b]	Moderate	Low	Low
Exsanguination[a]	Moderate	Moderate[b]	Low	High	Low
Electrocution[a]	High	Moderate[b]	High (equipment)	High	Low

[a] Recommended veterinarian only.
[b] Operator training required.

- Physical disruption of brain activity caused by direct destruction of brain tissue (e.g., gunshot, penetrating captive bolt)
- Drugs that directly depress the central nervous system (e.g., anesthetics, barbiturates) and induce death by hypoxia
- Agents that induce unconsciousness followed by mechanisms that induce hypoxia (e.g., narcotics followed by exsanguination).

The decision as to which is the most appropriate method to employ in a given situation (Table 14.1) is influenced by a number of considerations, including the following (AABP 1999):

- The welfare of the animal
- The amount and type of restraint available
- The possibility of salvage of the carcass (e.g., use of certain drugs may disqualify the carcass from human or animal consumption)
- Economics (euthanasia methods differ in cost, and also in their effects on salvageable product, as above)
- Availability of equipment, materials
- Human safety
- Practicality
- Skill of operator (e.g., effective use of the captive bolt requires training and skill)
- Esthetics
- Diagnostic considerations (e.g., importance of preserving intact brain tissue for rabies verification)
- Logistical considerations (euthanasia of a large number of animals within a short period of time, as in the face of mass pandemics, will preclude some of the options available for euthanasia of a tractable individual).

The following represent either acceptable, or conditionally acceptable, euthanasia methods or agents for ruminants, as determined by the American Veterinary Medical Association (accessed December 18, 2004, at http://www.avma.org/policies/animalwelfare.asp).

- Acceptable: Barbiturates, potassium chloride in conjunction with general anesthesia, penetrating captive bolt
- Conditionally acceptable: Chloral hydrate (intravenously after sedation), gunshot, electrocution.

The AABP (1999) recommends the following as being **unacceptable** for bovine euthanasia:

- Manually applied blunt trauma to the head
- Injection of chemical agents into the conscious animal (e.g., disinfectants, electrolytes such as KCL and MgSO4, nonanesthetic pharmaceutical agents)
- Air embolism (i.e., injection of a large air bolus into the blood stream)
- Electrocution using a 120-volt electrical cord.

For facilities and operations that regularly encounter the need for euthanasia of animals, it is recommended that personnel be appropriately trained and that facilities and materials are available to perform the task humanely and safely, with due regard to esthetic considerations.

Cloning by Nuclear Transfer

Advances in biotechnology allow the artificial creation of genetically identical individuals, with this process having become relatively commonplace in cattle. However, as discussed in chapter 10, the current process is arduous, expensive, and inefficient. Some questions are still unresolved about the viability and health of the offspring so obtained, and these have welfare implications including the necessity to deliver such calves via Cesarean section. Other problems include developmental abnormalities, decreased viability, premature ageing, and excessive weight gain. A useful report on the welfare implications of cloning in livestock has been prepared by the Farm Animal Welfare Council in the UK, which may be accessed at http://www.fawc.org.uk/reports.htm.

Cloning, or the process of somatic cell nucleus transfer, has been accomplished in a number of species (e.g., cattle, sheep, swine, goats). Apart from generally poor success rates, other adverse outcomes occur relatively frequently, most often during early development. Pregnancies may terminate spontaneously because of fetal abnormalities or difficulties with placentation. A common fetal abnormality in cattle is fetal overgrowth (or large offspring syndrome), which comprises a group of factors including large fetal size (>20% greater than normal), relative uterine size, and dystocia. Other calf problems, which may or may not be related to large offspring syndrome, include underdeveloped respiratory, cardiovascular, and renal systems. In addition, a number of calves are born with excess fluid within either the placenta or organs. Thus, surrogate dams carrying clone pregnancies are at increased risk of dystocia and of complications associated with hydrops. Other reported problems in neonate, cloned calves include contracted flexor tendons, respiratory failure, and large umbilici. Once through the critical neonatal period, cloned calves appear to grow and be as healthy as their conventional counterparts.

WELFARE CODES

Whereas a number of European countries have relied on governmental regulation to improve animal welfare in livestock systems, in the United States and Australia, those involved have looked more to the industries themselves for education and self-regulation. Here, the emergence of welfare questions about some traditional procedures has prompted cattle producers to articulate general principles and guidelines for the care and handling of cattle. These guidelines are produced by consensus among interested constituents and stakeholders, and are being steadily augmented as new knowledge arises and cultural changes allow. It is important for the livestock industries to be perceived as acting in good faith, and with due haste, on animal welfare issues. It is also important for such guidelines to be based on principles that reflect an objective balance between science, philosophy, and economic survival. Some examples of such guidelines are provided below.

National Beef Cattlemen's Association (NBCA; excerpts from "Statement of Principles," 2003):

- I believe in the humane treatment of farm animals and in continued stewardship of all natural resources
- I believe my cattle will be healthier and more productive when good husbandry practices are used
- I believe that my, and future, generations will benefit from my ability to sustain and conserve natural resources
- I believe that it is the purpose of food animals to serve mankind, and it is the responsibility of all human beings to care for the animals in their charge.

The NBCA has also developed a "Producer Code of Cattle Care" (NBCA 2003), as shown in Figure 14.1.

In Australia, codes of practice have been developed for different industries in a process of consultation among interested parties (Penson 1998). *The Code of Practice; Cattle*, for example, attempts to meet three general objectives as follows:

- To promote humane and considerate treatment of cattle and the use of good husbandry practices to improve the welfare of cattle in all types of cattle farming enterprises
- To provide information concerning their obligations and responsibilities to all people responsible for the care and management of cattle
- To set an industry standard by defining acceptable cattle management practices.

Agriculture Canada has published a *Recommended Code of Practice for the Care and Handling of Food Animals: Beef Cattle* (Hurnick 1991) which covers a wide spectrum of topics from shelter and housing to commercial transportation.

The Farm Animal Welfare Council in the United Kingdom has recognized the basic needs of livestock in terms of the "new" Five Freedoms:

Beef cattle producers take pride in their responsibility to provide proper care to cattle. The Code of Cattle Care below lists general recommendations for care and handling of cattle:

- Provide necessary food, water, and care to protect the health and well-being of animals.
- Provide disease prevention practices to protect herd health, including access to veterinary care.
- Provide facilities that allow safe, humane, and efficient movement and/or restraint of cattle.
- Use appropriate methods to euthanize terminally sick or injured livestock and dispose of them properly.
- Provide personnel with training/experience to properly handle and care for cattle.
- Make timely observations of cattle to ensure basic needs are being met.
- Minimize stress when transporting cattle.
- Keep updated on advancements and changes in the industry to make decisions based on sound production practices and consideration to animal well-being.
- Persons who willfully mistreat animals will not be tolerated.

Figure 14.1. Producer code of cattle care. (NBCA 2003)

- Freedom from hunger and thirst
- Freedom from discomfort
- Freedom from pain, injury, or disease
- Freedom to express normal behavior
- Freedom from fear and distress.

In the Unites States, the Federation of Animal Science Societies has produced a useful compendium that covers welfare considerations for a number of species including beef cattle (Federation of Animal Science Societies 1999). Other relevant publications include *Grazing Animal Welfare* (Moore and Chenoweth 1985) and *Food Animal Well-Being* (Baumgardt and Gray 1993), proceedings of a meeting hosted by the USDA and Purdue University in 1993. Readers interested in this topic might also consult relevant sections in a text edited by Temple Grandin (1998) and those by Bernard Rollin (1995) and Hemsworth and Coleman (1998).

A consortium of animal protection agencies, foundations, and individuals has been formed in the United States to define and promote the use of a "certified humane" label for products derived from livestock. This label, originally launched by Humane Farm Animal Care (http://www.certifiedhumane.com) would reassure purchasers (via a USDA-verified process) that the animals involved have been raised and handled in a humane manner. In addition, provisions would be made to ensure "wholesome" food products, which would include providing a nutritious diet without antibiotics or hormones and the raising of animals with shelter, resting areas, sufficient space, and the ability to engage in natural behavior. The quality standards also include provision for stress and disease prevention. Some U.S. producers have already been approved to carry the "certified humane" logo on their products. In time, there is little doubt that this label will become more apparent in supermarkets and restaurants. Although it is probable that these rules will have little effect on most cow/calf operators, extra costs will be borne by packers and processors who must provide verification for the standards. A similar program is in operation in the United Kingdom, where a "Freedom Food" label is promoted by the Royal Society for the Prevention of Cruelty to Animals.

Finally, the Office International des Epizooties recently established a provisional working group on animal welfare that has developed guiding principles and guidelines for the development of science-based welfare standards for animals involved in trade.

CRITICAL WELFARE CONTROL POINTS FOR COW/CALF OPERATIONS

1. Body Condition Score of Cows: Ninety percent of the cows should be maintained at a body condition score of greater than 2. The minimum acceptable body condition score is 2. Emaciated, skeletal animals must be euthanized on the premises. (A score of 2.75 on the Elanco dairy chart is equal to a beef 2.)
2. Euthanasia: Severely debilitated, nonambulatory, or emaciated animals must be euthanized on the ranch using the American Association of Bovine Practitioner's guidelines. Permitted methods are gunshot, captive bolt, or other approved method.

3. Non-ambulatory Animals: Dragging of nonambulatory animals is not permitted. Euthanasia on the premises is strongly recommended.
4. Cancer Eye: Any animal with ocular neoplasia (cancer eye) that has invaded the facial tissue must be euthanized on the ranch or stocker operation.
5. Calving Management: Per Federation of Animal Science Societies (1999) guidelines.
6. Cattle Handling: Use the scoring system on the feedlot audit form.
7. Castration: Per Federation of Animal Science Societies (1999) guidelines, as early as possible.
8. Dehorning: Horns should be removed before calves are 4 months old. Removal of horns from older animals requires local anesthesia
9. Branding: Should be avoided unless required by law. No face branding.
10. Ear Marking: Cutting the animal's ear or dewlap for identification purposes is not permitted. Small notches made with a punch are permitted.
11. Weaning: Calves should be preweaned and vaccinated 5–8 weeks before they leave the ranch of origin. Shipment of bellowing calves should be avoided. Low-stress, fenceline weaning is recommended.
12. Protection from Extreme Weather: Depending on the location and topography of the ranch, cattle may need shade, windbreaks, or other devices to protect them. In northern areas, ranchers should plan for how to protect their cattle from severe winter storms.
13. Spaying Heifers: Anesthetics are required for flank spaying of heifers. Other, less invasive methods of spaying that do not require a flank incision are recommended.
14. Acclimate Cattle to Handling: Calves destined for a feedlot should be acclimated to people on foot and on horseback, and to vehicles. This will help keep them calmer during handling at the feedlot and packing plant.

RESOURCES

Farm animal welfare guidelines have been issued by a number of organizations and groups, such as

- Farm Animal Welfare Council (http://www.fawc.org.uk).
- Food Marketing Institute and National Council of Chain Restaurants (http://www.fmi.org, http://www.nccr.net).
- Free Farmed certification (American Humane Association, http://www.americanhumane.org).
- Animal Welfare Information Center (http://www.nal.usda.gov/awic).
- Animal Welfare Institute (http://www.awionline.org).
- Federation of American Science Societies (http://www.fass.org) in conjunction with the American Registry of Professional Animal Scientists (http://www.arpas.org).
- United States Department of Agriculture, Food Safety and Inspection Service (USDA/FSIS) (http://www.fsis.usda.gov)
- Animal Agriculture Alliance (http://www.animalagalliance.org)
- Centers for animal welfare at various land grant universities (e.g., University of California Davis, Purdue University, Washington State University).
- The Humane Society of The United States (http://www.hsus.org)
- Farm Animal Care Training and Auditing, LLC, provides training for auditors and producers involved in ascertaining compliance with industry-established standards of animal welfare in the United States (http://www.factallc.com).
- The Department for the Environment, Food and Rural Affairs, formerly the Ministry for Agriculture, Fisheries and Food (UK) (http:www.defra.gov.uk/animalh/welfare/default.htm).
- Code of Recommendations for the Welfare of Livestock. Cattle. Defra Publications, Admail 6000, London, UK SW1A 2XX.

Additional information on many aspects of cow/calf welfare are available:

- Temple Grandin: http://www.grandin.com
- U.S. Department of Agriculture Animal Welfare Information Center: http://www.nal.usda.gov/awic

Humane Certification

- Humane Farm Animal Care: http://www.certifiedhumane.com
- Humane Society of the United States: http://www.hsus.org

Handling

- National Institute for Animal Agriculture: http://www.animalagriculture.org
- *Cattle Handling and Transportation* (video)

- *Livestock Handling Guide* (pamphlet)
- *Livestock Trucking Guide* (pamphlet)

Euthanasia

Practical Euthanasia of Cattle, American Association of Bovine Practitioners.

2000 Report of the AVMA Panel on Euthanasia. *Journal of the American Veterinary Medical Association* 218669-696

Practical Euthanasia of Cattle American Association of Bovine Practitioners (http://www.aabp.org)

Heat Stress

- NebGuide G00-1409-A (http://ianrpubs.unl.edu/beef); particularly aimed at feedlots
- Animal Welfare Information Center (http://www.nal.usda.gov/awic)

REFERENCES

American Association of Bovine Practitioners. 1999. *Practical euthanasia of cattle*. Rome, GA: American Association of Bovine Practitioners.

Aissat, D., J.M. Sosa, D.M. Avila, K.P Bertrand, J.J. Reeves. 2002. Endocrine, growth, and carcass characteristics of bulls immunized against luteinizing hormone-releasing hormone fusion proteins. *Journal of Animal Science* 80:2209–2213.

American Veterinary Medical Association. 1991. *The veterinarian's role in animal welfare*. Schaumburg IL: American Veterinary Medical Association.

Australian Association of Cattle Veterinarians. 1995. *The Veterinary Examination of Bulls*. Brisbane: Australian Association of Cattle Veterinarians.

Bath, G.F. 1998. Management of pain in production animals. *Applied Animal Behaviour Science* 59:147–149.

Baumgardt, W., H.G. Gray, eds. 1993. Food animal well-being. *Proceedings of the USDA and Purdue University Conference.* West Lafayette, IN.

Clark, R.H., R.W. Hewetson, B.J. Thomson. 1973. Comparison of the fertility of bovine semen collected by artificial vagina and electro-ejaculation from bulls with low libido. *Australian Veterinary Journal* 49:240–241.

Federation of Animal Science Societies. 1999. *Guide for the Care and Use of Agricultural Animals.* Savoy, IL: Federation of Animal Science Societies.

Fordyce, G., M.E. Goddard, R. Tyler, et al. 1985. Temperment and bruising in *Bos indicus* cross cattle. *Australian Journal of Experimental Agriculture* 25:283–288.

George, L. 2003. Pain control in food animals. In *Recent Advances in Anesthetic Management of Large Domestic Animals*, edited by E.P. Steffey. New York: IVIS.

Gonyou, H. 1994. Why the study of animal behavior is associated with the animal welfare issue. *Journal of Animal Science* 72:2171–2177.

Grandin, T., ed. 1998. *Genetics and Behavior of Domestic Animals.* San Diego: Academic Press.

Grandin, T. 1998. Review: Reducing handling stress improves both productivity and welfare. *The Professional Animal Scientist,* 14(1):1–10.

Hemsworth, P.H., G.J. Coleman. 1998. *Human-Livestock Interactions*. Wallingford: CAB International.

Hurnick, F. 1991. *Recommended Code of Practice for the Care and Handling of Farm Animals: Beef Cattle.* 1870/E. Ottawa: Agriculture Canada.

National Beef Cattlemen's Association. 2003. *Guidelines for handling and care of beef cattle.* NBCA Cattle Care Working Group.

National Cattlemen's Beef Association. 1999. *Executive Summary of the 1999 National Market Cow and Beef Quality Audit.* Englewood, CO: National Cattlemen's Beef Association.

Moore, B.L. and P.J. Chenoweth, ed. 1985. *Grazing Animal Welfare*. Indooroopilly: Australian Veterinary Association (Queensland Division).

Morrow-Tesch, J.L. 2001. Evaluating management practices for their impact on welfare. *Journal of the American Veterinary Medical Association* 10:1374–1376.

Mosure, W.L., R.A. Meyer, J. Gudundson, et al. 1998. Evaluation of possible methods to reduce pain associated with electro-ejaculation in bulls. *Canadian Veterinary Journal* 39:504–506.

Penson, P. 1998. The Australian model code of practice for the welfare of cattle. *Proc. XX World Buiatrics Congress*, Sydney, Australia 2:1067–1073.

Rollin, B.E. 1995. *Farm Animal Welfare*. Ames: Iowa State Press.

Singleton, E.F. 1970. Field collection and preservation of bovine semen for A.I. *Australian Veterinary Journal* 46:160–163.

Spire, M.F., J.P. Holly. 1995. Establishing culling criteria in beef cow/calf operations. *Veterinary Medicine* 90:693–700.

Stafford, K.J. 1995. Electroejaculation: a welfare issue? *Surveillance* 22:15–17.

Merriam-Webster. 1981. *Webster's Third New International Dictionary*. Springfield, MA: Merriam-Webster.

Underwood, W.J. 2002. Pain and distress in agricultural animals. *Journal of the American Veterinary Medical Association* 221:208–211.
Welsh, T.H. Jr., B.H. Johnson. 1981. Stress-induced alterations in secretion of corticosteroids, progesterone, luteinizing hormone, and testosterone in bulls. *Endocrinology* 109: 185–190.
Wythes, J.R., R.K. Kaus, G.A. Newman. 1985. Bruising in beef cattle slaughtered at an abattoir in southern Queensland. *Australian Journal of Experimental Agriculture* 25:727–733.

15
Environmental Aspects of Livestock Production

Glenn Nader, Gary Veserat, Valerie Veserat, and Lee Fitzhugh

INTRODUCTION

Livestock producers, by the nature of their operations, often use extensive holdings of land and water resources. Therefore, they are considered a major resource manager of the environment. This attribute can be positive or negative, depending on the differing perspectives of livestock producers, agencies, and the public. Because many view grazing as an impactive land use, many livestock producers are on the defensive about their use of these resources.

It has been said that the best environmentalist is one who has the disposable income to change adverse practices without affecting operational integrity. Therein lies the challenge for livestock producers. A western states study in the United States placed the return on investment for cattle operation at negative −2% to plus 3% (Torell et al. 2001). This clearly paints a picture of the average cattle producer as one who manages resources that are highly valued by an expanding urban population, but who has a limited income to address the issues that society is now presenting as additional costs of operation. Survival under these conditions is a matter of how livestock producers approach perceived problems. Some producers, tired of low-income livestock production, approach environmental challenges with a defensive attitude, whereas others look at them as an opportunity to improve the production of their operation. In this chapter, the environmental issues that press livestock producers will be discussed, and the methods used to deal with them will be provided. The topics to be covered include rangeland monitoring, water quality monitoring, the Endangered Species Act and endangered species, and conservation banking and easements.

Often, it is not one but many environmental issues that affect a livestock operation. Solutions may involve several agencies; however, the outcome can be invaluable as resources are improved and transformed into stronger operational assets. For example, in one case, a ranch facing both endangered species and water quality issues was able to propose a riparian pasture fencing project that regulated grazing in the riparian area. This riparian fencing raised the number of pastures from five to thirteen, which greatly increased the number of cattle grazed on the ranch. This producer was successful in receiving a grant that funded the project because the eroded or "entrenched" stream presented problems for an endangered fish habitat and sediment into the stream caused a water quality problem. However, it is important to note that nothing comes for free. The producer signed an agreement to maintain the project for 20 years. The producer was obligated to follow a specific grazing "prescription" in the project plan. The approach was to be proactive and address issues before a government agency limited the grazing operation.

The U.S. Department of Agriculture–Natural Resources Conservation Service (USDA-NRCS) can provide technical assistance and cost-share funding. Similar organizations and resources exist in a number of countries. Local Resource Conservation Districts in many cases can work with landowners to sponsor and obtain grant funding to address issues on ranches and farms. The local Cooperative Extension office may also be able to assist livestock producers and farmers to develop improvements that insulate the operation from agency action and can direct them to potential funding sources.

RANGELAND MONITORING

Monitoring or developing environmental information on the ranch can be the best method to educate government agencies or the interested public on the relationship between livestock management and perceived problems. Monitoring means systematically recording observations of processes or activities to detect changes in the rangeland over time. Monitoring can range from a simple collection of photos to recording detailed information about species and soils. Most livestock producers begin with simple photographic records. As interest, experience, and knowledge grows, monitoring develops to suit the livestock producer's goals and situation.

Monitoring information can be used in two ways. The first are "point-in-time" comparisons, which are used to evaluate short-term, annual site conditions. The second type is "trend monitoring," which is used to track changes over time. As records accumulate, trends in the conditions of rangelands become more apparent.

It is often said that a picture is worth a thousand words. Photographs taken consistently over the years provide an easy, convenient, and inexpensive method by which livestock producers can establish a visual representation of resource conditions. This is the simplest form of monitoring and can be more convincing than species- or site-quantitative data.

Landscape photographs document change over time. Here it can be useful to locate old family or ranch photographs that show historical vegetation conditions. These historical photos provide documentation of previous range conditions. Contemporary pictures from the same locations can immediately illustrate vegetation changes. This will vividly portray the condition and trend of the rangeland. One example is the issue of the effect of a juniper invasion on rangeland grazing capacity. When this issue was broached by the livestock producers, it was highly discounted by some wildlife agencies, which said that the rangeland in question had always been juniper woodland. The pictures in Figure 15.1 were taken in 1916 and 1993, and the difference in conditions spoke convincingly of the issue.

If historic photographs of rangeland sites are not available, take new ones. Sites where photographs could be taken may include riparian areas, upland areas, burns, revegetation areas, stream diversions such as rock dams, areas likely to be discussed in allotment or ranch management planning, and fencelines.

Photographs should be taken from the same designated point (marked by a steel post, global positioning system, or other type of permanent marker). The permanent marker may be used to direct by compass the position of the camera for the photo. Include a distinctive landmark in the background such as a peak, a rock outcrop, or a tree. Previous photographs are helpful when setting up photographs in subsequent years.

The following should be recorded on each photograph:

- date (a camera that dates pictures works well)
- photo point number (i.e., pasture or grazing allotment)
- actual use (i.e., cattle on or off pasture dates)
- any unusual event that occurred that year (i.e., drought, high rainfall, etc.)

WATER QUALITY MONITORING

In the United States, there is significant interest in the effect of Clean Water Act regulation on ranches. Water quality monitoring within the rangeland community reflects an awareness of nonpoint source pollution (NPS) water quality issues by livestock producers and rangeland managers. NPS occurs when water runs over land or through the ground, picks up pollution (sediment, nutrients, and organic and toxic substances originating from land use activities such as farming and grazing) and deposits them in surface waters or introduces them into groundwater.

Water quality monitoring can be conducted for both proactive and reactive purposes. Proactively, a livestock producer who cares about the environment and wildlife can monitor to provide a baseline reading for the operation. This may also become a protective measure should the producer need to react if the U.S. Environmental Protection Agency (EPA) or a similar agency believes the operation is impairing the quality of water leaving the producer's premises. Recording the pasture use by livestock can be combined with water quality monitoring to help clarify the effect of grazing management practices on bacteria and pathogen levels. This section of the chapter familiarizes livestock producers and veterinarians with basic monitoring methods and concepts. A livestock producer's water samples can augment samples taken by regulatory agencies; however, they need to conform to scientific design, as described in this section. This section will also provide the livestock producer and veterinarian with knowledge of the correct methods that regulatory agencies should follow.

Unless a water quality–monitoring program addresses specific objectives, much money and time can be spent with little return on the investment. Potential errors must be minimized during design and recognized during data interpretation. Properly designed

Figure 15.1. A. Photograph from West Valley grazing area circa 1916. **B.** Photograph from West Valley grazing area circa 1993.

rangeland water quality monitoring incorporates the concepts of NPS prevention.

Important Concepts of NPS

NPS is driven by meteorological and hydrological events on a watershed. Excess NPS is caused by human activity that alters natural processes. The occurrence and magnitude of NPS is directly linked to the hydrologic cycle. These facts lead us to four basic concepts that must be considered when developing or evaluating an NPS monitoring program. First, by definition, a watershed is the region draining into a body of water. It is also the basic land unit of the hydrologic cycle and thus is the source of nonpoint pollution generation and transport. Water quality at any point along a stream reflects all pollutants from all sources in the watershed above that point such as natural geological processes, grazed uplands, roads, housing developments, channel erosion, wildlife activities, campers, and so forth.

Second, because nutrient cycling and erosion are natural processes, there is some "background" level, or natural NPS, for each watershed. Background levels vary from watershed to watershed based on unique climatic, hydrologic, geologic, or soil and ecological factors that are usually unknown. The potential exists for background levels to exceed water quality standards during rare storm flow events such as 10-, 25-, and 100-year storms. If water quality standards have not been adjusted to reflect natural variation among watersheds, noncompliance may be caused by large storms.

Third, the amount of NPS from rangelands depends on storm intensity. Most NPS will be generated and transported during high-flow storm events, which represent a short period (hours or days) of high stream flow. In general, concentrations of nonpoint source pollutants such as suspended sediment, nitrogen, phosphorus, and pathogens increase as storm flow increases. However, pollutants such as calcium, magnesium, and pH that are controlled by groundwater discharge from springs and seeps to streams often decrease during a storm because groundwater inflows are being diluted by the higher stream flows.

Fourth, NPS generation and transport are unpredictable because they are driven by rainfall and depend on individual watershed characteristics. Interacting climatic, hydrologic, and geologic causes, or soil, vegetation, and land-use factors create a high level of natural variability in NPS generation and transport through time (such as duration of rainfall) and space (such as size of watershed).

Developing a Monitoring Program

Realistic Expectations of the Monitoring Program

Without a valid experimental design and use of appropriate statistical principles early in the development of a monitoring plan, it is difficult to identify cause and effect. Only a well-planned and well-implemented monitoring program will answer cause-and-effect questions. In most cases, monitoring simply represents observation of water quality variables over time.

Even with the "best" techniques and designs, detecting changes in NPS is difficult. Spooner et al. (1987) evaluated the sensitivity of grab sample–based NPS monitoring programs. They determined that a 30%–0% increase or decrease in the average pollutant concentration over a 5-year period is required to document a significant change in water quality caused by management changes.

Types of Water Quality Monitoring

Although there are many types of monitoring, two basic approaches are often used: cause-and-effect monitoring and compliance monitoring.

Cause-and-Effect Monitoring

Cause-and-effect monitoring tries to prove or disprove a cause-and-effect relationship between a specific land activity and water quality degradation. For instance, the question might be: does fencing the riparian zone on Ranch X reduce suspended sediment concentrations in the stream? (Remember that suspended sediment concentration depends on stream flow.) How then are the following two scenarios to be interpreted?

The stream is fenced. For the next 4 years, rainfall is low, and thus, stream flow and sediment concentration are below normal. Was the reduced sediment concentration caused by the fence or the low rainfall?

The stream is fenced. For the next 4 years rainfall is high, and stream flow and sediment concentrations are above normal. Was the fence ineffective, or was the increased sediment concentration caused by the high rainfalls?

Without adoption of a feasible method to quantify the effect of natural weather and stream flow variation, there is no way to determine the effectiveness of the fencing practice on Ranch X.

The key to the cause-and-effect monitoring design is adequate baseline or control information and an understanding and application of experimental design and statistical principles. There are three basic cause-and-effect experimental designs for documenting water quality problems or changes in water quality resulting from changes in land use or management (Spooner et al. 1985, EPA 1993). These designs attempt to separate natural geologic, weather, or upstream impacts from land management affects:

1. The "before-and-after" experimental design employs water quality monitoring before and after a change in management to determine whether the change alters water quality, and thus, whether a cause-and-effect relationship actually exists. However, without associated long-term monitoring of water quality, weather, and stream flow, this method will provide little insight. Of the three cause-and-effect designs discussed here, this method is the least useful for determining true cause and effect.
2. The "above-and-below" experimental design calls for sampling water quality over time immediately above and below a potential source of NPS, such as monitoring immediately above and below where a road crosses a stream. The primary advantage of this design over the

before-and-after design is that it allows for separation of nonpoint source pollutants contributed upstream above the ranch. This advantage can be lost if the monitoring sites do not isolate the source of interest from other inputs to the stream. In addition, changes to the channel may cause changes to upstream reaches. For example, a poorly designed bridge could cause bank erosion upstream.

3. The "paired watersheds" experimental design involves monitoring water quality on two or more watersheds over time. The watersheds are initially under the same management. After a sufficient pretreatment period (several years at a minimum), one watershed is selected as a control and the others are treated. The control watershed measures the year-to-year and seasonal climatic variation. This design is the most useful of the three methods for establishing cause-and-effect relationships. It is also the most technical and expensive method of those discussed.

Compliance Monitoring

Compliance monitoring evaluates whether water quality parameters are within (comply with) predetermined minimum or maximum chemical values. State or regional agencies have set water quality standards based on "beneficial use categories" for each stream, river, and lake. "Beneficial use" is a term used by the EPA to describe different functions of water. Most waters support several beneficial uses. The rationale for public regulation of water quality is to protect the existing and designated beneficial uses of water (EPA 1991).

Although the specific designated uses vary from region to region, they often include agricultural use, industrial use, domestic water supplies, recreational use, and the propagation of fish and wildlife. Each state (or region) determines which uses should be applied to the water bodies within their boundaries. Numeric parameters for water quality are assigned to each beneficial use and then become minimum criteria for water quality. The specific water quality parameters that are monitored will be dependent on the designated beneficial uses of the water.

Important Questions about Water Monitoring

There are six basic questions that, if specifically addressed, will help producers begin the process of identifying objectives and developing a meaningful, NPS-monitoring program.

Why Monitor Water Quality?

The most important step in developing a meaningful NPS-monitoring plan is to establish clear, explicit, and realistic objectives. What questions about water quality does a producer need to answer? What does a producer intend to do with the information he or she obtains from this monitoring? How confident does a producer need to be in the accuracy and precision of this information?

Once these monitoring objectives (the "why" questions) have been established, the "who," "what," "where," "when," and "for how long" questions can be systematically answered. The best monitoring program for each individual watershed and monitoring objective is unique. The objectives and the watershed set the protocol. Because it is impossible to monitor everything, everywhere, all the time, experience and judgment must be used to select the appropriate type and intensity of monitoring. A producer's objectives will help him or her define not only the type of monitoring to use but also the level and intensity of monitoring.

When to Sample?

When (during the day and also during the year) will water quality be monitored with regard to stream flow, and what relation does the timing of sampling have to the monitoring objective? Water temperatures will be affected by air temperatures. The daily stream temperature will vary with the stream size, shape, and amount of shade provided by the vegetation. Monitoring should occur in the afternoon if a critical maximum temperature is of concern. If a producer is trying to determine whether the stream is providing a proper fish habitat, he or she would want to sample the stream at times when temperatures are highest or lowest.

The proper time to sample water depends on the monitoring objectives. The time of year, time of day, and flow conditions are major considerations. For certain water quality properties (such as dissolved oxygen, temperature, and bacteria/pathogen levels), the time of day that water is sampled can affect the results. Temperature, dissolved oxygen, and bacteria/pathogen levels are often most critical during low flows.

A grab sample represents nonpoint pollution transport only at the time of sampling. For proper interpretation, each water sample should be accompanied by a stream flow measurement and an indication of whether the sample represents rising, peak, or falling flow.

Table 15.1. Beneficial uses.

	Warm/Cold Fishery	Domestic Water Supply	Recreation	Hydroelectric
Items suggested to monitor by U.S. Environmental Protection Agency	Temperature Dissolved oxygen Turbidity Channel characteristics Nitrogen Phosphorus Stream flow Biochemical oxygen demand	Sediment Bacteria/ pathogen Nitrate	Bacteria Pathogen Sediment Nitrogen Phosphorus Flow rate	Sediment Flow rate

Source: Environmental Protection Agency 1991.

What to Monitor?

Which water quality variables will be monitored, and what relation do the variables have to the monitoring objective? If a producer has trouble in establishing objectives that help determine what to monitor, he or she may want to consider the structure of compliance monitoring (Table 15.1), in which the important consideration is the beneficial use of the water either on or downstream from a producer's ranch. Once a producer has determined which beneficial uses are relevant to his or her situation, the producer can review the EPA (or other appropriate) monitoring suggestions. These are simply suggestions: What a producer decides to monitor must be based on his or her specific monitoring objectives, as discussed above.

Where to Monitor?

Where will water quality be monitored, and what relationship does the location have to the monitoring objective? Sampling site location can greatly influence temperature readings. For a typical stretch of stream, select sites from both shaded pools and open water for water temperature sampling. Stream characteristics can greatly influence water temperature. During midsummer, shade provided by plants and close overhanging banks can keep a stream 7°–12°F cooler than a stream exposed to direct sunlight.

Water in a stream mixes with air through exposure and turbulence at the surface and is influenced by the air temperature. Because of larger surface area, a wide, shallow stream will heat more rapidly than will a deep, narrow stream. Color and composition of a streambed also affect how rapidly stream temperature rises. A dark, bedrock channel will gain and pass to the stream more heat from solar radiation than will a lighter-colored channel. Similarly, solid rock absorbs more heat than does gravel.

The stream flow or volume of water in a stream also influences temperature. The larger a body of water, the more slowly it will heat. Rivers and large streams have more constant temperatures than do smaller streams.

The direction a stream flows also affects how much solar radiation it will collect. Southerly flowing streams receive more direct sunlight than do streams flowing north, because of the angle of the sun's rays. Eastward- or westward-flowing streams may receive shading from adjacent ridges, trees, and riparian vegetation.

Site location variability can be introduced into the data by the site of sample collection including the stream reach (i.e., pool vs. riffle) and a within-stream cross section (i.e., bottom vs. top). The section of the stream sampled is crucial and depends on the monitoring design. Considerations include the effects of inflows such as tributaries, springs, seeps, or irrigation return flows. It is important that the producer sample at a site where the waters are well mixed, not from a pool or in the middle of a riffle area, but somewhere in between. Stagnant water should never be sampled.

It is advisable to sample more than one spot in the stream cross section and to integrate water collection from the top to the bottom of the stream and across the stream. In general, samples should be collected in the main current, away from the riverbank, avoiding pool areas, riffle areas, and rock dams. In shallow stretches, this can mean carefully wading into the center current to collect the sample. It is important to try not to disturb the bottom sediment; in fact, the sample collected should not contain bottom sediment at all. It is helpful to collect the sample from in front, facing upstream. If wading is not possible, the sample bottle may be taped to an extension pole, or a boat may be used. The boat should be maneuvered into the center of the main current, with the water sample collected from the upstream side of the boat, as far from the side of the boat as safely possible.

Who will Monitor?

Who will conduct the monitoring, and what is his or her understanding of water quality monitoring and data interpretation? Because timing is an important part of water quality monitoring, the best person to collect water samples may be the livestock producer, as he or she can usually respond to weather changes faster than can a hired expert or agency. An automated sampling station can be programmed to sample on an hourly, daily, or other desired frequency; however, such stations are expensive and can handle a limited number of samples only.

How Long to Monitor?

How long will the monitoring last, and how will the producer know when it is time to stop? The longer a producer studies an area, the more accurately he or she can evaluate the situation, taking into account varying rainfall intensities and timing. The length of a study will also depend on how often sampling is conducted during the season.

How to Collect Samples and Related Information

Stream Flow

A livestock producer must know the stream flow amount to determine pollutant load. The pollutant load is calculated by the pollutant concentration (from the lab report) multiplied by the stream flow amount. Some streams have a U.S. Geological Service gauging station that records flow volumes. Producers can obtain this information from the U.S. Geological Service Web site (http://h2o.usgs.gov/). The data may be miles away from the property, and other streams may add (or a diversion may subtract) flow volume by the time the stream reaches the producer's property. Although the flow measurements may not be exact, they will allow the producer to estimate flow data for his or her stream.

Because stream flow is simply the volume of water passing a point on a stream per unit of time, the producer can estimate stream flow as the area of water within a cross section of the stream channel, multiplied by the current velocity. Stream flow is usually expressed as cubic feet per second (ft^3/sec). One ft^3/sec equals 7.5 gallons per second.

$$\text{Stream flow } (\text{ft}^3/\text{s}) = \text{cross-section area } (\text{ft}^2) \times \text{average velocity (ft/s)}$$

$$\text{Average depth} = 1' + 2' + 3' + 2' + 1' \,/\text{Five measurements.}$$

$$\text{Area} = \text{Width} \times \text{Average depth}$$
$$18 \times 10' \times 1.8'$$

Calculating the Area of a Channel Cross Section

Figure 15.2 illustrates the partition of a stream channel cross section into five divisions and the estimation of area within each division. The number of divisions in a cross section depends on the characteristics of the

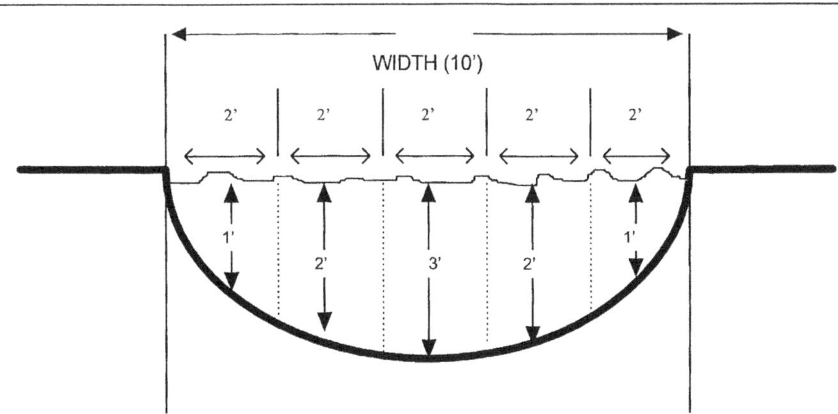

Figure 15.2. Calculating the area for a channel cross section.

cross section. The more irregular (rough) the cross section, the more divisions required.

Velocity is a measure of how fast something moves. Water velocity can be measured by timing how fast a floating object travels 50 ft in a stream deep enough to float the object. Wind can be a factor, so use an object that floats low in the water, such as an orange, fishing bobber, or stick. Calculate the velocity using the formula below. The average of three to five trials will give a good velocity figure.

$$V = 50\,\text{ft}/X\,\text{sec},$$

where X is the number of seconds it takes the object to float 50 ft and V is velocity. For example, if an object floats 50 feet in 25 seconds, velocity is

$$V = 50\,\text{ft}/25\,\text{sec} = 2\,\text{ft/sec}.$$

Completion of Stream Flow Calculation Example

$$V \times \text{Area} = \text{cubic feet/second};$$

$$2\,\text{ft/sec} \times ;1.8\,\text{ft}^2 = 3.6\,\text{ft}^3/\text{sec}.$$

Calculation of Load

$$\text{Concentration} \times \text{stream flow} = \text{Load}$$

$$10\,\text{oz sediment/ft}^3 \times 3.6\text{ft}^3/\text{sec.} = 36\,\text{oz sediment/second.}$$

Collection and Preservation of Samples

Before sampling, a producer should contact a laboratory that has a Quality Assurance/Quality Control state certification and request collection bottles and sampling and handling procedure information. The certification ensures that the lab complies with state procedural standards. When sampling, the collector should make sure that the lip of the bottle is not touched. The producer should also pay special care and attention to handling the bottle cap to prevent contamination. Ask the laboratory to explain the proper way to place the water sample into the bottle. Use an ice chest to keep the water samples below stream temperature during transportation. Try to get the samples to the lab within 24 hours. In general, the shorter the time between collection and analysis of a sample, the more reliable the results are. Table 15.2 shows the maximum storage times for various samples.

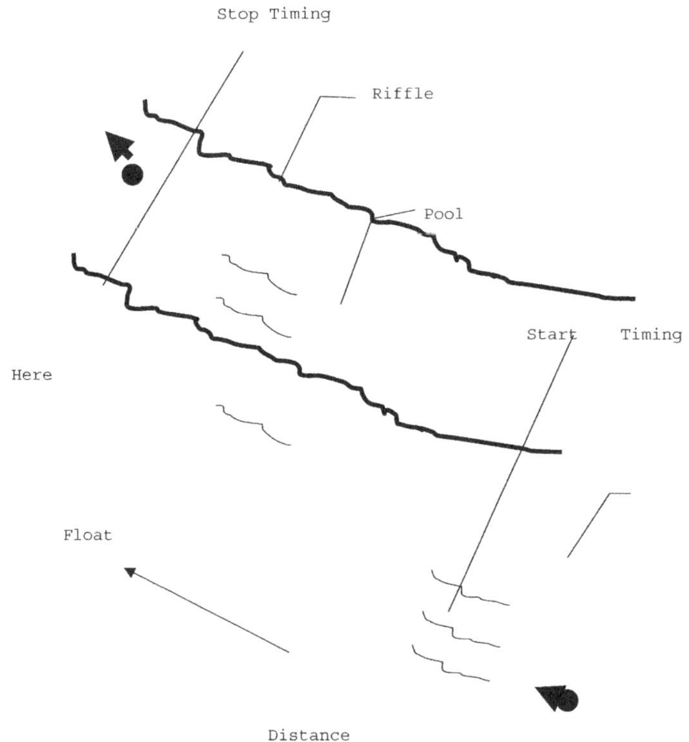

Figure 15.3. Simple method to determine stream velocity. The velocity can be measured by a velocity meter (pygmy) or by timing a floating object between two points.

Table 15.2. Summary of special sampling or handling requirements.

Determination	Container	Minimum Sample Size (mL)	Preservation	Maximum Storage Recommended Regulations[a] by (recommended /EPA maximum)	Method	Location (lab or field)
Nitrate	P,G	100	Add H_2SO_4 to pH < 2, refrigerate	48 h/48 hours		
Nitrate + nitrite	P,G	200	Analyze as soon as possible or refrigerate; or freeze	None/48 hours	Cadmium reduction withcompact or	Either
Total solids			Measure within 24 hours		Oven drying	Lab
					Meter	Either
Dissolved oxygen	G	300	Analyze immediately	0.5 h/1 hour	Meter	Either
					Winkler	Either
			Titration may be delayed after acidification	8h/8 hours		Lab
Phosphate	G	100	For dissolved phosphate, filter immediately; refrigerate or freeze	48h/8 hours	Ascorbic acid with color	Either
					Comparator spectro-photometer	Lab
Temperature			Measure in the stream		Thermometer or recording thermistor	Field
Turbidity	P,G		Analyze same day; store in dark up to 24 hours	24h/48 hours	Meter	Either
					Secchi disk	Field

Sources: Environmental Protection Agency 1995, American Public Health Association 1992.
Refrigerate = storage at 4°C (39°F), in the dark; P = plastic (polyethylene or equivalent); G = glass.
[a] Environmental Protection Agency handling guidelines.

Table 15.3. Maximum weekly average temperatures for growth and short-term maximum temperatures for the following fish °F).

Species	Growth	Maximum	Optimum or mean	Upper temperature[a]
Channel catfish	90	95	81	84
Largemouth bass	90	93	70	81
Smallmouth bass	84	63		

[a] The upper temperature for successful incubation and hatching reported for the species (ERL-Duluth 1976).
Source: EPA 1986.

Quantity of Water Sample

For most chemical or physical analyses, a two-quart or smaller sample should be sufficient. In some cases, larger samples may be required. Often, more than one type of bottle will be needed. Table 15.2 shows the volumes required for different types of analyses. The lab to which the producer plans to send his or her samples should be able to advise about how much water to collect for each analysis as well as provide the producer with the bottles required.

Make a record of every sample collected and identify every bottle, preferably by attaching an appropriately inscribed tag or label. Include the name of the collector, date, hour, exact location, and any other data that may be needed for correlation, such as weather conditions, water level, and actual or estimated stream flow. Fix sampling points by detailed description, by maps, or with the aid of stakes, buoys, global positioning system, or landmarks in a manner that will permit their identification by other persons without reliance on memory or personal guidance.

Water Sample Containers

Label the bag or bottle with the site number, date, and time. Fill in the bag or bottle number or site number on the appropriate field data sheet. This is important. It is the only way the lab coordinator will know which bottle goes with which site. Producers may choose between two container types, screw-cap bottles or bags. Consult the lab as to whether they have a preference.

Specific Water Quality Variables

Water Temperature

Water temperature can be the most limiting factor for fish production, as it greatly affects the amount of dissolved oxygen available in water. Most aquatic organisms assume the temperature of the water that surrounds them. Their metabolic rates are controlled by water. Trout generally need 50°–60°F for normal production. Temperatures above 78°F may be lethal. Temperature ranges for several species are shown in Table 15.3.

Water temperature thermometers come in two types. One type is a red, alcohol-filled thermometer. This is a cheap, simple, but labor-intensive method of monitoring temperature. If the producer is interested in obtaining less labor-intensive and more frequent samples, sophisticated electronic temperature recorders can automatically take up to several thousand readings. These devices can be placed in several sites in the stream and monitor temperature throughout the season. Data can easily be downloaded into a computer. (See Resources section for more information on available devices.)

Dissolved Oxygen

Dissolved oxygen controls aquatic organism growth and development, and it concerns most fish biologists. Livestock producers can monitor dissolved oxygen if there is a concern about the lack of stream vegetation, possible manure inflows, or high temperatures. Dissolved oxygen is measured in parts per million of oxygen to water. Dissolved oxygen levels are affected by water temperature, altitude, water agitation, types and numbers of aquatic plants, light penetration, amounts of dissolved or suspended solids that use oxygen (such as organic matter), and salinity.

The capacity of water to hold oxygen in solution (dissolved oxygen) is directly related to water temperature. Oxygen is more soluble at lower temperatures; thus, dissolved oxygen decreases as temperature increases. At higher altitudes (elevation), the dissolved oxygen saturation point is lower than when under the same conditions at lower altitudes.

Trout prefer having the dissolved oxygen near saturation. Water absorbs oxygen from the atmosphere as it contacts air at rapids, riffles, or waterfalls in a

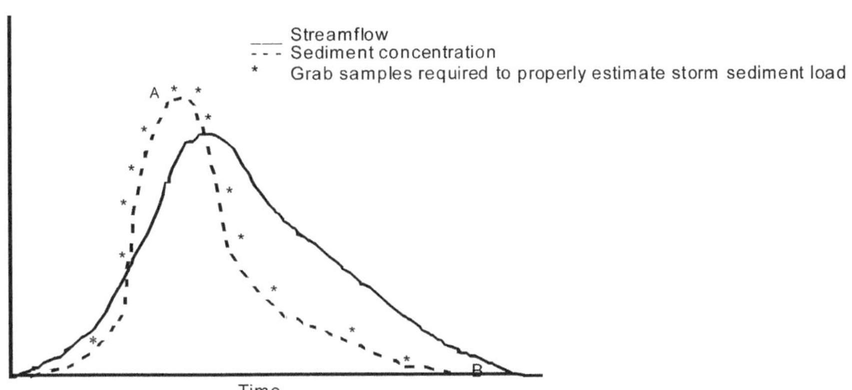

Figure 15.4. General relationship between stream flow and suspended sediment concentrations during a stormflow event.

stream. Oxygen can also be added to water by plant photosynthesis.

Dissolved oxygen can be measured in the field with an electric meter with a membrane sensor, or with a kit that can be purchased (see Resources section).

Sediment and Turbidity

Often turbidity is measured because it is easier and faster than measuring suspended sediment. Turbidity refers to the amount of light that is scattered or absorbed by a fluid. Turbidity is actually an optical property of the fluid. An increase in turbidity is visually described as an increase in cloudiness. Turbidity in streams is usually caused by suspended particles of silt and clay (eroded soils), but other materials such as colored organic compounds and microorganisms can add turbidity.

High sediment levels adversely affect fish. Very high concentrations of suspended sediments can irritate and actually clog gill filaments, causing fish to suffocate. Bedload sediments, such as gravel, create beds used by fish for spawning. Increased water sediment can reduce the amount of suitable spawning habitat. It fills the spaces between the rocks where the eggs can be deposited, or it covers deposited eggs, which reduces the amount of oxygen available to them by blocking water circulation; sediment also traps fry in the gravel. Sediment-caused changes in plant and insect composition can reduce the amount and types of food available during different stages of fish development.

Turbidity can be measured either by purchasing a turbidity meter (which is very expensive) or by sending a sample of water to a laboratory to be analyzed. The sample needs to be kept in a cool, dark place until it can be delivered to the lab. The sample should be delivered to the lab within 24 hours of collection.

Suspended sediments or turbidity sampling results depend on timing and frequency of sampling in relation to the stream flow. Figure 15.4 illustrates the general relationship between suspended sediment concentration and stream flow during a single storm event on a headwater's stream. Suspended sediment is the major NPS pollutant on rangelands and is used as an example. Most other nonpoint source pollutants on rangelands have a similar, if not more complex, relationship with flow.

In headwater streams, suspended sediment concentration increases rapidly with stream flow and typically peaks before stream flow does (Figure 15.4). Erosion is usually highest at the start of a runoff event, when sediments are readily available for transport by overland flow to the stream channel. Sediment concentrations drop rapidly as the supply of readily eroded sediment is diminished. Note the timing and frequency of grab samples required to adequately estimate total storm sediment load.

The relationship in Figure 15.4 will repeat throughout a wet season as storms come and go and stream flow rises and falls. The amount of suspended sediment per storm will be reduced throughout the season, as the readily available sediment supply in the watershed is depleted, but the relationship of suspended sediment concentration to flow will remain the same. There will be brief periods of sediment loading (i.e., high sediment concentrations and high flow). Each storm event would have to be sampled with timing and frequency similar to that illustrated in Figure 15.4 to estimate the total NPS loads in the stream.

Table 15.4. Water quality parameters for fish.

	General	Salmonids	Warm water fish
Dissolved oxygen (minimum)	6 mg/L	7 mg/L	6 mg/L
Conductivity (maximum)	300 mmhos/cm	300 umos/cm	500 umos/cm
Phosophorus[a]	50 (ppb)	50 (ppb)	50 (ppb)
Nitrogen (maximum)	300 (ppb)	300 (ppb)	300 (ppb)

Source: Oregon Cooperative Extension, Klamath County 1995.

[a] Values used by Oregon Department of Environmental Quality may not be universally appropriate. Check with a local source.

Phosphorus

Phosphorus may be of interest if erosion, sediment, fecal contamination, or applied phosphorus fertilizers are a concern. Phosphorus is usually measured as total phosphorus and orthophosphates. Orthophosphate is a chemistry-based term that refers to the dissolved inorganic form of phosphate found in aquatic environments. Total phosphorus measures include orthophosphates, condensed phosphate (inorganic), and organic phosphate. Phosphorus can negatively affect water quality, as it is one of the most important limiting nutrients for algae growth. When algae grow, respire, die, and decay, they lower dissolved oxygen levels. The analysis for phosphorus can be done in the field using a Hach kit or can be determined from samples sent to a laboratory. Phosphorus concentrations vary with the time of day because of fluctuations in temperature and biological activity.

Sample containers need to be glass, acid-washed plastic, or disposable Whirl-pak bags. The EPA analysis procedures are outlined in Table 15.2.

Nitrogen

Nitrogen may be of interest when erosion or sediment is of concern. Nitrates can be of major concern in excess amounts (along with phosphorus) because they can increase algae's growth, which will ultimately decrease dissolved oxygen. Nitrogen may also be important if fecal contamination or applied nitrogen fertilizers are at issue. Substances that are generally measured in nitrogen include nitrates (NO_3), ammonia, and total nitrogen. Nitrogen is found in the following forms: ammonium (NH^{+3}), nitrates (NO^{-3}), and nitrites (NO^{-2}). Total N includes these forms and the nitrogen bound into organic matter.

The natural levels of nitrate are typically low, less than 1 mg/L. Because nitrogen-monitoring equipment is expensive (nitrate electrodes and meters can cost $700–$1200), laboratory measurement is most common. Delayed or warm transport of samples can alter nitrogen levels.

Electrical Conductivity (Nitrogen and Phosphorus)

Electrical conductivity is a measure of the ability of water to pass an electric current. The basic unit of measurement of conductivity is micromhos/centimeter (μmhos/cm), or microSiemens/centimeter (mS/cm). Distilled water has conductivity and a range of 0.5–3.0 μmhos/cm. Rivers generally range from 50 to 1500 μmhos/cm. Streams that support fish range from 500 to 1500 μmhos/cm. Conductivity in water is affected by the presence of inorganic dissolved solids such as salts, nitrates, and phosphates (Table 15.4).

Bacteria/Pathogen

Contamination of streams by warm-blooded animals with microorganisms is monitored by measuring fecal coliform (FC) and fecal streptococci (FS) concentrations in the water. FC and FS are often referred to as "indicator bacteria." FC and FS do not generally cause disease themselves, but they may indicate the possible presence of other disease-causing bacteria and protozoan. The sources of FC and FS are warm-blooded animals including wildlife, domestic animals, and humans. Therefore, a sample of water that tests positive for FC and FS will not clearly identify which mammal, including humans, contaminated the water.

Allowing water samples to reach warmer temperatures may increase the bacterial counts and misrepresent the real situation in the stream. The laboratory needs to process the samples within 30 hours of sampling. Bacteria can be measured either with a simple field kit called "total bacterial count" or through laboratory analysis. The field kit will give a general idea of the number of bacteria present. Cryptosporidium is also an important pathogen that can be checked. For a more specific analysis, take samples to a laboratory. (See Resources section.)

U.S. ENDANGERED SPECIES ACT

Many livestock producers in the United States have concerns about the potential or present effect of the Endangered Species Act (ESA) on the viability of their operation. The state and federal ESAs were passed to prevent animals and plants from becoming extinct in the United States. These laws restrict the actions of people who would make changes in the landscape or engage in other activities that affect certain plants or animals. These restrictions can be costly, and in general the person or agency that wants to make the change in landscape or activity must bear the cost of compliance with the law. It also is possible to violate these laws without making any changes in landscape or activity, if a person's actions result in harm to a threatened or endangered (listed) species.

What Is a Livestock Producer's Interest in Endangered Species?

A few years ago, livestock producers were concerned about trespass, poaching, and the laws relating to those problems. Recently, more livestock producers are concerned about wildlife and plants that some government agency has officially "listed." This listing may create a legal status for the species that can influence production practices. The species may be endangered, threatened, rare, poorly known, or food for another animal that is in one of those categories. Livestock producers may be concerned because the endangered species threatens the continued use of the land. However, livestock producers may make money by selling development rights to a "conservation bank" or "mitigation bank." To find out whether a ranch or farm has animals or plants that are officially listed, see the Web site of the U.S. Fish and Wildlife Service (USFWS) field office that covers the livestock producer's area. The *Federal Register* reports contain information about the listed species and recovery as well.

How Do Livestock Producers Find Out Whether They Have "Critical" or Sensitive Habitats?

In some situations, the USFWS has identified designated "critical habitats" for the recovery of endangered or threatened species. Contacting the closest field office of the USFWS should help a producer learn whether any of these habitats are on his or her ranch or are nearby.

What Is the ESA?

The federal ESA is a visionary and powerful law aimed at protecting the natural diversity that has enriched our lives. The ESA's method is to force those who would change the landscape to think about the changes in terms of wildlife and plants, and in some cases to repair damages their plans may cause or replace the damaged resources. Some species that are federally listed are not state listed, and vice versa.

The ESA prohibits "taking," "harming," or "harassing" endangered wildlife. "Take" means to harass, harm, pursue, hunt, shoot, wound, kill, trap, capture, or collect, or to attempt to do so. "Harass" means any act or omission that is likely to "significantly" disrupt normal behavioral patterns of the animal. "Harm" means an actual injury to an animal or its habitat. With a few exceptions, a livestock producer cannot legally possess, sell, deliver, carry, transport, or ship endangered or threatened wildlife, or violate any regulation made regarding them. The latter phrase allows the USFWS to prosecute violations of minor laws that involve endangered species, using ESA's penalty provisions.

ESA prohibitions other than those mentioned in this chapter do not pertain to normal agricultural or management activities on private or public land. Violation of the ESA carries civil and criminal penalties of between \$12,000 and \$25,000 per violation, with criminal penalties running between \$25,000 and \$50,000, which may include a year in jail.

Incidental Take

The USFWS may allow some "taking" of listed species as "incidental" to other legal operations. An "incidental take permit" usually is part of a habitat conservation plan (HCP) for the species involved. It also may be obtained with a Section 10 "safe harbor agreement" or as an "incidental take statement" in connection with a Section 7 "biological opinion." The process by which someone obtains permission to incidentally take listed species varies under federal law depending on whether the producer encounters federal involvement ("nexus") in the project.

Federal "Nexus"

If action by a livestock producer, even on private land, will affect a listed species, someone is required to consult with the USFWS. If there is federal involvement in the action in question, then the federal agency involved does the consultation. This is called a "Section 7 consultation," after Section 7 of the ESA, which discusses agency consultations. If there is no federal involvement, then the private landowner or agent is required to consult under Section 10 of the ESA. The federal involvement that triggers a Section 7 consultation is called a "nexus."

What kind of involvement creates a federal nexus? If the action or the land involved is funded even in part by the federal government, there is a nexus. (This means that if a livestock producer is taking advantage of a federal farm program, there is a nexus.) If the federal government will help a livestock producer carry out the action, including providing technical assistance, there is a nexus. If a livestock producer will use pesticides in the project, because pesticides are permitted according to federal label restrictions, there is a nexus.

Once there is a federal nexus, part of the consultation will involve assessment of cumulative effects. These, by definition, involve other local or private actions that will predictably occur on the same land. So, if a livestock producer has a federal nexus for one action—for example, the use of pesticides—then everything else a livestock producer does on that land must be considered as part of the consultation for the pesticide use.

Can a livestock producer affect the listing decision?

A decision to list a particular species is based on the best available data. If a livestock producer can provide more data showing larger numbers of individuals or populations, or populations in formerly unknown areas, this information may be enough to prevent listing or to reduce the level of listing. In this situation, it will be necessary to document locations and provide evidence of the species' existence and numbers in these new areas.

Even if the numbers warrant listing, a recent policy may allow a livestock producer to delay or prevent the listing. The Policy for Evaluation of Conservation Efforts has guidelines for evaluating ongoing conservation efforts during the listing determination. If existing programs seem likely to improve the species' status without listing it, the listing decision may be delayed to give the conservation effort time to succeed. To be considered, the effort must have explicit objectives, steps, and procedures for accomplishing the goals, dates for accomplishment, and monitoring standards. Any laws or regulations that are necessary to the effort must be in place, and funding must be ensured.

Consulting with the USFWS

Once a livestock producer conceives of an action that might affect a listed species in any way, then he or she must consult with the USFWS and the appropriate Wildlife Agency. If a livestock producer has no federal involvement at all, then he or she must consult with the USFWS, or the livestock producer can hire someone to do it (see section on consultation). If there is a federal nexus, the livestock producer should ask the federal agency involved to enter into an "early consultation" with the USFWS under Section 7 of the ESA. The USFWS is required to do so, and the livestock producer is authorized by law to ask this agency to do so.

If a livestock producer will be assisting all relevant listed species, then he or she may be able to work through the required applications by himself or herself. If not, the producer may need to mitigate for harm to listed species, and the burden of technical preparation of the various plans and applications will be considerably larger and may require hiring a consultant with knowledge of the species. If the livestock producer decides to proceed with the project and do the conservation, he or she may want to hire a wildlife consultant to help with the application procedure.

One of the first things required will be an "effects determination." This is a report that determines whether the project will have an effect on the listed species. With federal involvement in the project, the agency will make this report. Without the federal nexus, the livestock producer or the designated agent must make the report. The USFWS will either approve or reject it. "Effects" are determined differently under federal and state law. The livestock producer will need a separate section of the report to determine effects to state-listed species, and his or her Wildlife Agency will approve or reject that part of the report. These reports need to be prepared thoroughly, because occasionally they are challenged in court. The effects report should address

- History and trends of listed species and grazing in the area
- Season, length, distribution, and stocking capacity of grazing
- A description of current or baseline conditions
- Population viability analysis for listed species
- Details, dates, and descriptions of any management activities
- Details, dates, and descriptions of any acts of God (e.g., fire, floods, drought)
- Description of the plans and any effects on listed species or habitat

- Measurable, noticeable, or detectable effects of the plans
- Identification of direct, indirect, and cumulative efforts
- Determination of whether efforts are beneficial or detrimental
- Effects of the project that may affect listed species beyond the precise project area.

The conclusion of the effects report must state

- "No effect"; for example, no consultation needed
- "Not likely to adversely affect" determination; for example, informal consultation should occur
- "Likely to adversely affect" determination; for example, formal consultation is required, including developing a HCP.

The USFWS will review the effects report and may disagree with the conclusions. If consultation is necessary, the livestock producer or agent must perform it unless there is a federal involvement in the project. If there is federal nexus, the agency must do the consultation, and they are restricted by law so that they must move the species toward recovery.

Is an "incidental take permit" a good option for a livestock producer? In some cases, it may be. Here are some other options. See Table 15.5 for a complete listing.

- Producers may be able to avoid a take or impact of endangered species under state and federal laws
- Some state laws provide a statutory exemption for normal and ongoing agricultural operations; there is no similar federal provision
- Producers may obtain an incidental take authorization for the project under certain conditions under both state and federal law
- Producers may obtain an incidental take authorization as part of a conservation plan under both state and federal law
- If a livestock producer wants to make improvements in habitat, he or she may obtain a safe harbor agreement under federal law that allows the producer to return the habitat to its "baseline" or beginning condition at the end of the agreement; safe harbor also provides for an incidental take authorization
- Federal law also allows for a "candidate conservation agreement" for improving habitat of candidate species to the ESA. This will provide protection in the event the species is listed.

Avoiding Take

If a livestock producer will do something that could possibly be considered by agencies to affect a listed species or its habitat, he or she may need to officially justify that no species take or habitat impact will occur as a result of that action. This is called "avoidance of take." The state and federal procedures for doing this are different, and a livestock producer should contact the USFWS and Wildlife Agency for details. Livestock producers will need to document the details of their project to demonstrate that there will be no effect on the listed species. If they are in a designated critical habitat for a listed species, the species need not be present for "taking" to occur. Documenting avoidance of take means a livestock producer need not devote further time and effort toward other applications under the ESA, but it provides no protection to the producer should he or she inadvertently "take" a listed species.

Statutory Exemptions

As an example, under California Fish and Game Code (section 2087), no permit is needed if accidental take of state-listed species occurs during routine and ongoing agricultural activities. Fish are not covered under this exemption, nor is land conversion. However, nonagricultural activities such as flood control, levee work, or diversions are not exempted. There is no federal counterpart to this California exemption, so the federal ESA may still apply.

Incidental Take Authorizations

Federal incidental take authorizations may be obtained in several ways. If there is an official determination that no effect will occur to listed species, a permit or an official incidental take statement will be issued. This usually occurs with a federal nexus, following a "no effect" determination or a "biological opinion" or "biological assessment." Without the federal nexus, the USFWS will issue incidental take permits following approval of a safe harbor agreement or a HCP.

Safe Harbor Agreements

Safe harbor agreements apply to situations in which a livestock producer plans to improve habitat for listed species. The agreement will need to describe the existing (preproject) habitat or population status of the species (the "baseline"), the anticipated management, and a description of the land on which it will occur. The agreement should include how specific habitat improvements and management will help

Table 15.5. Comparison of regulatory mechanisms providing incidental take authorization.

	Voluntary Local Program	Section 2081 Incidental Take Permit.	Section 2835 NCCP Act	Safe Harbor Agreement	Section 7 BO/ITS	Section 10 HCP
Federal or state	State	State	State	Federal	Federal	Federal
Listed or unlisted species	Listed (no fish)	Both	Both	Listed (no anadromous fish)	Listed	Both
Activities covered	Routine and ongoing agriculture	Private or public	Private or public	Private or public	Federal activities	Private or public
Covers restoration activities	Yes, on agricultural land	Yes	Yes	Yes	Yes	Yes
Applicant	Single farmer, group of farmers, or organization	Single landowner or organization, or land use agency	Land-use agency, or owner of large contiguous area of land	Private or public project proponent or landowner, or land-use agency	Federal agency, or private or public project proponent seeking permit or funding	Private or public project proponent or landowner, or land-use agency
Geographic scope	Small or large area; limited to agricultural land	Small or large area	Large area or ecosystem	Small or large area	Small or large area	Small or large area

Conservation required	Habitat enhancement; management practices to avoid or minimize take	Take must be minimized and fully mitigated.	Conserve and manage natural diversity on a landscape or ecosystem level; habitat reserves or equivalent; adaptive management	Net conservation benefit that contributes directly or indirectly to recovery of the species	Must not jeopardize the continued existence of listed species; must include reasonable and prudent alternatives, if necessary	Must minimize and mitigate take to the maximum extent practicable
Monitoring Required	Report of acreage benefited	Yes	Yes	Yes	Sometimes	Yes
Assurances to participants	None. Wildlife Agency may seek amendment to VLP if law changes	Long-term, as long as plan is being implemented properly	Long-term, as long as plan is being implemented properly	Long-term, if property does not move below baseline	None	Long-term, as long as plan is being implemented properly
Assurances to NonPartici-pating Neighbors	None	None	None	None	None	None
Length of time for approval	8–12 months	1–3 years	3–5 years	8–12 months	5–6 months	1–5 years
Environmental review	Certified regulatory program	EIR	EIR	None	EA or EIS	EA or EIS

Resources Law Group, LLP 2003 (http://www.resourceslawgroup.com/contact/contact.htm).

the species. The USFWS or a designated consultant probably will participate in designating the baseline. The livestock producer will draft an agreement and monitoring plan that will ensure that he or she will accomplish the project as described. The producer will submit a plan, along with an application for an "enhancement for survival" permit. The USFWS may modify these documents.

Following internal approval of the documents, USFWS will submit them according to environmental protection laws to publication in the *Federal Register* and public comment before USFWS can give final approval. Once the agreement is signed, the USFWS can revoke it if it is determined that the situation of the species has changed so that a producer's project causes jeopardy to the species. As long as it is in force, the agreement allows a producer to return his or her land to other uses at the end of the agreement period, as long as the producer's land does not become less hospitable to the species than the baseline conditions.

Candidate Conservation Agreements

A "candidate" species is one that has not yet been listed but has been proposed for listing and is under consideration, or will be considered later. A livestock producer may now obtain a "candidate conservation agreement with assurances" that will provide some predictability to future management decisions. This is a federal program only. The program is new as of this writing, and exact requirements are unknown at this time. However, once the agreement is signed by the USFWS, a producer agrees to improve habitat or other conditions for the candidate species, in an effort to prevent the need for listing. A livestock producer has assurance that if the species is listed, he or she will have no restrictions in addition to those already in the agreement. A producer also can include nonlisted species in an HCP, and the incidental take permit for them would be activated if they were listed. However, the incidental take permit and HCP do not necessarily preclude additional restrictions because of a new listing, as might happen, for example, when a critical habitat is designated.

Voluntary Local Programs

Voluntary local programs protect routine and ongoing agricultural practices and wildlife habitat restoration on agricultural lands as defined in an agreement a producer reaches with his or her Wildlife Agency.

HCP

What Is a HCP?

Do livestock producers avoid creating wildlife habitats because listed species might invade the habitat? Are they concerned that they might have problems using management practices because of endangered species? These are substantial and real concerns, but a potential solution exists. It is the HCP. Briefly, the HCP is a plan that provides for endangered species' conservation and continuing commercial operations in an area. Once approved by the USFWS and Wildlife Agency, it allows normal land management operations to go on according to the plan, without jeopardy from the ESA. Even if a producer kills a limited number of endangered species during normal management operations, the species loss is covered under the "limited take" provisions of the ESA, as defined in the HCP.

Will the HCP Cover All Endangered Species?

The HCP will cover only the species mentioned in the plan. In addition to listed species, a producer may also include species that are not presently listed but that may be listed in the future. A livestock producer may write a plan for one species or for several. The species need not be present when the plan is written, as new species may be added by amendment.

Will an HCP Allow a Livestock Producer to Plow Up New Ground or Change Management?

The plan may allow the destruction of a specified, limited number of an endangered species or some portion of its habitat for normal management practices if conservation criteria are in place and defined in the plan. The livestock producer must somehow replace the habitat destroyed, perhaps in another area. Producers can probably do this by purchasing land or paying for land improvements that will provide such habitat (i.e., mitigation via a land bank, discussed below). The other side of this is that a livestock producer may be able to sell conservation rights to others if he or she has a habitat that is in a high-quality area for the species and can improve it or permanently protect it from development.

How Large an Area Must Be Included?

The HCP may be prepared for a single ranch or for a larger area. Most of the recent HCPs now under preparation are for areas ranging from 6000 to 10,000,000 acres. A large HCP has several advantages. First, a large HCP will include a broader spectrum of conditions and future management scenarios, and it is more likely to include all the possible species and management options on a producer's land. Second, a large plan will have more latitude in identifying conservation areas, giving more flexibility to the producer. Finally, there is also the potential of sharing the cost of preparing the HCP among several producers.

What Is the General Procedure?

Before a livestock producer begins preparing a HCP, he or she should contact the USFWS and Wildlife Agency, and involve them in the process. All the affected parties should be encouraged to attend a meeting so they can learn of the proposed action and have an opportunity to become a part of it, or to exempt themselves. Affected parties may include local, state, and federal government agencies, private landowners, and special interest groups involved in wildlife and natural resource issues, as well as those involved in agricultural and water issues.

Once a producer and all the agencies agree on the scope and direction of the project, the HCP can be prepared. For a small project, discussions among the affected parties will provide the necessary agreements and resolve details in the HCP. For most HCPs, a consultant will be needed, and at considerable expense. Preparation of an HCP for a large area may take a year or more.

The USFWS may require more information, but at a minimum, the HCP must include

- Location, management actions, and people involved
- The likely effect on the included species from the planned activities
- Monitoring and reporting procedures
- Conservation measures and funding for them
- Criteria for determining success, including a baseline (preproject) biological assessment
- Alternatives considered and reasons why they were rejected
- Procedures to ensure performance and protect wildlife sites forever.

The USFWS cannot issue a permit if the activity will be to the disadvantage of the species. A livestock producer also must convince the USFWS that he or she will actually carry the plan to completion. The USFWS could require such things as bonding, contracts, and periodic reports on monitoring results and management actions. The producer will be expected to describe an adaptive management procedure in connection with monitoring results and to describe some restoration activities to benefit the listed species. He or she may include nonlisted species in the HCP, or they may be added later by amendment. If newly listed species are added later, much of the same process will be required as was necessary for the original species.

Once the HCP is prepared, the livestock producer must submit it to the USFWS, which may return it informally for changes before "accepting" the submission. Once it is "accepted," the USFWS is bound to a time limit for processing the approval. This limit, however, may be several years. The main time consumer in this process is the preparation of an environmental impact report, with its associated public meetings and requirements for advance notice in the *Federal Register*. In general, the USFWS will ask that the people who request the HCP also provide the environmental impact report.

A safe harbor agreement made in connection with an HCP will allow a livestock producer to return the landscape at the end of the safe harbor agreement to baseline conditions, and will allow any agreed-on changes made because of conservation. The agreement will apply to all species included in the HCP. A safe harbor agreement can be very complex. Before entering into one, producers should engage in an in-depth study of the ramifications.

"No Surprises"

The USFWS and National Marine Fisheries Service have a "no surprises" policy in connection with the HCP. The policy is that once the HCP is finalized and working, as long as it is being properly implemented, the government will not require additional conservation from the landowner, even if the listing status of the species changes, except where there are extraordinary and unforeseen circumstances.

For more information on the HCP process, a livestock producer can call the Habitat Conservation Planning Branch, the Wildlife Agency, or the USFWS field office that has jurisdiction over their area.

Another way to consider which state or federal programs are likely to be involved in a project is according to project size. For large, regional-scale projects, there usually will be federal involvement, and the appropriate approaches are the state Natural Community Conservation Plans and federal HCP (the state version of the federal HCP is the Natural

Community Conservation Plan). For medium to small projects, the Natural Community Conservation Plan and HCP also are options, but a state voluntary local program and federal safe harbor agreement may be sufficient on private lands only. Federal protection for small, private (no federal nexus) projects will involve a safe harbor agreement or HCP. If there is a federal nexus, small projects may receive federal protection through a biological assessment or biological opinion, with either one accompanied by an incidental take statement.

"CONSERVATION" AND "CONSERVATION BANKING"

Some HCPs include an idea known as "conservation banking," which gives credit for previous habitat protection or improvement not already used for conservation. Wetlands projects under the Clean Water Act may involve a similar concept called "mitigation banking." People who own endangered species habitats may be able to sell easements or otherwise improve the habitat in return for habitat credits. Some commercial conservation banks have been approved. They purchase improvements or easements done in advance and hold them for sale to people who need to mitigate damage done by a project. In some cases, the bank may simply act as a broker between the developer and the owner of conservation credits.

People have criticized the practice of conservation banking as trading dollars for the right to destroy habitats. That is an incomplete description of the process. The dollars traded are used to improve habitats or to protect other habitats for the same species whose habitat is in danger of destruction. Conservation banking is a voluntary process. A willing landowner offers to the "bank" an easement or management system to improve conditions for an endangered species. A developer who wants to destroy some endangered species habitat goes to the "bank" and essentially pays to have the other land improved to substitute for the damage proposed so that the development can continue.

Under what conditions can a landowner profit from these procedures?

PROPERTY IDENTIFICATION

Location

The property must be important to a threatened or endangered species. A critical habitat listing, recovery plans, and HCPs describe areas that are desirable for preservation or development for one or more endangered species. If any of these documents include a livestock producer's land, he or she may identify habitats important to the species of concern. For example, the producer's land may be in a core habitat area or in land proposed as a linkage between core areas. (Please note that a recovery plan is prepared by the USFWS and describes the steps necessary to protect a threatened or endangered species and improve its habitat. The plans are available on the Internet at http://endangered.fws.gov/recovery/#plans).

Potential for Improvement

The habitat for endangered species may be susceptible to improvement for listed species. Such improvement may take the form of planting appropriate plants, managing water, changing crops or pesticide use, or other active improvement measures appropriate to the endangered species to be favored. A livestock producer must provide maintenance of the improvement, if needed.

Protection

If the property is in danger of change that would adversely affect the endangered species, a permanent protection against that change may be an appropriate strategy. Protection of this sort is often done through permanent easements or deed restrictions. Sometimes, under prior agreement, protection can still allow continued compatible uses. Two examples might be continuation of a waterfowl-hunting club on land preserved from development to benefit the giant garter snake, or livestock grazing in areas used for nesting by bald eagles.

Cost-Effectiveness

The proposed change must be cost-effective for the landowner, who will be giving up future options and who may be committing to make improvements. The price received for doing the conservation should be sufficient to cover the cost of habitat enhancement, management into perpetuity, and value foregone. Values vary widely, but people have sold recent conservation credits for as much as 50% of fee value for land protection. Credits may exceed total land value when management is part of the agreement.

Firm Commitments

The USFWS will require firm commitments for any services required from outside parties, such as for water delivery. These commitments must span the duration of the project, and normally the duration would be "into perpetuity"—or permanent.

Acceptance by the "Bank"

The USFWS and Wildlife Agency, and perhaps other agencies, must approve the proposed action through the administrators of the conservation bank. A livestock producer and the agencies will sign a formal agreement in which they agree to perform the actions for the value received.

Indirect Benefits of Helping Listed Species

There may be indirect financial benefits from helping listed species. If a livestock producer has an agritourism business, the effort the producer has made can be an advertising bonus and may create an onsite attraction if human presence would not harm the listed species. Potentially, livestock producers can produce a product that can be labeled "environmentally friendly." Be sure to receive permission for such statements on food labels from the USDA authorities who approve these labels.

If a livestock producer owns land that potentially could support, or does support, endangered species, management of that land could be a headache, or it could mean additional income. If endangered species have made a livestock producer's land a lemon, maybe the producer can make lemonade: Perhaps a producer can even help remove an endangered species from the list.

Technical and Financial Assistance

Conversation assistance may be available from various agencies and organizations, including:

- USDA
- USDA-NRCS
- USDA-Farm Service Agency
- Resource Conservation Districts
- United States Forestry Service
- United States Department of Interior
- USFWS
- Land grant university in your state (Cooperative Extension Service)

Several incentive programs are available to help private landowners improve lands for wildlife and protect listed species:

- Conservation Reserve Program (USDA-FSA)
- Environmental Quality Incentives Program (USDA-NRCS)
- Farm Credit and partial debt cancellation (USDA-FSA)
- Forestry Incentives Program (USDA-NRCS)
- Forest Legacy Program (United States Forestry Service)
- Stewardship Incentive Program (United States Forestry Service)
- Wildlife Habitat Incentives Program (USDA-NRCS)
- Wetlands Reserve Program (USDA-NRCS)
- Conservation Easements (USFWS, Wildlife Conservation Board, Wildlife Agency, USDA-NRCS)
- Debt Restructuring Program (USFWS, USDA-FSA)
- Partners for Fish and Wildlife (USFWS)
- Private Stewardship Grants (USFWS)
- Cooperative Conservation Initiative Grants (USFWS)
- North American Wetlands Conservation Act Grants (USFWS)
- Land trusts
- Open space districts

Conservation Easements

Livestock producers use conservation easements as a method to raise cash or provide income to maintain their operation, or to reduce taxes. Livestock producers have a number of private rights that go along with the ownership of property. By granting a conservation easement, livestock producers can agree to give up some of those rights in trade for money or tax benefit.

A "conservation easement" is an agreement between a landowner and a qualified land trust, conservation group, or government agency regarding the future uses of private property. The conservation easement is recorded and becomes part of the deed to the property. It is generally designed to be appropriate to the specific circumstances and to address concerns of both the landowner and the easement holder. It can also be used to reduce the value of the land for estate planning purposes. Some landowners who need tax deductions can deduct the value of the donation of the easement from their taxes.

Most conservation easements involve selling or deeding the development rights either for a specified number of years, under a long term agreement or, most commonly, for perpetuity (forever). The terms of the easement usually involve restriction of housing developments and some set management guidelines for livestock grazing. The amount paid for a conservation easement is directly related to how much the owner is willing to give up in rights. Most easements require a management plan. Some groups require a plan that addresses the acceptable level of grazing. It is best to consider adaptive or flexible management guidelines, as the term of the easement in most cases is forever.

Given the legal and tax implications of conservation easements, it is imperative that these transactions be completed with the guidance of an attorney and a tax advisor. In addition, any outstanding debt secured by the property must be subordinated to the conservation easement; that is, any lenders must acknowledge that the property is encumbered by a conservation easement. The compliance by the producer to these terms is monitored by the easement holder (e.g., the land trust or agency) on an annual or semiannual basis. Most easement holders require an additional cash donation from the landowner to cover the cost of this monitoring into perpetuity. The terms of the easement can vary greatly depending on the agency or group that may purchase it. Because most easements are forever, much thought should go into any terms or conditions. Livestock producers can exclude public access from the conservation easement. For example, the California Rangeland Trust was formed by the California Cattlemen's Association to assist livestock producers in the construction and maintenance of conservation easements. They have an example easement agreement at http://www.rangelandtrust.org/.

Some of the general steps to begin the conservation easement process are

- Collect baseline inventory of the property to record the agricultural, scenic, historical, and wildlife values that the easement seeks to protect
- Obtain legal and tax advice
- Make contact with agencies or nonprofit organizations that provide funding for, or help producers with, securing conservation easements
- Work with the prospective easement holder to complete an appraisal of the easement value
- Secure the cooperation of any lenders
- Create an endowment to ensure that the easement is protected over time
- Ensure that the deed of easement is signed by both parties, notarized, and recorded with the county clerk and recorder

Before considering an easement, cattle producers should take a long time to digest the positive and negative aspects of entering into a conservation easement. Many livestock producers are not interested in selling the land use rights, as they are concerned that it limits their heirs, if cattle production is no longer viable sometime in the future. With urban development in the West, many livestock producers feel that the only real profitable enterprise in a cow/calf operation is the land resource. They see this as a potential asset in the future. Others have a deep ranching heritage that they feel a conservation easement will help to maintain. It is imperative that livestock producers realize that perpetuity is a long time. "Forever" or "infinity" is hard to contemplate or truly understand.

Some livestock producers are legitimately concerned about who will hold the conservation easement. This entity will now be a right holder with the landowner on the property. Because the conservation easement holder has title to certain rights on the property, it is important that the easement holder be a trusted organization that understands ranching. To address this concern, some livestock producers have formed their own easement-holding organizations. Others have included a condition in the conservation easement that it is nontransferable without the permission of the landowner. This also addresses the other concern that somewhere out in perpetuity the holder of the easement could sell or transfer it to another party that is less understanding of ranching.

The producer's vision—and the easement holder's—should match before the parties sign the easement contract. The producer should consider all future alternatives or diversification possibilities (gravel operations, logging, communication towers, farming, future ranch buildings, and so forth) before signing and make sure that future options that could potentially arise in perpetuity are well thought out and are provided for in the contract.

Finally, the future economic viability of ranching operations on lands that are under conservation easement is an important consideration. Instead of receiving a single, lump-sum cash payment for an easement, some livestock producers opt to receive payment over time. Others invest the money in an endowment that provides an annual payment to the owners of the ranch in perpetuity. Although a one-time, up-front payment may benefit the current owners of a ranch, these alternative payment and investment strategies help the future ranch owners realize income from the sale of the conservation easement.

RESOURCES

Endangered Species Resources

The USDA's Integrated Taxonomic Information System is a partnership of U.S., Canadian, and Mexican agencies, other organizations, and taxonomic specialists cooperating on the development of an

Table 15.a. Partial list of chemical and scientific equipment suppliers.

Fisher Scientific 711 Forbes Ave. Pittsburgh, PA 15219-4785 Tel. (800) 766-7000	**Millipore Corporation** 397 Williams Street Marlborough, MA 01752 Tel. (800) 225-1380	**Onset Instrument**[a] P.O. Box 2450 Pocasset, MA 02559-3450 Tel. (508) 563-9000 Fax (508) 563-9477
Hach Company P.O. Box 389 Loveland, CO 80539 Tel. (800) 525-5940	**Thomas Scientific** 99 High Hill Road at I-295 P.O. Box 99 Swedesboro, NJ 08085-0099	**Wildlife Supply Company** 301 Cass Street Saginaw, MI 48602 Tel. (517) 799-8100
Hydrolab Corporation P.O. Box 50116 Austin, TX 78763 Tel. (512) 255-8841	**Carolina Biological Supply Company** 2700 York Court Burlington, NC 27215-3398 Tel. (800) 334-5551	**YSI Incorporated** 1725 Brannum Lane Yellow Springs, Ohio 45387 Tel. (513) 767-7241
LaMotte Chemical Products P.O. Box 329 Chestertown, MD 21620 Tel. (800) 344-3100	**VWR Scientific** P.O. Box 2643 Irving, TX 75061 Tel. (800) 527-1576	

[a] Manufacturer of the temperature monitor

online, scientifically credible list of biological names, focusing on the biota of North America. ITIS is also a participating member of Species 2000, an international project indexing the world's known species; http://www.itis.usda.gov/plantproj/itis/index.html.

The USDA's Natural Resources Conservation Service Web page; links for regional and state offices are available; http://www.nrcs.usda.gov.

USDA-NRCS "field office technical guides" for cost-share projects; http://www.nrcs.usda.gov/technical/efotg.

USDA-NRCS National PLANTS Database. The PLANTS Database provides a single source of standardized information about plants. The database focuses primarily on plants of the USA and its territories. The PLANTS database includes checklists, distributional data, crop information, plant symbols, plant growth data, references, and other plant information. The PLANTS database reduces costs by preventing the duplication of efforts and by making information exchange possible across agencies and disciplines. These Web pages allow you to query the PLANTS database, download information, and view images of plants; http://plants.usda.gov.

- U.S. Department of Commerce, National Marine Fisheries Service Web page; http://www.nmfs.noaa.gov
- Endangered Species Program, U.S. Fish and Wildlife Service; http://endangered.fws.gov
- Private Landowners and the Endangered Species Act; http://endangered.fws.gov/landowner/index.html
- United States Department of the Interior-Fish and Wildlife Service, Region 1, ecological services page; http://news.fws.gov/newsreleases
- USFWS news releases and archive. The February 6, 2001 news release discusses grants for endangered species projects; http://endangered.fws.gov/esa.html
- An online copy of the Endangered Species Act; http://endangered.fws.gov/consultations/index.html
- An online copy of the Section 7 (agency) consultation handbook and the HCP handbook; http://www.epa.gov/ebtpages/ecosystems.html
- United States Environmental Protection Agency page to browse ecosystem topics; http://www.epa.gov/espp
- United States Environmental Protection Agency page about the Endangered Species Protection Program; http://edcwww.cr.usgs.gov/Webglis/glisbin/finder_main.pl?dataset_name=NAPP
- United States Department of Interior index to locations of orthophoto aerial photography

from the U.S. Geological Service, at various scales, which can be ordered by telephone or over the Internet; http://geography.usgs.gov/products.html#maps

- United States Department of the Interior official repository of domestic geographic place names. Links to sites providing map viewers are provided; http://www.gpoaccess.gov/fr/tips.html
- U.S. Library of Congress page of helpful hints for searching the *Federal Register* online via GPO access.

The U.C. Davis glossary of monitoring and water quality terms (http://animalscience.ucdavis.edu/extension/RangelandResources/pdfs/glossary.PDF) is an Internet address that may be useful in your efforts to answer questions about water quality.

A site with information about developing conservation trusts (http://www.rangelandtrust.org/land-conservation. html).

Water Quality Monitoring Resources

Table 15a is a partial list of chemical and scientific equipment supply companies from which to purchase equipment for water quality monitoring.

REFERENCES

American Public Health Association. 1992. *Standard Methods for the Examination of Water and Wastewater*, 18th ed. Washington, DC: American Public Health Association.

Mitchell, M., W. Stapp. *Field Manual for Water Quality Monitoring,* 8th edition. Available from 2050 Delaware Ave., Ann Arbor, MI 48103.

Oregon Department of Fish and Game. *The Stream Scene*. Available from Oregon Department of Fish and Game, P.O. Box 59, Portland, OR 97207.

Oregon State University Cooperative Extension, Klamuth County, 1996.

Resources Law Group. 2003. Available at http://www.resourceslawgroup.com/contact/contact/htm.

Spooner, J., R.P. Maas, S.A. Dressing, M.D. Smolen, F.J. Humenic. 1985. Appropriate designs for documenting water quality improvements from agricultural nonpoint source programs programs. In *Perspectives on Nonpoint Source Pollution*. USEPA 440/5-85-001. Washington, DC: U.S. Environmental Protection Agency, p. 30–34.

Spooner, J., R.P. Maas, M.D. Smolen, C.A. Jamieson. 1987. Increasing the sensitivity of nonpoint source control monitoring programs. In *Symposium on Monitoring Modeling and Mediating Water Quality*. Minneapolis, MN: American Water Resources Association.

Torell, L.A., N.R. Rimbey, J.A. Tananka, S.A. Bailey. 2001. *The Lack of a Profit Motive For Ranching: Implications for Policy Analysis*. Society for Range Management, Annual Meeting 2001. Lakewood, CO: Society for Range Management.

U.S. Environmental Protection Agency. *Quality criteria for water: 1986*. Washington, DC: U.S. Environmental Protection Agency, Office of Water Regulations and Standards.

U.S. Environmental Protection Agency. 1991. *Monitoring Guidelines*. Publication 910/9-91-001. Available from Region 10, NPS section (WD-139), 1200 Sixth Ave., Seattle, WA, 98101.

U.S. Environmental Protection Agency. 1995. *Volunteer Stream Monitoring: A Methods Manual*, p. 111. Washington, D.C.: GPO.

Index

Note: *Italicized* page locators indicate figures/tables.

C

I

M

N

Q

R

S

T

U

V

X

Y

Z

Printed and bound by CPI Group (UK) Ltd, Croydon, CR0 4YY

16/04/2025

14658424-0003